CRC SERIES IN ENZYME BIOLOGY

Editor-in-Chief
John R. Sabine, Ph.D.

3-HYDROXY-3-METHYLGLUTARYL COENZYME A REDUCTASE

Editor
John R. Sabine, Ph.D.
Reader in Animal Physiology
Department of Animal Sciences
Waite Agricultural Research Institute
University of Adelaide
Adelaide
Australia

PYRUVATE CARBOXYLASE

Editors
J. C. Wallace
Senior Lecturer in Biochemistry
D. B. Keech
Reader in Biochemistry
Department of Biochemistry
University of Adelaide
Adelaide
Australia

Pyruvate Carboxylase

Editors

D. B. Keech, Ph.D.
Reader in Biochemistry
Department of Biochemistry
University of Adelaide
Adelaide, South Australia
Australia

J. C. Wallace, Ph.D.
Senior Lecturer in Biochemistry
Department of Biochemistry
University of Adelaide
Adelaide, South Australia
Australia

CRC Series in Enzyme Biology

Series Editor-in-Chief
John R. Sabine, Ph.D.

CRC Press
Taylor & Francis Group
Boca Raton London New York

CRC Press is an imprint of the
Taylor & Francis Group, an **informa** business

First published 1985 by CRC Press
Taylor & Francis Group
6000 Broken Sound Parkway NW, Suite 300
Boca Raton, FL 33487-2742

Reissued 2018 by CRC Press

Library of Congress Cataloging in Publication Data

Main entry under title:

Pyruvate carboxylase.

(CRC series on enzyme biology)
Bibliography: p.
Includes index.
1. Pyruvate carboxylase. I. Keech, D. B.
II. Wallace, J. C. (John C.) III. Series.
QP619.P87P97 1985 574.19'254 84-23080
ISBN 0-8493-6552-X

A Library of Congress record exists under LC control number: 84023080

Publisher's Note
The publisher has gone to great lengths to ensure the quality of this reprint but points out that some imperfections in the original copies may be apparent.

Disclaimer
The publisher has made every effort to trace copyright holders and welcomes correspondence from those they have been unable to contact.

ISBN 13: 978-1-315-89704-2 (hbk)
ISBN 13: 978-1-351-07614-2 (ebk)

Visit the Taylor & Francis Web site at http://www.taylorandfrancis.com and the
CRC Press Web site at http://www.crcpress.com

DEDICATION

We wish, respectfully, to dedicate this volume on pyruvate carboxylase to the memory of the late Professor Merton F. Utter, with whom we had the privilege and pleasure of working at the Case Western Reserve University Medical School, Cleveland, Ohio.

D. B. Keech
J. C. Wallace
G. J. Barritt

SERIES PREFACE

The "CRC Series on Enzyme Biology" is a series of books, each one devoted to a single enzyme and each one endeavoring to draw together in a comprehensive and systematic manner all that is currently known about the chemistry, biochemistry, and physiology of that particular enzyme. Each volume, written or edited by one or more international enzyme specialists, draws together the latest information available on the occurrence, structure, function, role, and control of the biologically more important enzymes.

Each chapter or section of each volume is not written primarily for those specializing in that narrow area (innumerable specialist reviews do this), but rather to provide a coherent and integrated summary of that aspect for those working on other, quite different facets of the same enzyme. In this way several distinct classes of scientist, and student, will benefit from each volume, namely, those concerned with "type" situations of which that enzyme is an example, those working on the metabolic systems of which that enzyme is a key component, and those concerned with diseases in which the enzyme has an important role.

PREFACE

Pyruvate carboxylase has an interesting history. Its discovery in 1960 provided the answer to the vexing question of how pyruvate is converted to P-enolpyruvate during the gluconeogensis in the liver. Prior to this discovery, there had been no explanation of how the energy barrier was overcome for the conversion of pyruvate to P-enolpyruvate. In the 1960s, this was a most exciting development. Shortly thereafter, electron microscopy of chicken liver pyruvate carboxylase came into focus. It was one of the first enzymes to be subjected to structural analysis using the electron microscope. The resulting beautifully clear profiles showed a square planar tetramer which was an inspiration to others to use this procedure for study of the quaternary structure of enzymes. Much later it was found that the square planar structures were not pyruvate carboxylase, rather they were a highly visible minor contaminant of the pyruvate carboxylase preparations that overshadowed the more lightly staining pyruvate carboxylase which dissociated under the conditions used for staining. The enzyme has now been found to be a tetrahedron-like structure composed of two pairs of subunits with the planes orthogonal to each other. The mechanism of the pyruvate carboxylase reaction, likewise, has been a subject of lively interest. At first, based on steady state kinetics, the mechanism was considered to involve a "two-site" pingpong reaction. More recently, the reaction has been shown to occur by a sequential pathway.

These are but a few examples of the turbulent events that have occurred in the history of pryuvate carboxylase. The present book is the first comprehensive review to be writtten since these new discoveries with pyruvate carboxylase have come to light. Three of the authors of this book at one time or another worked directly with Merton F. Utter, the father of pyruvate carboxylase. All the authors write with first-hand authority concerning the various aspects of this remarkable enzyme and its role in metabolism in relation to living processes. This book is a welcome source of information on this enzyme which is vitally important to many forms of life.

Harland G. Wood

SERIES EDITOR

John R. Sabine, M.Agr.Sc., Ph.D. is Reader in Animal Physiology at the Waite Agricultural Research Institute of the University of Adelaide, Adelaide, Austrialia. Dr. Sabine obtained his Bachelor's and Master's degree in Agricultural Science from the University of Melbourne and then his Ph.D. in Animal Nutrition in the laboratory of Dr. B. Connor Johnson at the University of Illinois, Urbana. After several research appointments — Monash University, Australian National University, and the University of California, Berkeley, he was appointed to the faculty of University of Adelaide in 1967. At various times since then he has held Visiting Professorships at Brandeis University (Graduate Department of Biochemistry), the University of Stockholm (Wenner-Gren Institute), and the University of Oklahoma Health Sciences Center (Department of Biochemistry and Molecular Biology).

Dr. Sabine is a member of several scientific societies, including particularly the Australian Biochemical Society, the Nutrition Society of Australia, and the Australian Physiological and Pharmacological Society. He has presented his research findings at various national and international meetings, and has been chairman of a number of sessions at these meetings. He was co-convenor of, and leader of the Australian delegation to the U.S./Japan/Australia Cancer Conference (Hawaii, 1975) and the convenor and chairman of the unique international symposium "Lipids in Cancer", which was held on board the Indian-Pacific Express as a satellite meeting to the 12th International Congress of Biochemistry, 1982. In 1979 he delivered the 12th Patricia Chomley Oration to the Australian College of Nursing.

Dr. Sabine has published some 50 research articles and 12 invited reviews. He is on the Editorial Board for *Nutrition and Cancer*, and his book *Cholesterol* was the first comprehensive coverage of this important field to appear in 20 years. His major research interests revolve around cholesterol physiology, with particular reference to the control of its synthesis and to its role in membrane structure and function, and with emphasis upon the role of cholesterol in the etiology of cancer. He has further research interests in such diverse fields as the role of earthworms in biological resource recovery and the physiology of goats for meat and fiber production.

THE EDITORS

Bruce Keech, Ph.D., at the time of writing, Reader in the Department of Biochemistry at the University of Adelaide; he has since retired.

Dr. Keech received his early training at the University of Adelaide where he obtained a B.Sc. and 1st Class Honors in Biochemistry in 1953, and in 1957 he obtained his Ph.D. His post-doctoral career began when he won a C.S.I.R.O. oveseas scholarship and spent 2 years at the University of Oxford. In 1959 he moved to Western Reserve University, Cleveland, for a further 2 years of full-time research before returning to the University of Adelaide in 1961 to take up the position of Lecturer in the Department of Biochemistry. In 1964 he was promoted to Senior Lecturer and in 1968 he was appointed Reader.

Dr. Keech was co-discoverer of the enzyme, pyruvate carboxylase at Western Reserve University (now Case Western Reserve). His major research interest has been the study of the physical, chemical, and kinetic properties of the biotin-dependent carboxylases — in particular, pyruvate carboxylase. He has published over 50 research papers in this general area. In 1980 he was the recipient of the Lemberg Medal awarded by the Australian Biochemical Society.

John C. Wallace, Ph.D., is a Senior Lecturer in the Department of Biochemistry, University of Adelaide, Australia.

Dr. Wallace graduated B.Sc.Agr. with Honors in 1960 and Ph.D. in 1965 from the University of Sydney. He was a post-doctoral fellow in the laboratory of Professor Sir Hans Krebs, F.R.S., University of Oxford, England, 1965-1966 and of Professor M. F. Utter, Case Western Reserve University, Cleveland, Ohio, from 1967-1969. He returned to Australia as a Queen Elizabeth II Fellow in 1969 and was appointed in 1970 to the staff of the University of Adelaide where he has remained except for periods of study leave at the Universities of Cambridge, England, and of Wurzburg, West Germany.

Dr. Wallace has been a council member of the Australian Biochemical Society and of the Medical Sciences Club of South Australia. His current research interests include studies on the structure and function of the biotin-dependent carboxylases and of the polypeptide growth factors present in colostrum and milk.

CONTRIBUTORS

Paul V. Attwood, Ph.D.
Senior Teaching Fellow
Department of Biochemistry
University of Adelaide
Adelaide, South Australia
Australia

Gregory J. Barritt, Ph.D.
Reader in Clinical Biochemistry
Department of Clinical Biochemistry
Flinders Medical Centre
Bedford Park, South Australia
Australia

Simon B. Easterbrook-Smith, Ph.D.
Lecturer
Department of Biochemistry
University of Sydney
Sydney, New South Wales
Australia

D. Bruce Keech, P.h.D.,
Reader in Biochemistry
Department of Biochemistry
University of Adelaide
Adelaide, South Australia
Australia

John C. Wallace, Ph.D.
Senior Lecturer in Biochemistry
Department of Biochemistry
University of Adelaide
Adelaide, South Australia
Australia

Harland G. Wood, Ph.D.
Professor of Biochemistry
School of Medicine
Case Western Reserve University
Cleveland, Ohio
U.S.A.

TABLE OF CONTENTS

Chapter 1

HISTORICAL BACKGROUND

D. Bruce Keech

During the 1950s, considerable interest centered on the pathway of carbohydrate synthesis in liver tissue from noncarbohydrate precursors such as lactate and pyruvate, i.e., the process of gluconeogenesis was, in effect, a reversal of glycolysis; it was appreciated that the enzymic mechanism could not, for thermodynamic reasons, be a straightforward reversal of glycolysis. Krebs and Kornberg[476] pointed out that all but three steps of glycolysis were, in fact, readily reversed, but at these three loci, major energy barriers occur, viz.,

$$\text{Glucose} + \text{ATP} \rightarrow \text{Glucose,6-Phosphate} + \text{ADP} \quad (1)$$

$$\text{Fructose,6-Phosphate} + \text{ATP} \rightarrow \text{Fructose,1-6-Diphosphate} + \text{ADP} \quad (2)$$

$$\text{Phosphopyruvate} + \text{ADP} \rightarrow \text{Pyruvate} + \text{ATP} \quad (3)$$

They calculated that if all other reactants were present at equal concentrations, then a low ratio of ATP:ADP (about 10^{-6}) would be adequate for a reversal of the first two reactions, while a high ratio (about 10^{6}) would be required to reverse the reaction in Equation 3. Clearly, the two conditions could not be expected to coexist in cells actively synthesizing glucose from pyruvate. In the case of the first two reactions (Equations 1 and 2) the energy barrier in the gluconeogenic pathway is overcome by two enzymes, viz., glucose,6-phosphatase and fructose,1-6-diphosphatase, because both of these enzymes catalyze hydrolytic reactions rather than phosphotransferring reactions to form ATP. However, the energy barrier presented by a reversal of the third reaction (Equation 3) presents a more complex situation.

The experimental data available at that time was obtained primarily from isotopic distribution studies, and the two most significant observations were as follows:

1. When glycogen is synthesized in the liver from pyruvate (or lactate) in the presence of $H^{14}CO_3^-$, radioactivity appears in carbons 3 and 4 of the glucose moieties produced. Topper and Hastings[883] calculated that a minimum of 84% of the pyruvate molecules must have fixed CO_2 before they became part of the sugar molecule in the liver.
2. If pyruvate entered the glycolytic pathway by a direct reversal of the pyruvate kinase reaction, then isotope from [2−^{14}C]pyruvate or [2−^{14}C]lactate should appear in carbons 2 and 5 of the glucose molecule, while [3−^{14}C]pyruvate or [3−^{14}C]lactate should yield glucose with ^{14}C only in carbons 1 and 6. In fact, the isotopic label is uniformly distributed over carbons 1, 2, 5, and 6 of glucose irrespective of whether [2−^{14}C]- or [3−^{14}C]-labeled pyruvate or lactate is administered to the animal.[512,513,883]

Obviously then, the pathway of carbohydrate synthesis from pyruvate could not be via a reversal of the pyruvate kinase step. This realization prompted Krebs[471] and Utter and Kurahashi[903] to postulate a pathway which by-passed this enzyme and is shown in Scheme 1:

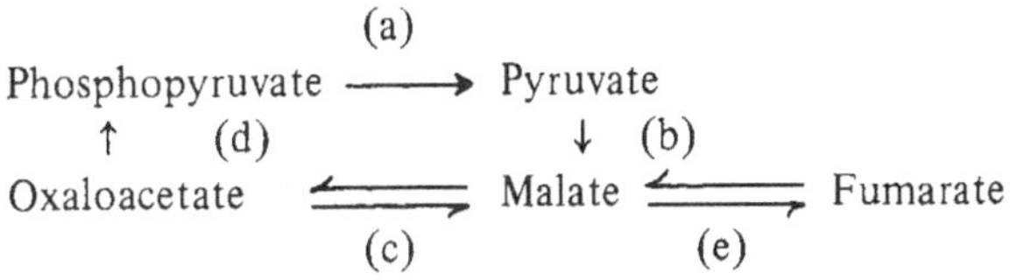

SCHEME 1. Dicarboxylic acid shuttle.

The enzymes involved in this pathway are: (a) pyruvate kinase, (b) malic enzyme, (c) malate dehydrogenase, and (d) phosphoenolpyruvate (PEP) carboxykinase. The equilibrium between malate and fumarate (e) is catalyzed by fumarase.

This sequence of events could account for the experimental data available. In the first step, pyruvate is carboxylated by the malic enzyme to form malate. If the malate then equilibrates with fumarate, a symmetrical molecule, then a redistribution of radioactivity would occur, and the appropriate labeling patterns in the synthesized glucose residues would be obtained. However, in terms of the overall energetic properties of this scheme, there is little to distinguish this pathway from a reversal of the pyruvate kinase reaction, i.e., reversal of (a) in Scheme 1.

Probably the most significant observation in this area was made by Utter[891,892] in 1959, when studying the intracellular distribution of the enzymes involved in the dicarboxylic acid shuttle shown in Scheme 1. Utter found that mitochondria isolated from chicken liver, although capable of synthesizing phosphopyruvate from pyruvate, contained only trace amounts of pyruvate kinase and malic enzyme. Since the first of these enzymes is responsible for the pyruvate kinase pathway and malic enzyme is an essential part of the shuttle (illustrated in Scheme 1), these findings suggested that neither pathway was correct. On the basis of these results, a further search of the enzymic mechanisms capable of carrying out this interconversion in chicken liver mitochondria was carried out.

Reexamination of chicken liver mitochondria revealed another very active CO_2-fixing enzyme which could form oxaloacetate from pyruvate[898] and catalyzed the following reaction:

$$\text{Pyruvate} + HCO_3^- + \text{ATP} \rightarrow \text{Oxaloacetate} + \text{ADP} + \text{Pi} \qquad (4)$$

Within a short time, the enzyme responsible for this reaction (later referred to as pyruvate carboxylase) was concentrated 100-fold from chicken liver extracts and the general properties of the reaction[899] and enzyme[439] were described. The discovery of this enzyme prompted a revision of the dicarboxylic acid shuttle to an abbreviated form as shown in Scheme 2:

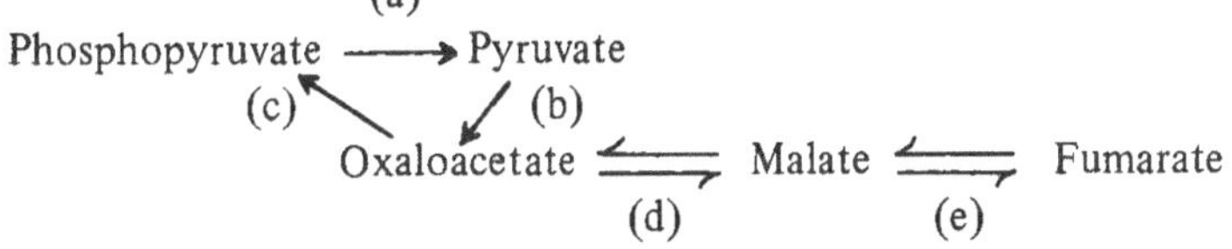

SCHEME 2. Abbreviated dicarboxylic acid shuttle.

The enzymes involved in this pathway are: (a) pyruvate kinase, (b) pyruvate carboxylase, (c) PEP carboxykinase, (d) malate dehydrogenase, and (e) fumarase.

The energetic relationships for the synthesis of phosphopyruvate (later referred to as phosphoenolpyruvate) from pyruvate via the action of pyruvate carboxylase and PEP carboxykinase is favorable. Furthermore, experiments with mixtures of the two purified enzymes showed that synthesis of phosphopyruvate was accomplished and no coupled reacions or special conditions were required.[439]

At the time of its discovery, perhaps the most significant observation concerning this enzyme was its complete dependence on acetyl-CoA for its activity,[898] although this substance participated only in a catalytic role. The control of the enzymic activity and the role of acetyl-CoA in the reaction mechanism will be discussed elsewhere, but it is appropriate that the question of how acetyl-CoA was shown to be involved in activating the enzyme should be mentioned. Following the observation by Utter that mitochondria isolated from chicken liver were capable of synthesizing phosphopyruvate from pyruvate, a search for the enzyme responsible in mitochondrial extracts was initiated. In the absence of an adequate assay system, the investigation was begun using an isotopic exchange reaction between pyruvate and oxaloacetate. This procedure proved to be inadequate in crude extracts mainly because of the instability of oxaloacetate at neutral pH. A number of alternate methods were considered, assuming that oxaloacetate was the product of the reaction. For example, coupling the reaction to malate dehydrogenase in the presence of NADH was considered, but was discarded on the grounds that there was sufficient lactate dehydrogenase in mitochondrial extracts to remove most of the pyruvate and all of the NADH before much CO_2 fixation could occur. Eventually, a coupling system involving acetyl-CoA and the enzyme, citrate synthetase, to convert the unstable oxaloacetate to acid-stable citrate was used.

At that preliminary stage, the reaction solution contained pyruvate, $H^{14}CO_3^-$, acetyl-CoA, citrate synthetase, and a crude extract of acetone powder of chicken liver mitochondria. Mg^{2+} and ATP were added for thermodynamic reasons and under these conditions, significant amounts of radioactivity were fixed into acid-stable products after a 5-min incubation period. Strangely, an equal amount of radioactivity was present in acid stable products in the control lacking citrate synthetase and it was thought that there must have been enough citrate synthetase in the mitochondrial extract to account for the coupling reaction. However, whenever acetyl-CoA was omitted from the assay solution, no ^{14}C was fixed, and as soon as $(NH_4)_2SO_4$ fractionation of the mitochondrial extract was attempted, it was realized that the enzyme system had an absolute requirement for acetyl-CoA. The results of those early experiments are presented in Reference 898 and since that time, the role of acetyl-CoA in controlling the activity of pyruvate carboxylase has continued to capture the interest of investigators in this field.

Chapter 2

DISTRIBUTION AND BIOLOGICAL FUNCTIONS OF PYRUVATE CARBOXYLASE IN NATURE

J. C. Wallace

TABLE OF CONTENTS

I. INTRODUCTION

With pyruvate, phosphoenolpyruvate (PEP), and oxaloacetate occurring at one of the major crossroads of intermediary metabolism, it is not surprising that the enzymes catalyzing their interconversions have been the subject of numerous investigations in a wide range of life forms.

Most cells have a combination of only three of these enzymes (viz., I, III, and V, see Figure 1) with which to negotiate this central triangle of metabolism. None has yet been reported to have enzymes catalyzing all six reactions, but some have four (e.g., *Salmonella typhimurium*[866] and *Escherichia coli*[31] have III, IVa, V, and VI; *Azotobacter vinelandii*[758] has I, II, IVa, and V); *Pseudomonas citronellolis*[612] has five (I, II, IVa, V, and VI).

Although as described in Chapter 1, pyruvate carboxylase was first sought and recognized as an enzyme of the gluconeogenic pathway in liver, its occurrence in many nongluconeogenic tissues as well (and even in microorganisms cultured in glucose-containing media) clearly indicates its wider biological roles. Thus, whenever intermediates of the tricarboxylic acid cycle are withdrawn for biosynthetic purposes, e.g., oxaloacetate for gluconeogenesis or glyceroneogenesis, citrate for fatty acid and cholesterol synthesis, citrate and 2-oxoglutarate for neurotransmitter synthesis, succinyl-CoA for porphyrin synthesis, or oxaloacetate and 2-oxoglutarate for amino acid synthesis, there is a need for regeneration of these by an anaplerotic reaction sequence.[460]

Anaplerotic synthesis of C_4 compounds can occur independently of CO_2-fixing enzymes in some cells by operation of the glyoxylate cycle.[461-463] In higher plants and some bacteria this function is also fulfilled by the enzyme PEP carboxylase (EC4.1.1.31) and in a few other species by PEP carboxytransphosphorylase (EC4.1.1.38); for a review, see Utter and Kolenbrander.[902]

This chapter surveys those situations in which this anaplerotic function is served by pyruvate carboxylase.

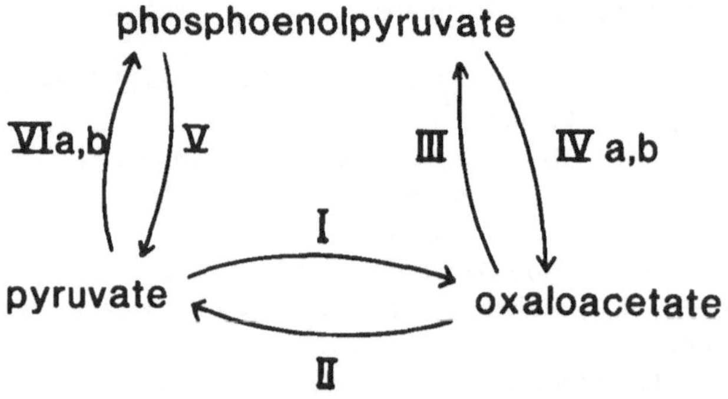

FIGURE 1. Enzymes Catalyzing the Interconversion of Pyruvate, PEP, and Oxaloacetate.

No.	Enzyme name and reaction catalyzed	Ref
I	Pyruvate carboxylase [pyruvate:CO_2 ligase (ADP); E.C.6.4.1.1]	899
	$$\text{Pyruvate} + \text{ATP} + HCO_3^- \xrightleftharpoons{Me^{2+}\ (\text{acetyl-CoA})} \text{Oxaloacetate} + \text{ADP} + \text{Pi}$$	
II	Oxaloacetate decarboxylase [oxaloacetate carboxy-lyase, E.C.4.1.1.3]	469
	$$\text{Oxaloacetate} + H_2O \rightleftharpoons \text{Pyruvate} + CO_2$$	
III	PEP carboxylase [GTP:oxaloacetate carboxylase transphosphorylating; E.C.4.1.1.32]	903
	$$\text{PEP} + CO_2 + \text{GDP} \xrightleftharpoons{ME^{2+}} \text{Oxaloacetate} + \text{GTP}$$	
IVa	PEP carboxylase [orthophosphate:oxaloacetate carboxylyase (phosphorylating); E.C.4.1.1.31]	65
	$$\text{PEP} + CO_2 + H_2O \xrightarrow{Mg^{2+}\ (\text{acetyl-CoA})} \text{Oxaloacetate} + \text{Pi}$$	
IVb	PEP carboxytransphosphorylase [pyrophosphate:oxaloacetate carboxylyase (transphosphorylating) E.C.4.1.1.38]	794
	$$\text{PEP} + CO_2 + \text{Pi} \xrightleftharpoons{Mg^{2+}} \text{Oxaloacetate} + \text{PPi}$$	
V	Pyruvate kinase [ATP:pyruvate phosphotransferase, E.C.2.7.1.40]	594
	$$\text{PEP} + \text{ADP} \xrightleftharpoons{Mg^{2+}, K^+} \text{Pyruvate} + \text{ATP}$$	
VIa	PEP synthetase [ATP:pyruvate 2-*O*-phosphotransferase (AMP) (E.C.2.7.1.dd)]	205
	$$\text{Pyruvate} + \text{ATP} \rightarrow \text{PEP} + \text{AMP} + \text{Pi}$$	
VIb	Pyruvate phosphate dikinase [ATP:pyruvate, orthophosphate phosphotransferase, E.C.2.7.9.1]	262, 383, 677
	$$\text{Pyruvate} + \text{ATP} + \text{Pi} \xrightleftharpoons{Mg^{2+}} \text{PEP} + \text{AMP} + \text{PPi}$$	

II. PYRUVATE CARBOXYLASE IN BACTERIA

A. Species Distribution

Aerobic organisms must possess at least one anaplerotic pathway to grow on hexoses or glycolytic intermediates. Such enzymes generate oxaloacetate to allow continued flux through the tricarboxylic acid cycle while its intermediates are continually removed for lipid, amino acid, and porphyrin synthesis.

It was at first thought[902] that the occurrence of either pyruvate carboxylase or PEP carboxylase was mutually exclusive of the other. However, these two enzymes have now been reported in *Azotobacter vinelandii*,[502,758] *Pseudomonas fluorescens*,[404] *Ps. citronellolis*,[612] *Brevibacterium lactofermentum*,[885] *Thiobacillus novellus*,[542] and some of the *Chromatiaceae*.[719,999] Among the *Chromatiaceae*, Sahl and Truper[719] found considerable variation between species as to whether both pyruvate carboxylase and PEP carboxylase occurred together and in the way their activities were affected by the growth conditions.

Both PEP carboxylase and pyruvate carboxylase have been found, albeit in reduced amounts, even under conditions of autotrophic growth in *Thiobacillus novellus*,[542] where ribulose-1,5 bisphosphate carboxylase (EC4.1.1.39) is the primary catalyst of CO_2 fixation in the reductive pentose phosphate pathway.

Nevertheless, the importance of this anaplerotic function, even to an autotroph, is demonstrated by the observation that a mutant strain of *Rhodopseudomonas spheroides* lacking pyruvate carboxylase failed to grow either autotrophically or heterotrophically on a glucose or lactate medium unless it was supplemented with malate.[639] Other genetic studies have similarly demonstrated very convincingly the importance of the anaplerotic function of pyruvate carboxylase in a variety of heterotrophic species (viz., *Bacillus subtilis*,[146,237] *Pseudomonas aeruginosa*,[648] *Rhizobium trifolii*,[701] and *Agrobacterium tumefaciens*[181]).

As is evident in Table 1, the reported occurrence of pyruvate carboxylase is quite widespread among the various groups of bacteria. Although so far only 17 of more than 200 bacterial genera listed by Bergey[414] are represented, many of the entries in Table 1 are relatively recent and this listing can be expected to grow as metabolic studies are undertaken on further species.

The distribution of pyruvate carboxylase appears to be influenced more by the particular niches the organisms occupy than by close adherence to generic relationships. There are apparently many interesting situations yet to be investigated, and clearly much work still to be done even with known bacterial sources of pyruvate carboxylase to bring our knowledge of all of these to a comprehensive state in molecular and genetic terms.

Pyruvate carboxylase is, however, notably absent from the Enterobacteriaceae,[31] from *Salmonella* spp.,[866] from some *Arthrobacter* species,[480] and from various methane-oxidizing organisms (viz., *Methylosinus trichosporium, Methylomonas methanica, Methylobacter chroococcum, Hyphomicrobium vulgare,* and *Pseudomonas methylica*), though present in others.[508]

B. Effect of Growth Conditions on Activity of Pyruvate Carboxylase

As can be seen in Table 1, light, temperature, the nature of the carbon source, and the availability of biotin are all factors which can influence the amount of pyruvate carboxylase present during growth of a particular organism. The specific effects of the growth conditions and the degree of their interactions vary considerably among the species studied.

1. Carbon Source

Although in some species the presence of malate, aspartate, or succinate in the

Table 1
PYRUVATE CARBOXYLASE IN PROCARYOTES

Organism	Substrate[a] (mM)	Specific activity[b]	Acetyl-CoA dependence[c]	Effect of aspartate[d]	Ref.
Rhodopseudomonas spheroides	Glucose (12)	15[e]	100%	–[g]	639
	Pyruvate (20)	19[f]			
	Acetate (25)	19[e]			
	DL-Malate (20)	41[f]			
	Lactate (20) + DL-Malate (5)	7[e]			
		20[f]			
		15[e]			
		17[f]			
		11[e]			
		35[f]			
Rhodospirillum rubrum	DL-Malate (50) + Glutamate (13)	+[h]	–	–	92
Chromatium (strain D)	DL-Malate	+	≥40%	–	297—298
C. okenii	Na_2S (3)	2[e]	60%	–	889
C. vinosum	Na_2S (3)	66[e]	–	–	719
	DL-Malate (3)	37[f]			
Thiocapsa roseopersicina	Na_2S (3)	0.5[e]	–	–	719
	DL-Malate (3)	1.1[f]			
T. pfennigii	Na_2S (3)	0.3[e]	–	–	719
Ectothiorhodospira mobilis	Na_2S (3)	13[e]	–	–	719
	DL-Malate (3)	1.9[f]			
E. shaposhnikovii	Acetate (25)	2.8[e]	–	–	999
		1.9[f]			

Table 1 (continued)
PYRUVATE CARBOXYLASE IN PROCARYOTES

Organism	Substrate[a] (m*M*)	Specific activity[b]	Acetyl-CoA dependence[c]	Effect of aspartate[d]	Ref.
Pseudomonas aeruginosa	Glucose (50)	5	0	–	671
	Glucose (25)	24	0	30% (5 m*M*)[e]	648
	Glycerol (30)	28			
	Pyruvate (30)	21			
	DL-Lactate (40)	35			
	Succinate (40)	13			
	Acetate (50)	7			
	Casamino acids (0.3%)	12			
	Casamino acids (0.3%) + glucose (10)	16			
Ps. fluorescens	Glucose (55)	49	0	0	404
	Glutamate (52)	19			
	Acetate (61)	4			
Ps. oleovorans	Methanol (625)	8	14%	–	508
Ps. citronellolis	Citronellol (10) + acetate (33)	13(22°C)	0	–	772
Ps. citronellolis	Glucose (55)	44 (25°C)	0	0	858
	DL-Lactate (54)	31			
	Acetate (61)	15			
	Citrate (39)	12			
	Glutamate (52)	9			
	Aspartate (56)	2			
	L-Malate (60)	0			
Azotobacter vinelandii	Sucrose (29)	12 (25°C)	0	0	758
	Glucose (55)	8			
	Glycerol (109)	12			
	Pyruvate (91)	32			
	Lactate (89)	25			
	Acetate (122)	11			
	Succinate (62)	0			
	Succinate (31) + Pyruvate (91)	68			
	Succinate (31) + Lactate (89)	52			

Rhizobium trifolii	Glucose (22)	8.4[j](28°C)	77%	–	701
Agrobacterium tumefaciens	Glucose (110)	10 (27°C)	100%	–	181
	Casamino acids (1%) + sucrose (59)	74	(Ka = 11 μM)		
Desulfovibrio desulfuricans (Teddington R strain, NCIB8312)[k]	Fumarate (50)	3.4 (25°C)			501
	Lactate (100)	3.8 (25°C)	–	–	
Thiobacillus novellus	Thiosulphate (40)	6	–	–	542
	Glucose (16.6)	101	83[178]	0[178]	
Thiobacillus A.2	Pyruvate (36)	45	95%	0	178
Streptococcus faecalis (var. liquefaciens)	Glucose (5)/casamino acids (1%)/	+	10% (@ 400 μM)	25% (@ 6.7 mM)	377
	yeast extract (1%)/brain-heart infusion broth (Difco)	+	40% (@ 35 μM)	35% (@ 3 mM)	483
S. faecium (var. durans)	Brain-heart infusion broth (Difco)	+	≥90% (Ka ≃ 15 μM)	95% Ki ≃ 0.4 mM)	483
S. lactis C10	Tryptone/yeast extract/lactose	+	33% (@ 375 μM)	100% (Ki ≃ 2.9 mM)	405
Micrococcus cerificans	Hydrocarbon	+	66%	–[e]	499
Bacillus megaterium NCIB7581 (biotin-deficient)	Glycerol (105)30°C	22	86%	50% @ 0.6 mM	10
	(105)37°C	9			
	(105)				
	+ biotin (40)30°C	24			
	Glycerol (105) + biotin (40)37°C	25			
B. subtilis M22 (wild type)	Tryptone (Difco, 1%)[m] + yeast extract (0.5%)	120 (45°C)	0	–	146
		50 (30°C)			
RB1950	Tryptone (Difco, 1°) + yeast extract (0.5%)	5 (45°C)			
		17 (30°C)			
B. subtilis 168 (strain 60015)	Glucose (50)	18	100%	–	237
	L-Malate (50)	22			
B. licheniformis	glucose (20)	4	100% Ka = 86 μM	Ki = 1 mM @ 20 μM acetyl-CoA	680
	glycerol (20)	4			
	Pyruvate (20)	4		Ki = 2.5 mM @ 50 μM acetyl-CoA	
	Malate (20)	3			
	Peptone/yeast extract	3		Ki = 6.5 mM @ 200 μM acetyl-CoA	

Table 1 (continued)
PYRUVATE CARBOXYLASE IN PROCARYOTES

Organism	Substrate[a] (m*M*)	Specific activity[b]	Acetyl-CoA dependence[c]	Effect of aspartate[d]	Ref.
B. licheniformis	Glucose (25)	22	99%	–	838
	Glucose (25) + biotin (0.4)	22			
	Succinate (50)	22			
	Succinate (50) + biotin (0.4)	22			
B. stearothermophilus[n]	Glucose (25)	22 (45°C)	≃ 100%[p]	Ki = 1.9—7.4 m*M* depending on [acetyl-CoA] (1—10 μM)[503]	838
	Glucose (25) + biotin (0.4)	160 (45°C)	Ka = 1.4—15 μM as [aspartate] varied 0—10 m*M*[503]		837
	Succinate (50)	2 (45°C)			
	Succinate (50) + glucose (25)	14 (45°C)			
	Succinate (50) +biotin (0.4)	160 (45°C)			
	Acetate (10—100)	6—3 (45°C)			
	Acetate (50) + biotin (0.4)	160 (45°C)			
Corynebacterium aquaticum	Glucose (55)	15	76%	–	84
Nocardia erythropolis	Glucose (55)	24	96%	–	84
Brevibacterium ammoniagenes	Glucose (55)	3	42%	–	84
B. lactofermentum AJ 3445	Pyruvate (220)	167	0	0	885
Arthrobacter globiformis	Lactate (89)	89 (25°C)	96% Ka = 25 μM	–[q]	350
	Methylamine	2	66%	–	508

[a] The substrate indicated is the main carbon source for growth, except where indicated. For other details of the growth media see original reference.

[b] The specific activity is the pyruvate carboxylase activity of a cell-free extract of the organism expressed as nanomoles CO_2 fixed per minute per milligram protein at 30°C, except where indicated in parentheses. This temperature then applies to all assays reported from that reference.

[c] The acetyl-CoA dependence is the percentage of the specific activity figure dependent on the presence of acetyl-CoA in the assay. The activator constant (Ka) under the conditions of the assay is noted also for the relatively few studies for which such details have been reported.

[d] The effect of aspartate is given as the percentage inhibition of activity at the concentration of aspartate indicated in parentheses.

[e] Photosynthetic growth conditions, anaerobic.

[f] Aerobic growth in the dark.

[g] The "–" indicates the effect was not investigated.

[h] The "+" indicates that clear evidence for pyruvate carboxylase activity was presented, but insufficient experimental details were given to enable a specific activity to be calculated.

[i] When the pyruvate concentration was near or below the Km value of 0.2 m*M*.

[j] Similar value with glycerol as sole carbon source. Pyruvate carboxylase required for growth on all carbon sources tested (viz., glucose, fructose, galactose, lactose, sucrose, sorbitol, mannitol, ribose, and glycerol) except succinate.

[k] Specific activities of 2.8 to 3.3 on lactate (100 m*M*) medium ± glutamine, glutamate, asparagine, or aspartate (each 10 m*M*) or yeast extract (1%). Essentially similar results reported by the same authors for *D. desulfuricans* strain Byron (NCIB 8458), also a fumarate dismuter, and for *D. vulgaris* strains Hildenborough (NCIB 8303) and Woolwich (NCIB 8457), two strains unable to dismute fumarate.

[m] Activity in organisms grown at 30°C.

[n] This organism was described originally[838] as *B. coagulans*, but was later characterized as *B. stearothermophilus.*[163] More recent data, especially from DNA-hybridization studies, have suggested, however, that it may not belong to the species *stearothermophilus.*[503]

[o] The letter "o" was skipped to avoid confusion with other symbols.

[p] For a detailed study of the activation of the purified pyruvate carboxylase from this thermophilic *Bacillus*, see References 163 and 503.

[q] The inclusion of aspartate in the assay mix has no effect on V, but raises the Ka for acetyl-CoA, up to 200 μ*M* at 4.5 m*M* aspartate.

growth medium partially (e.g., *Pseudomonas aeruginosa,*[648] *Ps. fluorescens,*[404] *Bacillus stearothermophilus,*[838] and *Chromatium vinosum*[719]) or even completely (e.g., *Azotobacter vinelandii*[758] and *Pseudomonas citronellolis*[501]) suppresses pyruvate carboxylase activity compared with that expressed in glucose- or lactate-containing media, in others (viz., *Rhodopseudomonas,*[639] *Thiocapsida,*[719] *Ectothiorhodospira,*[719] *Desulfovibrio,*[501] *Bacillus subtilis,*[237] and *Bacillus licheniformis*[680]) this effect is not evident. Where this means of regulating pyruvate carboxylase has been studied in detail,[858] it has been shown to operate at the level of polypeptide synthesis rather than at the stages of biotinylation of or the assembly of protomers into the holoenzyme.

2. Light

Similarly, there is considerable variation between those species yet examined in their responses to light as a factor affecting the specific activity of pyruvate carboxylase. The effect seen is also influenced by the nature of any metabolizable substrate present. Thus, with *R. spheroides* growing on glucose or malate as sole carbon source there was little difference in the specific activity of pyruvate carboxylase whether conditions were dark and aerobic or light and anaerobic. Conversely, if acetate, pyruvate, or lactate plus malate were the carbon sources, there was at least a 100% increase in specific activity when growth conditions were dark and aerobic.[639]

This, however, is not a general pattern of response to light as the level of pyruvate carboxylase in various members of the Chromatiaceae is affected quite differently by growth under either autotrophic or heterotrophic conditions.[719]

Furthermore, in most cases, the effect of light is compounded with a switch from aerobic to anaerobic conditions, and thus, further experiments are required to clarify this situation.

3. Temperature

So far, temperature as a factor determining the synthesis of pyruvate carboxylase has not received wide attention, but already two distinct genotypes have been recognized: in *Bacillus megaterium* NBCIB 7581 the reduction in pyruvate carboxylase activity seen at 37°C compared with growth at 30°C was shown by the restorative effect of biotin addition (10 μg/ℓ) to the medium to be a consequence of a temperature-sensitive defect in biotin synthesis.[10]

However, in a mutant strain of *Bacillus subtilis* unable to grow on a minimal glucose medium at 45°C unless supplied with utilizable intermediates of the tricarboxylic acid cycle (or their precursors), the primary metabolic lesion was shown by detailed genetic analysis to be the product of a heat-sensitive pyruvate carboxylase.[146] At an assay temperature of 30°C the specific activity of the mutant strain (RB 1950) was 33% that of its parent (M22), whereas at 45°C it was only 4%. The resultant deficit in oxaloacetate production at 45°C led to a decreased synthesis of *meso*-diaminopimelic acid, an amino acid unique to peptidoglycan, the major cell wall structural polymer.

4. Biotin

The addition of biotin to the medium at up to 500 μg/ℓ has been shown to markedly increase the level of pyruvate carboxylase activity in some organisms (*Rhodospeudomas spheroides,*[456] *Brevibacterium lactofermentum,*[885] *Bacillus coagulans,*[838] and some specific mutant strains of *Bacillus subtilis*[146] and *Bacillus megaterium* [10]) while in others (*Bacillus licheniformis,*[838] *Desulfovibrio vulgaris,*[501] and *Desulfovibrio desulfuricans*[501]) it is without effect.

Pyruvate carboxylase activity has been demonstrated in a number of organisms (*Pseudomonas aeruginosa,*[648,671] *Ps. citronellolis,*[772,858] *Azotobacter vinelandi,*[758] *Agrobacterium tumefaciens,*[181] *Thiobacillus novellus,*[542] *Corynebacterium aquaticum,*[84] *Nocardia erythropolis,*[84] and *Brevibacterium ammoniagenes*[84]) cultured on apparently biotin-free media. However, here, as with other organisms where biotin was included in the medium, there appear to be no data directly revealing the effect, if any, of biotin addition on the activity of pyruvate carboxylase or any other biotin-dependent enzyme.

The formation of biotin-dependent carboxylases from their inactive apoenzymes and (+)-biotin was established previously for propionyl-CoA carboxylase [EC6.4.1.3],[283,465-468,541,783,784] for acetyl-CoA carboxylase [EC6.4.1.2],[517] for transcarboxylase [EC2.1.3.1],[486,487,984] and for β-methylcrotonyl-CoA carboxylase [EC6.4.1.4].[420,453]

In a two-step mechanism,[486,487] the holoenzyme synthetases [(+)biotin-apocarboxylase ligase (AMP); EC6.3.4.1] first catalyze the formation of biotinyl-5′-AMP from biotin and ATP with the release of pyrophosphate[486,541] (see Equation 1), and then mediate the covalent attachment of the biotinyl moiety to a specific lysyl ε-amino group on the acceptor apoenzymes with the release of AMP[283,420,486,541,783,784] (see Equation 2).

$$\text{Biotin} + \text{ATP} \xrightarrow{Mg^{2+},\ \text{Synthetase}} \text{Biotinyl-5}'\text{-AMP} + \text{PPi} \tag{1}$$

$$\text{Biotinyl-5}'\text{-AMP} + \text{Apoenzyme} \xrightarrow{\text{Synthetase}} \text{Holoenzyme} + \text{AMP} \tag{2}$$

Kornberg and colleagues[160-162,839,840] demonstrated that this mechanism also applied to the formation of active pyruvate carboxylase from its apoenzyme, but that in the case of the thermophile *Bacillus stearothermophilus,* this process showed an almost complete requirement for acetyl-CoA as an allosteric activator. As with the expression of catalytic activity by this carboxylase,[163] the effect of acetyl-CoA was counteracted by L-aspartate. Three lines of evidence led to the conclusion that the actions of these allosteric effectors were on the apoenzyme and not on the synthetase. First, acetyl-CoA and aspartate affected the formation of holocarboxylase from synthetic biotinyl-5′-AMP and apoenzyme in the same way as they did the overall process from ATP and biotin, whereas they had no effect on the holoenzyme synthetase-catalyzed formation of biotinylhydroxamate from biotin, ATP, and hydroxylamine. Secondly, the apoenzyme was protected against thermal denaturation by acetyl-CoA and to a lesser extent by L-aspartate, whereas the holoenzyme synthetase was not markedly protected by acetyl-CoA and not at all by aspartate. Third, the kinetic parameters of the action of these two effectors on holocarboxylase formation were remarkably similar to the values obtained for their effects on holocarboxylase catalytic activity.[163]

Whereas Birnbaum[103] found that the total quantity of acetyl-CoA carboxylase (i.e., apoenzyme plus holoenzyme) increased approximately fivefold because of biotin deficiency in *Lactobacillus plantarum,* no such release from "coenzyme repression" was evident for pyruvate carboxylase in the thermophilic *Bacillus* grown under biotin-deficient conditions.[161]

C. Properties

1. Physical Properties

Only a few of the known bacterial pyruvate carboxylases have been purified to the extent necessary for a description in physiochemical terms of their molecular properties. These enzymes (viz., from *Pseudomonas citronellolis, Azotobacter vinelandii, Bacillus stearothermophilus,* and *Arthrobacter globiformis*) are discussed in detail in Chapter 3.

2. Effect of Acetyl-CoA

Much more widely studied have been the responses of the various pyruvate carboxylases to the addition of possible metabolic effectors such as acetyl-CoA and aspartate (see Table 1). In some species (viz., *Rhodospeudomonas spheroides,*[639] *Rhizobium trifolii,*[701] *Agrobacterium tumefaciens,*[181] *Thiobacillus novellus,*[542] *Thiobacillus A.2,*[178] *Streptococcus faecium* var. *durans,*[483] *Bacillus megaterium,*[10] *B. subtilis,*[146,237] *B. licheniformis,*[680,838] *B. stearothermophilus,*[838] *Arthrobacter globiformis,*[350] *Corynebacterium aquaticum,*[84] and *Nocardia erythropolis*[84]) pyruvate carboxylase activity is largely, if not wholly, dependent on the presence of acetyl-CoA. Only in a few of these, however, has the concentration dependence of this activation been reported in detail (see Table 1). In these cases the activator constant (Ka) has varied, depending on species and assay conditions, from 1 to 86 μM — a range remarkably similar to that reported for pyruvate carboxylases in vertebrate and other species.

Few attempts appear to have been made yet to relate this figure to the intracellular concentrations of acetyl-CoA found in the specific organisms. In the case of *Bacillus licheniformis,* Renner and Bernlohr[680] measured cellular acetyl-CoA levels during growth and sporulation, and noted that in 6.5-hr post-exponential phase cells this activator was present at a concentration only a quarter of the Ka value of that pyruvate carboxylase for it, suggesting the possiblility of a limitation to the synthesis of spore-specific compounds from pyruvate at that stage. However, as discussed by Scrutton and Griffiths[750] and Keech and Wallace,[440] the relationship between this parameter determined in vitro and the effective concentrations of acetyl-CoA and related compounds in the immediate environment of the enzyme is a complex one and in need of a great deal more examination in a variety of situations before firm conclusions can be reached.

With regard to the specificity of activation of bacterial pyruvate carboxylases by acyl-CoA compounds, it appears from the most intensively studied example so far (viz., the pyruvate carboxylase of the thermophilic *Bacillus*[503]) that its regulatory properties are a combination of features observed previously for this enzyme purified from other sources. Thus, at high concentrations of MgATP, HCO_3^-, and pyruvate, it can catalyze a rate of oxaloacetate synthesis in the absence of acetyl-CoA equal to about 20% of that observed in the presence of saturating concentrations of acetyl-CoA. These results are similar to those reported previously for sheep[29] and rat[545,764] liver pyruvate carboxylases. Only weak activation occurred on addition of CoA-SH or adenosine 3′,5′-bisphosphate. On the other hand, as with yeast pyruvate carboxylase, all alkylacyl derivatives of CoA tested were activators of the pyruvate carboxylase of this thermophilic *Bacillus.* Although acetyl-CoA was the most effective, certain long-chain acyl-CoA species were also very potent.

Another interesting facet of the influence that acetyl-CoA has on the pyruvate carboxylase of this thermophile is the effect of this ligand on the rate of biotinylation of the apo-enzyme[160-163,840] (see Section II. B.4).

Conversely, there are some species (viz., *Pseudomonas aeruginosa,*[648,671] *Ps. fluorescens,*[404] *Ps. citronellolis,*[772,858] *Azotobacter vinelandii,*[758] some strains of *Bacillus subtilis* [M22 and RB1950],[146] and *Brevibacterium lactofermentum*[885]) in which pyruvate carboxylase activity has been shown to be completely unaffected by the addition of acetyl-CoA.

3. Effect of Aspartate

As discussed by Scrutton,[747] regulatory inhibition of microbial pyruvate carboxylases by L-aspartate is an eminently appropriate means by which to control this area of anabolic metabolism, since this amino acid serves as a precursor of up to five other amino acids, of pyrimidines, and as a nitrogen donor for purine biosynthesis. However, as is evident in Table 1, there are a number of species in which this potential regulatory property has not been examined yet. Among those species in which the effect of aspartate has been investigated, the inhibition is generally relieved by increasing concentrations of acetyl-CoA in a quasi-competitive manner. There is evidence, however, that this relationship does not arise merely by competition between the two ligands for portions of the same binding site. Thus, for the enzyme from a thermophilic *Bacillus,* incubation with the lysyl group-specific reagent, trinitrobenzene sulfonic acid, causes specific loss of sensitivity to activation by acetyl-CoA, but is without effect on the inhibition of this enzyme by aspartate.[503] Similarly, the different ways in which their effects on enzyme activity vary with temperature, and the fact that they can act synergistically to stabilize the enzyme at elevated temperatures, are consistent with there being separate binding sites on the enzyme molecule for these two compounds.[163]

Further discussion of the regulation of pyruvate carboxylase activity by dicarboxylic amino and oxo-acids is presented in Section III. C and in Chapter 5, Sections III.B and C.

4. *Effects of Other Potential Regulatory Effectors*

Although the NAD:NADH ratio has been shown to influence oxaloacetate synthesis by PEP carboxylase in *Ps.* MA[596] and by yeast pyruvate carboxylase[159] (see, however, Chapter 5, Section III.B), there is as yet no such report of the activity of pyruvate carboxylase being so influenced in bacteria.

As is the case with the pyruvate carboxylases of eukaryotes, ADP has been shown to inhibit this enzyme purified from *Bacillus stearothermophilus*[163] and from *Ps. fluorescens,*[570] with a Ki comparable with the Km for ATP. However, no evaluation of this possible regulatory mechanism appears to have been made under physiological conditions in whole cells (cf., Chapter 5, Sections I.C.2 and III.B)

Similarly, although there are some reports quantitating the activating effects of divalent metal ions (mainly Mg^{2+})[163,350,671,772] over and above the amount required to complex with ATP^{4-} and of monovalent cations,[163,350,570,758] in general, the regulatory potential of these ions remains unexplored among prokaryotes.

III. PYRUVATE CARBOXYLASE IN FUNGI

Although somewhat fewer species are represented in this group (see Table 2), several (viz., *Saccharomyces cerevisiae*[304,514] and *Aspergillus*[111]) are among the longest known and most extensively studied examples of this enzyme.

A. *Saccharomyces cerevisiae*

1. *Effect of Growth Conditions*

a. *Carbon Source and Gas Phase*

Pyruvate carboxylase has been shown to be a constitutive component of yeast,[156,355,710] though the level of its activity and its actual contribution to anaplerotic carbon dioxide fixation can vary depending on the growth conditions.[619]

The highest activity was found in glucose-grown cells under anaerobic conditions (Table 2) and although only a slight level of aeration (1% oxygen in the sparging gas) was sufficient to decrease pyruvate carboxylase activity considerably, further increases in aeration intensity did not result in any specific trend in the activity of this enzyme.[356] In one nonrespiratory variant (CBS strain 2355), pyruvate carboxylase activity was low under both anaerobic and aerobic conditions,[356] whereas it had been shown previously[156] that the "petite" yeast[259] grown under aerobic conditions retained 60% of the pyruvate carboxylase activity of the wild type. A detailed genetic analysis of these two mutants should provide a relatively efficient approach to the controlling elements in the expression of this enzyme in *S. cerevisiae.*

Table 2
PYRUVATE CARBOXYLASE IN FUNGI

Organism	Substrate[a]	Specific activity[b]	Acetyl-CoA dependence[c]	Effect of aspartate[d]	Ref.
Saccharomyces cerevisiae	—[e]	8.3	—[g]	Ki = 1.9 m*M*	514, 627
	—[e]	2	66%	–	710
					204
	—[e]	44	50%	Ki = 1.4 m*M* 0 m*M* acetyl-CoA	157, 158
			Ka = 9 μM, 0 m*M* ASP Ka = 36 μM, 5 m*M* ASP Ka = 91 μM, 10 m*M* ASP	Ki = 3.4 m*M* 153 μM Acetyl-CoA	
S. cerevisiae wild-type "petite"	Glucose (110)	48	–	–	156
	(110)	31	–	–	
S. cerevisiae (Harden and Young strain)	Lactate (27)	70	–	–	996
S. cerevisiae (strain 306, Alko, Finland)	Glucose (280)[f]	116 (28°C)	–	–	619
	(280) + aspartate (28)[f]	19	–	–	
	Glucose (280)	56	–	–	
	(280) + aspartate (28)	28	–	–	
Rhodotorula glutinis	Glucose (110)	28	–	–	710
	Pyruvate (364)	18			
	Malate (144)	8			
	Aspartate (28)	9			
	Acetate (731)	11			
Hansenula anomala	Glucose (110)	20	–	–	710
	Acetate (731)	10			
Schizosaccharomyces pombe	Glucose (55)	15—40[i]	–	–	240
Candida tropicalis	Glucose (55)	20—50[i]	–	–	240
C. tropicalis	Glucose (80)	21	–	–	591
C. lipolytica	Glucose (80)	12	–	–	591

C. boidinii	Methanol (200)	40[j,k]	–	–	147
	(160)	70[j]—140[k]			
	+ $NaHCO_3$ (50)		–	–	
C. methylica	Methanol (160)	4	68%	–	508
Aspergillus niger	Sucrose (25)	41[m]	0	Ki = 1.9 m*M*	269
	(290)	4[m]			
A. nidulans	Sucrose (20)	100	–	–	795
	Acetate (100)	80	–	–	
A. nidulans	Glucose (20)	20	0[n]	Ki = 0.6 m*M*[p]	615
Penicillium camem-berti	Glucose (56)	1	50%	90% (5 m*M*)	818
P. isariiforme	Sucrose (90)	60	17%	–	330
Verticillium albo-atrum (ATCC	Glucose (28)	15	–	–	378
26289)	Glucose (28)	18	0	28% (5m*M*)	379
	Pyruvate (45)	19			
	Acetate (61)	13			
	Malate (30)	8			
	Xylose (33)	11			
	Glycerol (55)	11			
	Aspartate (28) + glucose (1)	16			
Fusarium culmorum	Glucose (194)	0.4	–	–	490
Helminthosporium cynodontis (ATCC 24938)	Glucose (110)	0.12	–	–	187
	Succinate (124)	0.13			
Neocosmospora vas-infecta	Glucose (166)	6 (25°C)	93%	–	141
	Acetate	14	87%	–	
Phycomyces blakes-leeanus	Glucose (166) + asparagine (22)	5.5 (25°C)	73% (Ka = 18 μ*M*)	Ki = 2 m*M*	723
Rhizopus nigricans	Glucose (275)	25	100% (Ka = 10 μ*M*)	–	620
Mucor rouxii	Glucose (110)	40 (28°C)	38%	80% (5 m*M*)	154
Agaricus bisporus (ATCC 382581) (Spores)	–	27[q]	55%	–	672

Table 2 (continued)
PYRUVATE CARBOXYLASE IN FUNGI

Note: For footnotes a to d and g and h, see Table 1.

[e] Commericaly available bakers' yeast.
[f] Anaerobic growth conditions.
[i] Specific activity varies during growth period.
[j] Assayed with 5 mM Mg^{2+}, 400 μM acetyl-CoA.
[k] Assayed with 3.3 mM Mn^{2+}, 400 μM acetyl-CoA.
[m] Assayed at room temperature.
[n] When assayed in absence of aspartate (or 2-oxoglutarate; see footnote "p").
[o] The letter "o" was skipped to avoid confusion with other symbols.
[p] 2-Oxoglutarate also inhibits (Ki = 5.5 mM) values for ASP and 2-OG measured in the absence of acetyl-CoA.
[q] Specific activity is nanomoles product formed per minute per milligram dry weight spores.

Batch-wise growth of yeast under aerobic conditions on glucose as the carbon source may be considered as a glucose-ethanol diauxie:[537] the second growth phase is based on the utilization of ethanol produced during the preceding fermentative growth phase. Pyruvate carboxylase activity remains constant during the glycolytic growth phase,[156,240,355] and then in some cases increases.[355] This increase in the latter phase, however, is somewhat at variance with the levels of pyruvate carboxylase activity present in yeast grown aerobically in chemostat culture with a particular carbon source as the growth-limiting factor. Oura et al.[619] found specific activities (micromoles oxaloacetate formed per minute per milligram protein) of 0.056 on 0.28 *M* glucose, 0.032 on 0.22 *M* glycerol, 0.033 on 0.45 *M* pyruvate, and 0.042 on 0.54 *M* ethanol. These results are generally consistent with the extent and pattern of $^{14}CO_2$ fixation into different chemical fractions of the yeast grown aerobically in a chemostat on these substrates.[619] The proportions of yeast carbon derived from CO_2 during growth on glycerol or glucose are similar, while the increased amounts of radioactivity found mainly in the carbohydrate (i.e., "neutral") fraction when pyruvate or ethanol was the substrate correlated closely with the marked increases in phosphoenolpyruvate (PEP) carboxykinase activity and the development of glyoxylate cycle function (as indicated by isocitrate lyase activity) on these latter two carbon sources.

Even more clear cut, however, are the contrasts between growth anaerobically on glucose (0.28 *M*) or on glucose (0.28 *M*) plus aspartate (0.028 *M*) and growth aerobically on glucose (0.28 *M*). Here the proportions of yeast carbon derived from CO_2 are 5.5, 1.6, and 2.6%, respectively, closely reflecting the relative specific activities (micromoles oxaloacetate formed per minute per milligram protein) of 0.115, 0.019, and 0.056, respectively, for pyruvate carboxylase in cells grown on these three media. The presence not only of aspartate,[355,619] but also of asparagine,[356] glutamate, or glutamine[354] in the growth medium reduces the activity of pyruvate carboxylase in yeast. The effect is maximal under anaerobiosis and has been shown not to be mediated at the level of holoenzyme formation from apoenzyme, since no free apoenzyme could be found in cells grown in the presence of aspartate in biotin-sufficient (i.e., $\geqslant$ 100 μg/ℓ) medium.[357] Although direct measurement of the amount of enzyme present has yet to be made by immunoprecipitation, it seems most unlikely that the reduction in pyruvate carboxylase activity in yeast grown on aspartate-containing media would be attributable to abnormally high levels of aspartate (or other metabolites) in the cell extracts, since these had been dialyzed prior to assaying for enzyme activity.[355] This suggests a direct role for aspartate (or its metabolites) in the repression of pyruvate carboxylase at the level of *de novo* synthesis. If this proves to be the case, it would be in direct contrast to the situation in *Pseudomonas citronellolis* for which it has been shown that the synthesis of pyruvate carboxylase is induced by the presence of lactate (or a metabolite of it) rather than repressed by the presence of aspartate (or a metabolite of it).[858]

However, it has been shown with cell-free extracts[839] that aspartate can inhibit the holoenzyme synthetase-catalyzed formation of yeast pyruvate carboxylase from the partially purified apoenzyme, (+)-biotin, and ATP, as has also been found under similar conditions for the carboxylase of *Bacillus stearothermophilus*.[161] If the concentration of aspartate required to inhibit the formation of holoenzyme from apoenzyme is the same as that required to inhibit the activity[158] of the holoenzyme, as is the case with pyruvate carboxylase in the thermophilic bacillus,[162,163,840] then one might expect some effect of this amino acid given that its normal intracellular concentration in yeast cells is 5 to 20 m*M*,[411] and that it was at least 30 m*M* in the medium used by Haarasilta et al.[357] Some possible explanations could include:

1. At high levels of biotin in the medium (i.e., $\geqslant$ 100 μg/ℓ, the rate of holoenzyme formation from apoenzyme is not rate limiting in the expression of pyruvate car-

boxylase activity in comparison with the rate of *de novo* synthesis of apoenzyme.
2. The process of *de novo* synthesis of pyruvate carboxylase apoenzyme is much more sensitive to inhibition by aspartate.
3. The intracellular concentration of aspartate is not readily influenced by the concentration of this amino acid in the medium.

Clearly, further work is required to fully elucidate the mechanism of aspartate regulation of pyruvate carboxylase expression in vivo.

b. Biotin Content of Medium

In yeast grown under conditions of biotin limitation, pyruvate carboxylase activity is low,[156,572,573] and the cells contain substantial proportions of the apoenzyme.[240,353,357,572] However, the biotin deficiency is not compensated for by an increased amount of pyruvate apocarboxylase.[357] Similar results were noted for pyruvate carboxylase in *Bacillus stearothermophilus* grown under biotin-deficient conditions.[161]

When yeast were grown aerobically in a chemostat[357] with glucose (0.22 *M*) as the carbon source, and at biotin concentrations of 10 μg/ℓ, there was a low but detectable activity of pyruvate carboxylase present. This activity rose rapidly (some five- to six-fold) and reached a maximum when the biotin content of the medium increased to 50 μg/ℓ. Conversely, the cellular content of apocarboxylase fell as the biotin concentration in the medium increased, until at 100 μg/ℓ no apoenzyme was detected (see Figure 2). At low biotin concentrations (2 μg/ℓ), the yeast were able to utilize only a part of the available glucose, and in spite of intense aeration their metabolism was fermentative. Indeed, only when the biotin concentration of the medium was above 10 μg/ℓ was all the glucose consumed and ethanol no longer produced. It seems likely, but remains to be proved, that the limiting biotin levels restricted the production by pyruvate carboxylase of oxaloacetate, the acceptor of acetyl-moieties in the tricarboxylic acid cycle, and hence directed the catabolism of glucose via pyruvate to ethanol, rather than to CO_2, despite the aerobic conditions.

Biotin may be readily incorporated into the apoenzyme to form the active holoenzyme in permeabilized cells *in situ*[353] and in cell-free extracts[572,839] when supplied with MgATP. The enzymatic basis for this was reviewed briefly in Section II.B.4 and is considered to be essentially identical for all biotin-containing enzymes, except that in the case of the microbial pyruvate carboxylases studied to date, biotinylation of the apoenzyme is affected by its interaction with the allosteric effectors acetyl-CoA and aspartate. Thus, the omission of acetyl-CoA (usually present at 500 μ*M*) from the pyruvate holocarboxylase synthesis incubation led to a 60% reduction in the formation of active enzyme, while inclusion of L-aspartate (1 m*M*), even in the presence of acetyl-CoA (500 μ*M*), resulted in a 50% inhibition of the synthesis of the holoenzyme.[839] Of the biotin analogs and precursors tested (viz., biocytin, biotin sulfone, biotin-D-sulfoxide, biotin-L-sulfoxide, dethiobiotin, diaminobiotin, and 7,8-diaminopelargonic acid), only biocytin was able to produce active pyruvate carboxylase when incubated with MgATP and apoenzyme-rich permeabilized yeast cells.[353] The slow rate achieved with biocytin suggested that this may have resulted from the hydrolysis of biocytin to biotin and lysine, since formation of biotinyl holoenzyme requires a free carboxyl group on the valeryl side chain of the biotin.[468]

When tested in the presence of biotin, DL-dethiobiotin, even at a concentration tenfold in excess of the biotin, had no effect on the amount of active pyruvate carboxylase formed, suggesting that this analog may not be bound by either the holoenzyme synthetase or the apoenzyme. However, as D-dethiobiotin is equivalent to biotin as a growth factor for bakers' yeast,[841] this analog must be able to be converted to biotin by enzymes in intact cells, thus enabling them to support active holocarboxylase synthesis.

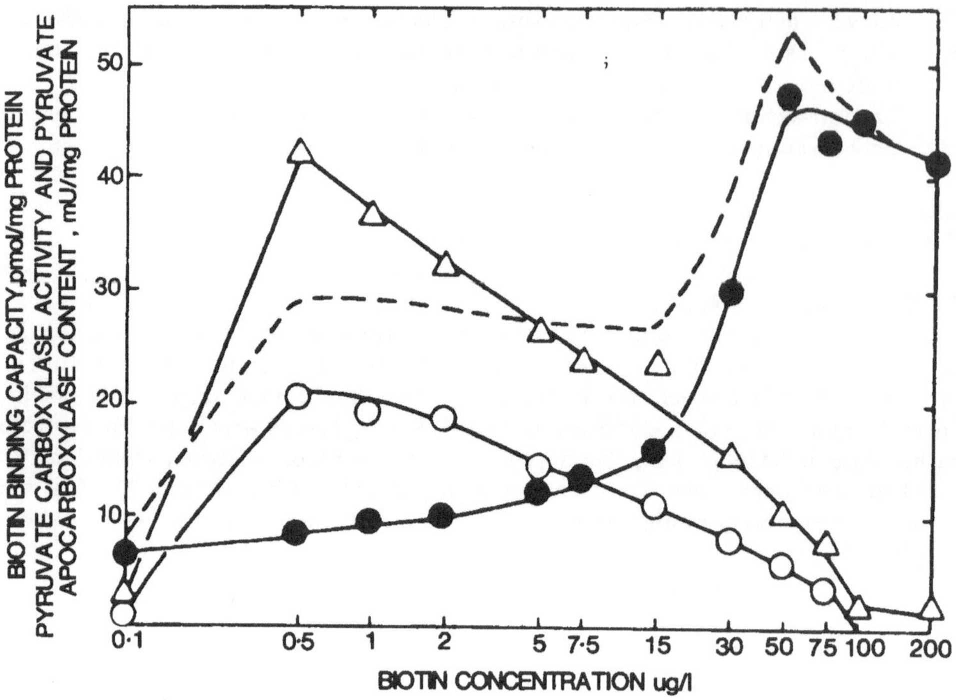

FIGURE 2. The effect of biotin concentration on the pyruvate carboxylase activity and pyruvate apocarboxylase content (expressed as units of active enzyme) of baker's yeast. The pyruvate carboxylase activity (●), apocarboxylase content (○), and biotin binding capacity (△) were assayed *in situ* in cells grown at different biotin concentrations in a chemostat under aerobic conditions with glucose as the carbon source. The total content of pyruvate carboxylase protein (the broken line) is calculated as the sum of the carboxylase activity and apocarboxylase content. (From Haarasilta, S., Oura, E., and Suomalainen, H., *Arch. Microbiol.*, 122, 121, 1979. With permission.)

Another exceptional aspect of the formation of pyruvate holocarboxylase is the almost complete lack of cross-reaction between the *Saccharomyces cerevisiae* and *Bacillus stearothermophilus* systems,[839] in contrast with the ready interchangeability reported for other carboxylase systems derived from different organisms.

2. Intracellular Location of Pyruvate Carboxylase

The weight of evidence favors an exclusively mitochondrial location for pyruvate carboxylase in most vertebrate tissues (see Section V.B). However, when Haarasilta and Taskinen[354,358] used sucrose density gradient centrifugation to fractionate yeast spheroplast lysates, they found 96% of the recovered pyruvate carboxylase activity in fractions of density less than 1.055 g/cm^3. Thus, pyruvate carboxylase was sedimented along with PEP carboxykinase [EC 4.1.1.32], hexosediphosphatase [EC 3.1.3.11], and isocitrate lyase [EC 4.1.3.1], and was clearly separated from the mitochondrial membrane marker enzyme, cytochrome *c*-oxidase, all of which was found in fractions of density 1.15 g/cm^3 or above. Although it is evident from the variation in the amount of sedimentable malate dehydrogenase and catalase in different fractionations (see Figure 3) that some breakage of mitochondria and microbodies has occurred, it is not possible to evaluate accurately the recovery of intact mitochondria and microbodies, since the proportion of malate dehydrogenase in mitchondria and of catalase in microbodies in yeast in vivo is not known. To this end it would have been of greater value to have used citrate synthase [EC 4.1.3.7] as a marker enzyme of the mitochondrial matrix since in all other situations that is its exclusive location. Malate dehydrogenase,

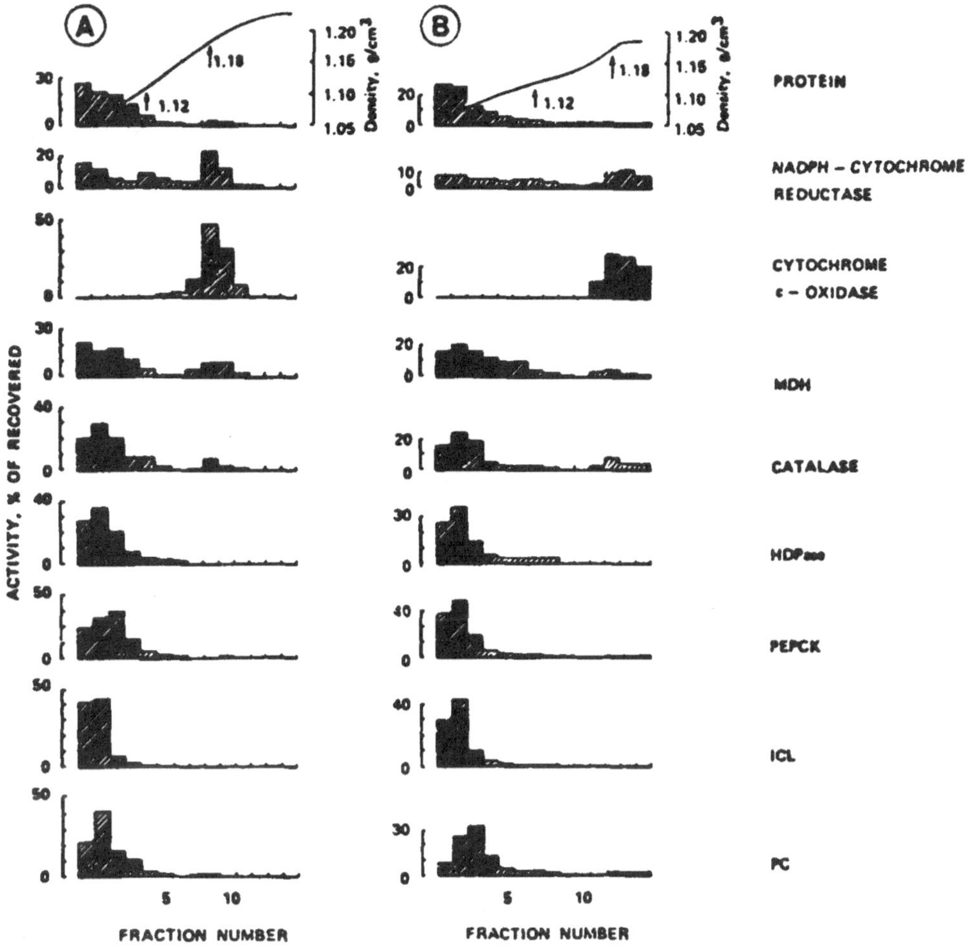

FIGURE 3. Distribution of protein and enzymes from yeast spheroplast lysates in sucrose density gradients. Spheroplast lysates in sucrose density gradients. Spheroplasts, prepared as described by Haarasilta and Taskinen,[358] were lysed in a Teflon-glass hand-homogenizer. Partially digested cells, intact spheroplasts, and nuclei were removed by centrifugation for 10,000 g × min. The supernatant, designated as the spheroplast lysate (2 m*l*, corresponding to 0.6 to 0.8 g of original wet yeast), was applied to a density gradient of 10 to 45% sucrose (w/w) in 25 m*M* Tris-HCl buffer (pH 7.2). The gradient was layered on top of a cushion (1.5 m*l*) of 60% (w/w) sucrose. After centrifugation at 4×10^6 g × min (A) or 9×10^6 g × min (B) at the sample zone in the Beckman SW25.1 rotor, 14 fractions of 2 m*l* were collected and analyzed as described by Haarasilta and Taskinen.[358] MDH, malate dehydrogenase; HDPase, hexose diphosphatase (EC 3.1.3.11); PEPCK, phosphoenolpyruvate carboxykinase; ICL, isocitrate lyase; PC, pyruvate carboxylase. (From Haarasilta, S., Oura, E., and Suomalainen, H., *Arch. Microbiol.*, 122, 121, 1979. With permission.)

on the other hand, is well known[64] to have two distinct isoenzymes, one in the mitochondrial matrix and the other in the cytosol. Nevertheless, had pyruvate carboxylase been entirely of mitochondrial origin it seems unlikely that only 3 to 4% of its total activity would remain within the mitochondria when at least 10% (Figure 3B) and up to 25% (Figure 3A) of the total (i.e., cytoplasmic plus mitochondrial) malate dehydrogenase activity was coincident with the cytochrome *c* oxidase activity. Hence, it seems that pyruvate carboxylase activity has a mainly cytosolic location in yeast. This must have important implications for the organization of lipid and amino acid biosynthesis in this organism, and raises the question of how the anaplerotic requirements of yeast mitochondria are fulfilled.

Furthermore, it is of interest to establish whether the 3 to 4% of pyruvate carboxylase activity that co-sediments with the mitochondria is

1. Merely an adventitious contaminant accounted for by the presence in those fractions of other typical cytosolic enzymes. Despite the relatively high sedimentation coefficient (16 to 17S) of pyruvate carboxylase,[768,856] it would seem, in view of the lack of PEP-carboxykinase,[358] hexosediphosphatase,[358] or isocitrate lyase[354] activity accompanying the carboxylase, that contamination is not the most likely explanation. However further detailed fractionations and the use of other cytosolic "marker" enzymes, preferably of high molecular weight, will be needed to substantiate this.
2. A residual amount of this enzyme that has bound specifically to the external surfaces of the outer membranes of the mitochondria in a manner analogous to that exhibited by the association of hexokinase with brain mitochondria.[971] Under other conditions, involving perhaps specific metabolic effectors, this "ambiquitous" behavior[972] may become more pronounced, and the proportion of pyruvate carboxylase bound to and hence co-sedimenting with the mitochondria might be increased, as has been observed for brain hexokinase.[973] Thus, increased levels of mitochondrially bound hexokinase have been found to result from conditions causing glycolytic stress.[454,455] Conversely, decreased levels of the bound form are found during anesthesia, a state in which glycolytic activity is depressed.[100] Other physiologically significant examples of this phenomenon of enzyme adsorption to the cellular substructure have been recognized.[535] It would not be surprising, in view of the close relationship between the anaplerotic role of pyruvate carboxylase and the activities of the citric acid cycle, and of the integral relationship of this enzyme with mitochondria in vertebrate tissues (see Section V.B), if yeast pyruvate carboxylase exhibited "ambiquitous" behavior in its intracellular distribution. This may well provide a sensitive mechanism for regulating the expression of the activity of this cytoplasmic enzyme in a manner attuned to the needs of the mitochondria.
3. Attributable to a distinct mitochondrial form of the enzyme, present in the mitochondrial matrix at only a fraction of the activity of the cytoplasmic form, and reminiscent of the situation for PEP carboxykinase in rat liver.[604]

A distinction between possibilities 2 and 3 should be readily achieved by utilizing immunochemical techniques, as was done for PEP carboxykinase,[53] and by fractionation of the mitochondria into outer membrane, intermembrane space, inner membrane/cristae, and matrix.[735] A distinct mitochondrial form of pyruvate carboxylase would be expected to be sedimented in the mitoplasts and be released with other matrix enzymes upon sonication of the mitoplasts. On the other hand, depending on the tightness of the binding of the enzyme to the outer membrane, it may either be sedimented with the membrane fragments or be released to appear, artifactually, with the marker enzymes of the intermembrane space. In neither case would it be sedimented with the mitoplasts, nor would it be readily distinguishable by immunochemical techniques from the unbound form in the cytosol. Genetic techniques should also be useful in differentiating between possibility 3 and the other two possibilities.

3. The Regulation of Yeast Pyruvate Carboxylase by Metabolic Effectors

Some details of the effects of acetyl-CoA and aspartate on the activity of yeast pyruvate carboxylase are summarized in Table 2. Discussion of the regulation of this enzyme by these and other effectors is presented in Chapter 5, Section III.B.

B. Pyruvate Carboxylase in Other Yeasts

As is evident from the data summarized in Table 2, pyruvate carboxylase activity has been found to occur in several other yeasts (viz., *Schizosaccharomyces pombe*,[240] *Rhodotorula glutinis*,[710] *Hansenula anomala*,[710] and four species of *Candida*[147,240,508,591]) at much the same level as in *Saccharomyces cerevisiae*. Growth on aspartate- or malate-containing media reduced the pyruvate carboxylase activity of *Rhodotorula glutinis*[710] as it did for *Saccharomyces cerevisiae*,[355,357,619] while inclusion of bicarbonate (50 m*M*) in the medium increased the activity of the enzyme two- to threefold in *Candida boidinii* growing on methanol.[147] In neither case was the molecular basis of this change in activity reported.

Genetic techniques have not yet been employed to demonstrate the essential function of this anaplerotic enzyme in these organisms, but in general, its activity, as a function of growth time, exhibits a pattern that is typical of an anabolic enzyme.[240] Furthermore, higher levels of activity are associated with an increased production of biomass.[240] Similarly, reduced levels of pyruvate carboxylase activity in *Candida tropicalis*, brought about by culturing this yeast on biotin-deficient, glucose-containing media, were also associated with a marked (55%) reduction in growth.[591] Biotin deficiency in *Candida tropicalis* also resulted in reduced activity of another biotin-containing enzyme, acetyl-CoA carboxylase, as well as a massive (8- to 13-fold) increase in the activity of isocitrate lyase, a key enzyme of the glyoxylate cycle. However, while the effect of biotin deficiency on cell growth and isocitrate lyase activity was evident in cultures utilizing the three hexoses tested (viz., glucose, fructose, and mannose), it was not seen in cells cultured on *n*-alkane. This contrast, together with the three-fold elevation of pyruvate concentration and a similar drop in oxaloacetate level in biotin-deficient cells, suggests an important role for the anaplerotic activity of pyruvate carboxylase vis-a-vis the activity of the glyoxylate cycle, an alternative source of C_4-compounds.

C. Pyruvate Carboxylase in *Aspergillus* Spp.

A pyruvate carboxylating system requiring ATP was first observed in *Aspergillus niger* by Woronick and Johnson,[985] and the biotin-containing enzyme responsible for this was subsequently shown[111] to be basically similar to that occurring in extracts of avian and beef liver,[439,898,899] except that, like the pyruvate carboxylase of *Pseudomonas*,[772] it did not require, nor was its activity stimulated by, acetyl-CoA. The activity of this enzyme is essential for citrate formation from sugars in *Aspergillus niger*,[392] but surprisingly, is not affected by the level of citrate in the medium.[269] However, when *Aspergillus niger* was grown in a high (0.29 *M*) sucrose medium (which should favor the accumulation of citrate,) there was only 10% of the pyruvate carboxylase activity that was found in mycelia cultured on low (0.025 *M*) sucrose media.[269] Furthermore, two other *Aspergilli*, *A. flavus* and *A. ustis*, which do not accumulate citrate, had a fairly high constant activity of pyruvate carboxylase at either concentration of sucrose.[269] Clearly, above some minimal value the level of pyruvate carboxylase activity is not directly related to the citrate accumulating ability of these various strains of *Aspergillus*.

Genetic analysis of the related fungus, *Aspergillus nidulans*, has revealed mutants (pyC A) lacking pyruvate carboxylase that are unable to grow on hexoses unless supplied with a source of C_4 tricarboxylic acid cycle intermediates.[549,795] Like the pyruvate carboxylase of *Aspergillus niger*,[111] the enzyme purified from *Aspergillus nidulans* appeared initially not to be activated by acetyl-CoA, but as is the case for a number of microbial pyruvate carboxylases, it was inhibited by aspartate (see Table 2). However, closer examination has revealed that acetyl-CoA and some of its longer-chain homologs do activate the pyruvate carboxylase of *Aspergillus nidulans* in the presence of either L-aspartate or 2-oxoglutarate, but not in their absence under normal assay conditions.[615] An apparently competitive relationship exists between the effects of acetyl-CoA and L-aspartate on catalysis by this enzyme, with each ligand increasing the concentration of the other required to achieve an effect. Further analysis of this situation has shown that partial activation by acetyl-CoA can be demonstrated at a pH on the acid side of the pH optimum, but only in the presence of nonsaturating concentrations of the substrates ATP and pyruvate.[617] Further discussion of the allosteric regulation of this pyruvate carboxylase from *Aspergillus nidulans* is presented in Chapters 3 and 5.

Another interesting recent report on the *Aspergillus nidulans* pyruvate carboxylase has been the observation that 98% of its activity is found in the cytosol.[618] Whereas pyruvate and 2-oxoglutarate dehydrogenases showed 99 to 100% latency, and citrate synthetase showed 15 to 20% latency when this parameter was assessed by assay in the presence or absence of 0.05% Non-Idet P40, pyruvate carboxylase and pyruvate kinase activities were fully expressed. Furthermore, while 95% of citrate synthetase and 100% of pyruvate, succinate, and 2-oxoglutarate dehydrogenase activities were sedimented by centrifuging at 12,000 g for 15 min, less than 1% of the pyruvate carboxylase and kinase activities were found in that pellet. These data demonstrate that pyruvate carboxylase is a cytosolic enzyme in *Aspergillus nidulans,* as is largely the case also in yeast.[358] A cytosolic location for pyruvate carboxylase has important implications for the organization of lipid and amino acid biosynthesis in these lower eukaryotes.

D. Other Fungal Pyruvate Carboxylases

As summarized in Table 2, pyruvate carboxylase activity has been demonstrated in a variety of other fungal species for which it appears to be the major CO_2 fixation enzyme present. Indeed, in some cases (viz., *Penicillium isariiforme,*[330] *Neocosmospora vasinfecta,*[141] and *Rhizopus nigricans*[620]), it has been shown to be solely responsible for this activity. Its function is clearly related to citric acid accumulation in *Penicillium isariiforme*[330] and is essential to fumaric acid production in *Rhizopus nigricans.*[620] In several species CO_2 fixation plays a role in the control of morphogenesis (e.g., *Penicillium isariiforme*[330] and *Mucor* sp.[83]) and of spore germination (e.g., *Agaricus bisporus*[672]), while it has also been proposed to be of nutritional importance in phytopathogenic fungi.[571] Since CO_2 is required for the growth of *V. albo-atrum* on glucose or glycerol, but not on acetate or succinate, and the concentration of CO_2 is generally considered higher than ambient in the vessels of higher plants, it seems likely that anaplerotic CO_2 fixation is significant in the pathogenic phase of vascular wilt caused by organisms such as this.[378] Even in those species (viz., *Fusarium culmorum,*[490] *H. cynodontis,*[187] and *Neocosmospora vasinfecta*[141]) for which CO_2 fixation has not yet been proven essential, there is nevertheless a substantial incorporation of carbon from $H^{14}CO_3^-$ into both nucleic acids and proteins. Not surprisingly, most of the radioactivity incorporated by *H. cynodontis* was found in aspartyl residues (and to a lesser degree in glutamyl and cysteinyl residues) as well as in the pyrimidines, cytidine and uridine, for part of which aspartate is a precursor.

The intracellular location of pyruvate carboxylase in these organisms is not clear; in some (viz., *V. albo-atrum*[378]) it has been found almost entirely in the cytosol, as indeed appears to be the case also in *Neocosmospora vasinfecta.*[141] However, an unspecified proportion of the citrate synthase activity was also found in the post-mitochondrial supernatant in this latter species,[141] so that a precise definition of location must await further studies.

There is considerable variation among these organisms as to the extent of dependence of their pyruvate carboxylases on acetyl-CoA for activity (see Table 2). Some show complete (viz., *Rhizopus nigricans*[620]) or nearly complete dependence (viz., *N. vasinfecta*[141] and *Phycomyces blakesleeanus*[723]), others only partial dependence (viz., *Aspergillus bisporus,*[672] *Phycomyces camembertii,*[818] *Phycomyces isariiforme,*[330] *H.*

cynodontis,[187] and *M. rouxii*[154]), and yet others are unaffected by this ligand (viz., *F. culmorum*[490] and *V. albo-atrum*[379]) — at least in the absence of aspartate. Both coenzyme A and acetyl-CoA are effective in activating pyruvate carboxylase in *Neocosmospora infecta,*[141] and in reversing the inhibitory effect of aspartate on the pyruvate carboxylase of *V. albo-atrum.*[379]

Wherever it has been tested among these fungi (viz., *P. camembertii,*[723] *V. albo-atrum,*[137] *Ph. blakesleeanus,*[723] and *M. rouxii*[154]) aspartate has been shown to be an inhibitor of pyruvate carboxylase. However, much remains to be investigated with regard to the control of the activity of this enzyme by aspartate and other potential regulatory effectors (e.g., NADP and 2,3-diphosphoglycerate[379]). In particular, their significance to the metabolic regulation of this enzyme will only become clear when their concentrations in vivo in the appropriate cellular compartment are also known. The partially purified pyruvate carboxylase of *P. camembertii* has been found to show an almost linear dependence on the energy charge of the adenylate pool.[818]

Similarly, the effect of the carbon source in the growth medium on the level of pyruvate carboxylase activity in these fungi has not received much attention as yet. Growth on acetate resulted in 1.4 to 3 times as much measurable pyruvate carboxylase activity in extracts of *N. vasinfecta* compared with its growth on glucose, but the rate of $^{14}CO_2$ fixation in vivo by the acetate-grown mycelia was only about a quarter that of the glucose-grown mycelia, apparently because of the limited availability of pyruvate in the acetate-grown mycelia.[141]

Conversely, pyruvate carboxylase was produced at high levels in *R. nigricans* during the period of high glucose utilization preceding fumarate accumulation.[620] However, the levels of pyruvate carboxylase in *V. albo-atrum* grown on glucose, acetate, malate, xylose, glycerol, or aspartate differed only slightly.[379] In *Ph. blakesleeanus,* growth on a glucose-asparagine medium in the light led to approximately twice the specific activity as growth in the dark on the same medium.[724]

IV. PYRUVATE CARBOXYLASE IN INVERTEBRATES

A. Protozoa

1. Marine Dinoflagellates

While the reactions of the Calvin cycle, particularly ribulose bisphosphate carboxylase (EC 4.1.1.39), are responsible for the majority of net CO_2 fixation in plants and marine phytoplankton, there is nevertheless a significant incorporation of $^{14}CO_2$ directly into oxaloacetate, especially in those species having the C_4 pathway.[382] This CO_2 fixation is mostly attributable to PEP carboxylase (EC 4.1.1.31), which is ubiquitous among higher plants, although the levels of this enzyme in C_4 plants is as much as 15-fold higher than in C_3 plants.[875] In some marine diatoms, however, it has been shown that an ADP-dependent PEP carboxykinase (EC 4.1.1.49) is responsible for this β-carboxylation activity.[412,478]

As yet there have not been any reports of pyruvate carboxylase in plants, but recently, in a survey of some ten marine phytoplankter, there has been a well-documented report of this enzyme in two dinoflagellates, (viz., *Amphidinium carterae* and *Gymnodinium* sp.[17]). This pyruvate carboxylase, which resembles those of *Ps. citronellolis, Ps. fluorescens,* and *Azotobacter vinelandii* in being unaffected by both acetyl-CoA and aspartate (see Table 1), is the only β-carboxylation enzyme (cf., PEP carboxylase and PEP carboxykinase) present in these species (see Table 3), and is present at an approximately constant specific activity throughout the growth period.

2. Parasitic Protozoa

The fixation of carbon dioxide by these organisms is well established, but the failure to detect in long-term cultures of *Trypanosoma mega* and *T.b. brucei* any significant

Table 3
PYRUVATE CARBOXYLASE AND SOME OTHER ENZYMES OF β-CARBOXYLATION IN INVERTEBRATES

Organism	Pyruvate carboxylase	PEP Carboxykinase[a,b]	Malic enzyme[c]	Ref.
Activities in nmol $^{14}CO_2$ fixed/min/mg Protein[d]				
Protozoa				
Marine dinoflagellates				
Amphidinium carterae	6	0[e]	—[f]	17
Gymnodinium sp.	5	0	—	17
Trypanosomatids				
Trypanosoma cruzi	0	46[a] (30°C)	4	155
T. mega	2	12[a]	8	445
T. brucei brucei				
Cultured	0	12[a]	36	445
Bloodstream	0	1[a]	2	445
Crithidia fasciculata	1	16[a]	4	445
C. fasciculata	45	45[a]	0	43
Leishmania	3	0	—	94
braziliensis	1	—	0	533
L. donovani	7	0	—	94
	3	—	0	533
L. mexicana	4	—	—	533
L. tropica	2	—	—	533
Helminths				
Moniezia expansa	9	86[b]	9	470
Mollusks				
Bivalves				
Mytilus edulis				
Posterior adductor muscle	2	128[b]	0.1	233
Mantle	0.6	48[b]	<0.1	233
Hepatopancreas	0.9	37[b]	<0.1	233
Placopecten magellanicus				
Muscle	0	49[b] (10°C)	21	613
Mantle	0	37[b]	17	613
Cephalopods				
Symplectoteuthis oualaniensis				
Mantle muscle	1.5[m]	0.6[b,m]	1.5[m]	407
Activities in μmol/min/g fresh wt. tissue[d]				
Arthropods				
Crustacea				
Crab				
Cancer magister				406
Hypodermis				
Intermolt	+g	0.15[b](20°C)	—	
Freshmolt	+	0.23[b]	—	
Gill				
Intermolt	+	0.08[b]	—	
Freshmolt	+	0.12[b]		

Table 3 (continued)
PYRUVATE CARBOXYLASE AND SOME OTHER ENZYMES OF β-CARBOXYLATION IN INVERTEBRATES

Organism	Pyruvate carboxylase	PEP Carboxy-kinase[a,b]	Malic enzyme[c]	Ref.
Carcinus maenus				315
hepatopancreas	7.4[h]	–	–	
Paralithodes camtschatica				90
skeletal muscle	15.2	8.8[b]	–	
Crayfish				
Astacus leptodactylus (intermolt, male)				
Antennal gland	0.186 ± 28(3)	0.297 ± 63(3)[b]	0.086 ± 8 (3)	912
Abdominal muscle	0.045 ± 0.008	0.005 ± 0.002[b]	0.017 ± 0.002	912
Heart	0.050 ± 0.011	0.025 ± 0.003[b]	0.131 ± 0.015	912
Hepatopancreas	0.064 ± 0.013	0.156 ± 0.014[b]	0.035 ± 0.004	912
Insects				
Locust (*Locusta migratoris*)	1.7[i]	<0.1[b]	–	211
(*Schistocerca gregaria*)				
Flight muscle	1.5	<0.1[b]	–	211
Hind leg muscle	<0.2	0.6[b]	–	211
Fat body				
Mitochondrial	0.03—0.300[j]	–[f]	–	211
Cytoplasmic	0.066—0.006[j]	–	0.212—0.028[j]	926
Cockroach (*Periplaneta americana*)				
Flight muscle	0.6[i]	<0.1[b]	–	211
Coxal muscle	<0.2	0.9[b]		
Waterbug (*Lethocerus cordofanus*)				
Flight muscle	2.1	<0.1[b]	–	211
Foreleg femoral	<0.2	0.7[b]	–	
Wasp (*Vespa vulgaris*)[k]	12	<0.1[b]	–	211
Honeybee				
Apis mellifera[k]	4.8	<0.1	–	211
Bumble bee				
Bombus pratorum[k]	19	<0.1[b]	–	211
Bombus agrorum[k]	17	<0.1[b]	–	211
Queen bumblebee				
Bombus terrestris[k]	12	–	–	211
Cuckoo bee				
Psithyrus vestalis[k]	18	–	–	211
Buff-tip moth				
Phalera bucephala[k]	5.2	<0.1[b]	–	211
Herald moth				
Scoliopterix libatrix[k]	4.7	–	–	211
Yellow underwing moth				
Noctua pronuba[k]	2.3	<0.1[b]	–	211
Blowfly				
Calliphora vomitoria[k]	10	<0.1[b]	–	211

[a] Values with this superscript are for the activity of the ADP-dependent form of this enzyme (EC 4.1.1.49).

[b] Values with this superscript are for the GDP- or IDP-dependent form of this enzyme (EC 4.1.1.32).

[c] L(−)Malate:NADP oxidoreductase (decarboxylating) EC 1.1.1.40.

[d] Enzyme activities were determined at 25°C unless otherwise indicated, in which case that temperature applies to all other values attributed to the same reference.

[e] Values of "O" indicate no measurable activity.

[f] The entry "–" indicates the value for this activity was not determined in that particular study.

[g] The entry "+" means that authors indicate that this activity was demonstrable but not quantifiable due to contaminating enzyme activities.

Table 3 (continued)
PYRUVATE CARBOXYLASE AND SOME OTHER ENZYMES OF β-CARBOXYLATION IN INVERTEBRATES

[h] Value at saturating concentrations of substrates; value is 2.6 at in vivo concentration of substrates.
[i] Authors comment that "since a systematic study of factors such as season, diet, size, age, and sex of animals was not attempted in this work, precise quantitative considerations based on the results must be made with caution".
[j] Value depends on age after the final ecdysis.
[k] Activities are for flight muscle.
[m] Activity is given as micro moles of product per minute per gram fresh weight of tissue at 25°C.

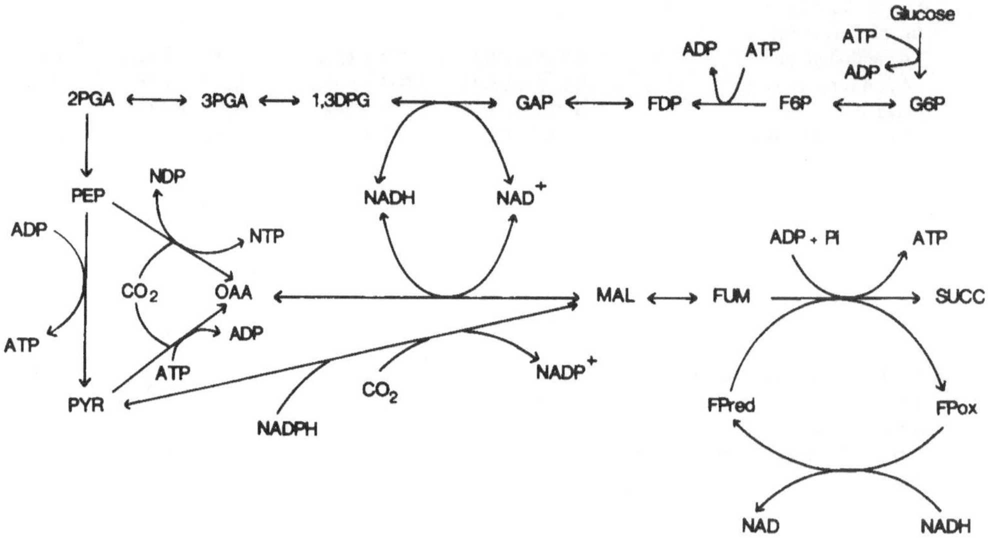

FIGURE 4. Some aspects of the pathways of anaerobic intermediary metabolism involving CO_2-fixation in invertebrate facultative anaerobes. *N.B.* As drawn, this scheme is not in redox-balance, nor does it account for the formation of alanine, propionate, or other end products of metabolism in these organisms. For a comprehensive review of this topic, see de Zwaan et al.[232] G6P, glucose-6-phosphate; F6P, fructose-6-phosphate; FDP, fructose 1,6 diphosphate; GAP, glyceraldehyde-3-phosphate; 1,3DPG, 1,3-diphosphoglycerate; 3PGA, 3-phosphoglycerate; 2PGA, 2-phosphoglycerate; PEP, phosphoenolpyruvate; PYR, pyruvate; OAA, oxaloacetate; MAL, malate; FUM, fumarate; SUCC, succinate.

amount of ^{14}C-label in amino acids compared with the overwhelming proportion (≥90%) present in succinate[445] suggests that the function of this process is less important in these organisms in a familiar anaplerotic pathway than as a basis for the regeneration of NAD for glycolysis.

Thus, as indicated in Figure 4, the formation of oxaloacetate from either PEP or pyruvate provides a means, in the presence of an active malate dehydrogenase, for reoxidizing NADH arising from glyceraldehyde-3-phosphate dehydrogenase (EC 1.2.1.12) in the glycolytic pathway. As these organisms are dependent on glucose as a substrate but lack the enzyme lactate dehydrogenase (EC 1.1.1.27),[266,533] the importance of this pathway is clear, particularly under anaerobic conditions. Excretion of large amounts of succinate is characteristic of anaerobically grown trypanosomes.[120,714] This pathway (Figure 4) of succinate formation via the carboxylation of PEP or pyruvate and the utilization of NADH for the subsequent reductions of both oxaloacetate and fumarate enables these organisms to recycle their pyridine nucleotides for further glycolysis.

As summarized in Table 3, there is surprising diversity even among the species examined so far in the possible means of effecting this CO_2 fixation. In some, e.g., *T. cruzi* and cultured *T.b. brucei,* there is a high level of activity of an ADP-dependent PEP carboxykinase (EC 4.1.1.49) accompanied by a variable level of malic enzyme (EC 1.1.1.40) activity, and no evidence for the presence of pyruvate carboxylase.[155,445] Although PEP carboxykinase is usually considered to function mainly in the decarboxylation of oxaloacetate to PEP, thereby beginning the gluconeogenic pathway from citric acid cycle intermediates,[874,903] its role as a CO_2-fixing enzyme has been proposed in invertebrates, such as mollusks[792] and helminths,[408] which share with trypanosomatids the capacity to ferment glucose to succinate. In others (see Table 3), e.g., *T. mega* and *Crithidia fasciculata,* there is pyruvate carboxylase activity present also in various proportions to the PEP carboxykinase activity,[43,445] while in *Leishmania* sp. it appears that pyruvate carboxylase is principally responsible for CO_2 fixation.[94,533] In those species having significant activities of more than one CO_2-fixing enzyme, it may be misleading to take the specific activities determined in in vitro assays as necessarily indicating their relative importance in vivo, without also taking into account the intracellular concentrations of relevant metabolites. The development and selection of mutant strains of these organisms would seem to be an attractive approach in analyzing this situation. Certainly the naturally occurring bloodstream form of *T.b. brucei* S42, which lacks carboxylating activity, has long been known to excrete large amounts of pyruvate,[713] and to reoxidize glycolytic NADH by a cytoplasmic α-glycerophosphate oxidase,[331] rather than use succinate excretion as an NAD-regenerating mechanism.

No details of any properties of these forms of pyruvate carboxylase have been reported as yet.

B. Helminths

In helminths, fixation of CO_2 and the ultimate appearance of carbon from this source in excreted succinate and volatile acids appears to be a rather general phenomenon.[135,408,731,792,915] As described in Section IV.A.2 and Figure 4, the importance of this pathway under anaerobic conditions is clear. The pathway of succinate formation via carboxylation provides an alternative means of reoxidizing NADH generated in the oxidation of glyceraldehyde-3-phosphate during anaerobic glycolysis. Whether a particular species of this class of organism excretes lactate or succinate under anaerobic conditions appears to depend on its relative activities of pyruvate kinase (EC 2.7.1.40) and PEP carboxykinase (EC 4.1.1.32/49).[142,915]

Pyruvate carboxylase activity has been demonstrated in the helminth, *Monezia expansa,* though its level was only a little more than 10% of the PEP carboxykinase activity.[135] This latter enzyme is present mainly in the cytosol in all helminths studied so far, and appears to be the major CO_2-fixing enzyme in such organisms. No pyruvate carboxylase activity could be detected in *Fasciola hepatica*[660] or in *Hymenolepis diminuta.*[659] The functional significance of CO_2 fixation and the regulation of respiratory metabolic pathways in parasitic helminths have been reviewed by Bryant.[136]

C. Mollusks

1. Bivalves

In bivalves, CO_2 fixation is involved in two major pathways: glycogen synthesis from pyruvate and related 3-carbon compounds, and anaerobic glycolysis when this stored glycogen is utilized during shell closure.[234] In a survey of nine marine mollusks, Simpson and Awapara[791] found that the most widely distributed and most active carboxylating enzyme was PEP carboxykinase. This enzyme was present in most of these species at high levels even in comparison with rat liver, while in one (viz., *Rangia cuneata*) it occurred at over ten times the level found in chicken liver. A role for PEP carboxykinase in glycolytic fermentation in bivalves is suggested by this enzyme being present at its highest specific activities in glycolytic tissue (viz., adductor muscle).[791] However, none of these organisms exhibited any pyruvate carboxylase activity, and only one had some malic enzyme activity. Similar results for other bivalve species have also been reported.[180,613]

Nevertheless, all three of those enzymes are present in the sea mussel, *Mytilus edulis*[233] (see Table 3). In this organism, pyruvate carboxylase activity is found exclusively in the mitochondrial fraction of the three tissues assayed (viz., the adductor muscle, the mantle, and the hepatopancreas), but is much lower (1/40 or less) than the activity of PEP carboxykinase, which is essentially cytoplasmic in its occurrence. For reasons of redox balance it is necessary for oxalacetate to be available to the cytoplasmic malate dehydrogenase. Thus, the involvement of pyruvate carboxylase in the formation of succinate from glycogen under anaerobic conditions seems quite unlikely.[232]

Whether the great disparity in activity indicates that the role of pyruvate carboxylase is mainly in gluconeogenesis, whereas PEP carboxykinase may function bidirectionally, i.e., it may participate both in succinate production during anaerobic glycolysis and in gluconeogenesis, as appropriate to the needs of the cell, remains to be established. The wide range of tissue types in which pyruvate carboxylase is found is consistent with the fact that the whole body of the sea mussel is involved in the accumulation of glycogen.[235]

2. Cephalopods

The mantle muscle of the fast-swimming squid, *Symplectoteuthis oualaniensis,* utilizes stored polysaccharide in an obligatorily aerobic manner to support both its sustained and burst activities.[407] The presence of pyruvate carboxylase activity in this tissue (Table 3) suggests a possible anaplerotic role for this enzyme during the transition from low to high respiratory activity. However, other potential sources of oxaloacetate for sparking the citric acid cycle include aspartate and the exceptionally high concentrations (about 150 m*M)* of proline in squid mantle. In common with other mollusk tissues that do not produce lactate during anaerobic stress, the squid's mantle muscle does accumulate succinate, and to a lesser extent, malate, albeit rather slowly. It would appear, however, that this tissue has a relatively limited potential for anaerobic metabolism in comparison with its considerable aerobic work capacity. This suggests the modest PEP carboxykinase activity in this muscle, along with the more or less comparable level of pyruvate carboxylase activity, may participate in glyconeogenesis from amino acids arising from the protein content of this predator's diet.

D. Arthropods

Gluconeogenesis in arthropods functions to supply hexose units for glycogen and chitin synthesis, as well as precursors for blood glucose, trehalose, and mannose. Although it follows basically the same pathway as in mammals, it occurs in a wider range of tissues, hepatopancreas, gill, fat body, epidermis, and muscle.[406]

1. Crustacea

Pyruvate carboxylase has been detected in several crustacean tissues (Table 3), but as yet has not been widely quantitated nor characterized in terms of regulatory or other molecular properties. However, the activity of *Carcinus* pyruvate carboxylase has been shown to be markedly dependent on the concentration of acetyl-CoA present, with an approximate Ka of 45 μM.[315]

In the various tissues of the crayfish, *Astacus leptodactylus,* the activities of pyruvate carboxylase, PEP carboxykinase, and malic enzyme are of a similar order, and individually are equal to only about 10% of the citrate synthase activities of the tissue and 1 to 2% of their pyruvate kinase activities.[912] These results suggest that CO_2 fixation is of secondary importance to glycolysis and the citric acid cycle, and that in this organism glycolyzable substrates are metabolized via PEP to pyruvate for subsequent oxidation in the citric acid cycle via acetyl-CoA.

Nevertheless, the demonstration of significant activities of both pyruvate carboxylase and PEP carboxykinase in the antennal gland and hepatopancreas of *A. leptodactylus* suggests that gluconeogenesis may be active in these tissues.

There is considerable evidence available to indicate that CO_2 fixation plays a significant role in pyruvate metabolism in lobster nerve.[174] Although isotope incorporation data clearly implicate pyruvate carboxylase in this CO_2-fixation activity, there appear not to have been any specific assays to determine the activity of this enzyme directly, nor any study of its properties in relation to the specific metabolic requirements of this tissue.

2. Insects

Pyruvate carboxylase is present in the flight muscle of insects from the orders of Hemiptera, Hymenoptera, Lepidoptera, Diptera and Orthoptera,[211] and in the fat body of the adult male desert locust, *Schistocerca gregaria*.[927] All insect flight muscles so far investigated possess pyruvate carboxylase, but none exhibits any PEP carboxykinase activity (see Table 3).[211] Conversely, the reverse is true of some other insect muscles not involved in flight.

Clearly, this distribution of these enzymes is not consistent with them having a role in carbohydrate synthesis from lactate and other substrates, as appears to be the case in some crustacean species.[90] However, it has been suggested that pyruvate carboxylase may provide a mechanism for increasing the levels of tricarboxylic acid cycle intermediates, and hence permit a higher (as much as 50-fold) rate of ATP synthesis when the insect takes flight.[211]

The role of pyruvate carboxylase in the insect fat body would seem to be analogous to that in vertebrate liver, where it participates in both gluconeogenesis and lipogenesis. It has been suggested[927] that the pyruvate carboxylase activity found in the cytoplasm of the fat body cells may play a part in providing oxaloacetate for lipogenesis early in the life of the locust, whereas the mitochondrial pyruvate carboxylase, which has its maximum activity later at a time when the locust is utilizing the lipid reserves in its fat body, may be involved in gluconeogenesis. However, further detailed studies are required to establish whether there are actually two different forms of this enzyme, and if so, what factors control their occurrence and functions.

The properties of the locust flight muscle pyruvate carboxylase, which is mainly (at least 80 to 90%) mitochondrial,[211] are better known[708] and the function and regulation of this enzyme are discussed in Chapter 5, Section III.A.

V. PYRUVATE CARBOXYLASE IN VERTEBRATES

A. Some Reported Occurrences of Pyruvate Carboxylase in the Tissues of Vertebrates

1. Quantitative Estimates

a. Introduction

Summarized in Table 4 are the results of numerous studies on the activity of pyruvate carboxylase in various tissues (mainly liver) of a wide range of vertebrate species. Although these values of enzymic activity have been determined in vitro under V_{max} conditions that certainly do not occur in vivo, nevertheless, they do provide an indirect measure of the concentration of pyruvate carboxylase in the various tissues and of its potential activity therein. No attempt has been made to express the results in Table 4 in a single uniform mode (e.g., micro moles oxaloacetate formed per minute per gram wet weight of tissue) corrected to an arbitary temperature, since in most cases there was insufficient specific information available to permit this to be done with an accuracy commensurate with the precision of the various authors' determination of en-

Table 4
PYRUVATE CARBOXYLASE IN VERTEBRATES

Species	Condition[a]	Tissue	Activity[b]	Ref.
Lamprey				
(*Mordacia mordax*)	Spawning migration (fasted)	Liver[m]	0.48 ± 0.1 (6) (25°C)	650
(*Lampetra fluviatilis*)	Spawning migration (fasted)	Liver	0.12 (30°C)	730
		Skeletal muscle	0	
		Heart	0	
Loach (*Misgurnus fossilis*)	Fully grown	Oocyte	1.9 ± 0.18[c] (25°C)	260
	2—20 hr	embryo	1.8 ± 0.2[c]	
Channel catfish (*Ictalurus punctatus*)	Lipid-free diet (22 weeks)			
	+ Biotin (8ppm)	Liver	74 ± 4[d] (15) (30°C)	691
	− Biotin		63 ± 10[d] (15)	
	+ Avidin		59 ± 1[d] (15)	
	8% Lipid diet (22 weeks)			
	+ Biotin (8ppm)	Liver	75 ± 6[d] (15)	
	− Biotin		48 ± 5[d] (15)	
	+ Avidin		52 ± 9[d] (15)	
Eel				
(*Anquilla australis occidentalis*)	Fasted (3—4 weeks)	Liver[m]	0.52 ± 0.06 (6) (25°C)	650
(*Anguilla japonica*)	Fasted (3—4 days)	Liver[m]	5.5 ± 2.5 (3) (37°C)	384
Plaice				
(*Pleuronectes platessa*)	Fed	Liver[m]	0.19 ± 0.02 (6) (15°C)	574
		Red muscle	0.005 ± 0.001 (6)	
		White muscle	0	
	Fasted (4 months)	Liver[m]	0.12 ± 0.03 (4) (15°C)	574
		Red muscle	0.01 (4)	
		White muscle	0	
Flathead (*Platycephalus bassensis*)	Fed	Liver[m]	0.109 ± 0.027 (9) (20°C)	577
		Red muscle	0.012 ± 0.001 (7)	
		White muscle	Trace (2)	

Table 4 (continued)
PYRUVATE CARBOXYLASE IN VERTEBRATES

Species	Condition[a]	Tissue	Activity[b]	Ref.
Snapper (*Chrysophrys auratus*)	Fed	Liver[m]	0.095 (2)	577
		Red muscle	0.016 (2)	577
		White muscle	0.001 (2)	577
Warehou (*Seriolella brama*)	Fed	Liver	0.082 (1) (20°C)	577
Pilchard (*Sardinops neopilchardus*)	Fed	Liver	0.25 (1) (20°C)	577
		Red muscle	1.42 (1)	577
		White muscle	0 (1)	577
Rainbow trout (*Salmo gairdneri*)	200 g body wt fasted 18 hr	Liver	0.79 ± 0.03 (6) (15°C)	209
		Kidney	0.11 ± 0.01 (6)	
		Gill	0.04 ± 0.01 (6)	
		Muscle	0.01 ± 0.001 (6)	
	Fed	Liver[m]	0.94 ± 0.13 (4) (25°C)	650
Brook trout (*Salvelinus fontinalis*)	Basal diet	Liver	182 ± 3[d] (4) (30°C)	657
	+ biotin	Liver	217 ± 12[d] (4)	657
	+ fat	Liver	162 ± 6[d] (4)	657
	+ biotin + fat	Liver	222 ± 5[d] (4)	657
Frog (*Rana catesbeiana*)		Gastric mucosa[m]	9.3[e] (30°C)	164
Toad (*Bufo marinus*)	Fasted	Gastric mucosa[m]	0.09 ± 0.02 (8) (30°C)	279
	Fed	Liver	0[f]	650
Axolotl (*Siredon mexicanum*)	Fed	Liver[m]	0.16 ± 0.02 (4) (25°C)	650
Lizard (*Trachydosaurus rugosus*)	Fed	Liver[m]	0.43 ± 0.01 (3) (25°C)	650

Chicken				
(*Gallus domesticus*)				
White leghorn — De Kalb strain	Embryo (9 days)	Liver[m]	30 ± 3 (4) (25°C)	393
	(12 days)	Liver[m]	44 ± 2 (4)	
	(17 days)	Liver[m]	45 ± 1 (11)	
	(21 days)	Liver[m]	38 ± 2 (7)	
	(17 days)	Kidney	5.0 (25°C)	393
	(17 days)	Adipose (s.c.)	0.42	
	(17 days)	Brain	0.50	
	(17 days)	Duodenum	0.26	
	(17 days)	Heart	0.20	
	(17 days)	Yolk sac	0.38	
	(17 days)	Skeletal muscle	0	
New Hampshire × Columbian	Embryo (20 days)	Liver[m]	15 ± 1 (8) (37°C)	777
	Chick (4 weeks)	Liver[m]	9.5 ± 0.4 (4)	
	Embryo (20 days)	Kidney[m]	1.7 ± 0.4 (16)	
	Chick (4 weeks)	Kidney[m]	4.8 ± 0.4 (8)	
	Embryo (12—20 days)	Kidney	(0.6 ± 1.2) (37°C)	775
	Chick (2—58 days)	Kidney	(3.6 ± 4.0)	
	Chick (6 weeks, fed)	Liver	7.4 ± 0.6 (4) (37°C)	776
	(fasted 2 days)	Liver	7.2 ± 0.7 (4)	
	(fasted 4 days)	Liver	7.1 ± 0.2 (4)	
	(fasted 4 days + refed 4 days)	Liver	4.8 ± 0.2 (4)	
	Chick (6 weeks, fed)	Kidney	3.0 ± 0.2 (4) (37°C)	776
	(fasted 2 days)	Kidney	2.8 ± 0.2 (4)	
	(fasted 4 days)	Kidney	4.2 ± 0.2 (4)	
	(fasted 4 days + refed 4 days)	Kidney	2.6 ± 0.2 (4)	
Chicken	Female (4 weeks, fed)	Liver	6.2 ± 0.7 (5) (37°C)	418
	(fasted 6 hr)	Liver	6.6 ± 1.3 (4)	
	(fasted 12 hr)	Liver	8.1 ± 0.8 (8)	
	(fasted 18 hr)	Liver	7.7 ± 0.5 (8)	

Table 4 (continued)
PYRUVATE CARBOXYLASE IN VERTEBRATES

Species	Condition[a]	Tissue	Activity[b]	Ref.
Marshall strain	Chicks (14 days)	Blood	13.2 ± 1.19 (10) (38°C)	951
	(28 days)	Blood	10.3 ± 0.99 (10)	
	(42 days)	Blood	9.4 ± 0.59 (5)	
			9.6 ± 0.79 (5)	
Turkey				
Broad-breasted white strain	Poults (21 days)	Liver	19.0 ± 1.8 (4) (30°C)	40
B.U.T. triple 6 strain	Male poults (14 days)	Blood	12.5 ± 1.69 (7) (38°C)	951
	Female poults (14 days)	Blood	12.5 ± 1.59 (3)	
	Male poults (28 days)	Blood	9.0 ± 0.79 (3)	
	Female poults (28 days)	Blood	8.3 ± 0.79 (3)	
Pigeon	3—6 month old (fed)	Liver	7.6 ± 1.3 (5) (30°C)	805
	(starved 92 hr)	Liver	11.4 ± 1.3 (5)	
Rat	Male (200 g body wt)	Liver[m]	15.5 ± 2.1 (5) (37°C)	57
(Hooded Wistar)		Kidney[m]	14.4 ± 1.0 (7)	
		Brain	0.88 ± 0.18 (4)	
		White adipose[m]	1.92 ± 0.29 (5)	
	New born	Brown adipose[m]	4.06 ± 0.59 (5)	
	Male (200 g body wt)	Adrenal gland	5.30 ± 1.30 (4)	
	Female (lactating 2 days)	Mammary gland[m]	5.00 ± 0.51 (8)	
	Male (200 g body wt)	Intestinal mucosa	0.14 ± 0.05 (4)	
		Testis	0.61 ± 0.04 (4)	
		Heart	0.59 ± 0.09 (4)	
		Skeletal muscle	0.12 ± 0.02 (4)	
(Wistar II and Sprague Dawley)	Fed (250—300 g body wt)	Liver[m]	11.1 ± 0.9 (6) (30°C)	118
	Fed (300 g body wt)	Liver[m]	8.7 ± 1.3 (10)	
(Wistar)	Female (150—200 g body wt)	Liver[m]	2.80 ± 0.5 (6) (30°C)	717
	Fed	Kidney cortex[m]	3.10 ± 0.1 (6)	
		Adipose[m]	146.5 ± 15.0[k] (6)	
(Buffalo strain)	Male (60—90 days, fed)	Liver[m]	8.21 ± 0.24 (7) (25°C)	650
(Wistar)	Male (220—300 g, starved 12 hr)	Liver[n]	5.1 ± 1.6 (7) (30°C)	397
	(starved 72 hr)	Liver[n]	10.0 ± 2.3 (6)	
	(starved 12 hr)	Kidney cortex[n]	3.8 ± 1.0 (5)	

	(starved 72 hr)	Kidney cortex[n]	6.1 ± 1.1 (6)	
(Wistar)	adx[j])	Liver[n]	3.2 ± 0.7 (6) (30°C)	397
	(adx[j] + cortisol[k])	Liver[n]	4.3 ± 0.7 (11)	
	(adx[j])	Kidney cortex[n]	4.3 ± 0.8 (7)	
	(adx[j] + cortisol[k])	Kidney cortex[n]	5.9 ± 0.9 (11)	
(Wistar)	Male (230—260 g, fed)	Liver	4.64 ± 0.22 (5) (30°C)	805
	(230—260 g, starved 48 hr)	Liver	11.67 ± 1.77 (5)	
(Sprague-Dawley)	(10 weeks, fed)	Liver	4.36 ± 0.59 (4) (37°C)	227
	(10 weeks, fasted 24 hr)	Liver	8.46 ± 0.34 (4)	
(Sprague-Dawley)	Fed	Liver	4.43 ± 0.22 (6) (30°C)	947
	Diabetic (acute)[q]	Liver	5.20 ± 0.28 (7)	
	(chronic)	Liver	9.31 ± 0.39 (8)	
(Wistar)	Female (300—400 g, fed)	Liver	3.51 ± 0.24[r] (5) (30°C)	278
	(300—400 g, fasted[p])	Liver	4.19 ± 0.32[r] (5)	
	(300—400 g, diabetic[q])	Liver	5.24 ± 0.48[r] (4)	
	(300—400 g, dexamethasone)	Liver	5.29 ± 0.15[r] (3)	
	(300—400 g, fed)	Kidney	4.95 ± 0.27[r] (5)	
	(300—400 g, fasted[p])	Kidney	6.28 ± 0.39[r] (5)	
(Hebrew Univ. albino)	Male (150—200 g, fed)	Liver	3.20 ± 0.25[r] (9) (30°C)	325
	(150—200 g, fasted 24 hr)	Liver	4.43 ± 0.40[r] (5)	
	(150—200 g, + triamcinolone[bb])	Liver	3.33 ± 0.25[r] (5)	
	(150—200 g, diabetic[cc])	Liver	9.15 ± 2.20[r] (7)	
	(150—200 g, diabetic + insulin[dd])	Liver	4.21 ± 0.82 (4)	
(Sherman)	Fed	Liver	1.38 (6) (30°C)	286
	Fasted 48 hr	Liver	2.13 (6)	
	Diabetic	Liver	2.22 (6)	
	Diabetic + insulin	Liver	1.28 (6)	
(Wistar)	Male (fed)	Liver	3.1 ± 1.0 (8) (30°C)	661
	(diabetic)	Liver	6.4 ± 0.9 (8)	
	(diabetic + insulin)	Liver	2.6 ± 0.64 (8)	
(Wistar)	Male (220—300 g, fed)	Liver	1.18 ± 0.07 (12) (25°C)	472
	(220—300 g, fasted, 24 hr)	Liver	1.21 ± 0.05 (4)	
	(220—300 g, low carbohydrate	Liver	0.94 ± 0.04 (6)	
	diet)	Liver	1.26 ± 0.03 (7)	
	(220—300 g, diabetes [alloxan])			
(Wistar II)	180—200 g (fed)	Liver	2.95 ± 0.60 (8) (30°C)	827
	(fasted 12 hr)	Liver	4.83 ± 0.60 (8)	

Table 4 (continued)
PYRUVATE CARBOXYLASE IN VERTEBRATES

Species	Condition[a]	Tissue	Activity[b]	Ref.
	(fasted 12 hr + 5 mg corti-	Liver	3.68 ± 0.91 (8)	
	sol)	Liver	4.96 ± 0.83 (8)	
	(fasted 12 hr + 0.5 mg be-			
	tame-	Liver	4.21 ± 0.58 (8)	
	thasone)	Liver	5.91 ± 0.73 (8)	
	(fasted 12 hr, Adx[j])			
	(fasted 12 hr, Adx[j] + 5 mg cor- tisol)			
(Hooded Norway)	Female (3 months old, nonpregnant)	Liver	3.38 ± 0.32 (6) (37°C)	797
	Pregnant (7 days)	Liver	1.74 ± 0.5 (6)	
	(16 days)	Liver	1.25 ± 0.25 (6)	
	(21 days)	Liver	2.25 ± 0.25 (6)	
	Lactating (2 days)	Liver	1.95 ± 0.44 (6)	
	(14 days)	Liver	1.85 ± 0.23 (6)	
	Post-weaning (3 days)	Liver	2.92 ± 0.47 (6)	
(Sprague-Dawley)	Adult (female)	Liver[m,n]	6.76 ± 0.35 (5) (37°C)	172
	Fetus (19 days)	Liver[m,n]	0.38 ± 0.03 (5)	
	Pup (4—5 hr)	Liver[m]	1.9 ± 0.12 (3—12)	
	(5 days)	Liver[m]	7.11 ± 0.5 (3—12)	
	(21 days)	Liver[m]	4.68 ± 0.42 (3—12)	
(Wistar)	Fetus (17 days)	Liver[n]	0.36 (37°C)	52
	(21 days)	Liver[n]	6.28	
	Pup (2 days)	Liver[n]	12.75 ± 0.4	
	(7 days)	Liver[n]	14.95 ± 0.5	(993)
	(14 days)	Liver[n]	10.8 ± 1.0	
	(21 days)	Liver[n]	12.4 ± 2.0	
	Adult male	Liver[n]	12.3 ± 1.3	
	fa/− (lean) fed *ad lib*	Liver	0.167 ± 0.013 (8) (25°C)	812
	fa/fa (obese Zucker) fed *ad lib*	Liver	0.133 ± 0.017 (8)	
	fa/fa (obese Zucker) limit-fed = fa/ −(lean)	Liver	0.333 (8)	

	fa/− (lean fed *ad lib*	White adipose	0.084 (8)	
	fa/fa (obese Zucker) fed *ad lib*	White adipose	0.179	
	fa/fa (obese Zucker) limit-fed = fa/−(lean)	White adipose	0.336 (8)	
(Hebrew Univ. albino)	Male (150—200 g, fed)	White adipose	30.5 ± 1.8[r] (8) (30°C)	325
	(150—200 g, fasted 24 hr)	White adipose	32.2 ± 3.2[r] (5)	
	(150—200 g, + triamcinolone)	White adipose	11.2 ± 1.5[r] (5)	
	(150—200 g, diabetic[cc])	White adipose	16.9 ± 1.5[r] (5)	
	(150—200 g, diabetic + insulin[dd])	White adipose	16.1 ± 1.2[r] (4)	
(Wistar)	Male (350—400 g, chow-fed)	White adipose	0.22 ± 0.02 (6) (37°C)	51
	(350—400 g, fasted 96 hr, refed chow 96 hr)	White adipose	0.24 ± 0.03 (6)	
		White adipose	0.16 ± 0.01 (6)	
	(350—400 g, fasted 96 hr, refed high fat 96 hr)			
(Wistar II)	Fed (190 g)	White adipose	0.5 (30°C)	118
(Wistar)	Male (130—150 g, fed)	White adipose[m]	304 ± 37[h] (8) (25°C)	532
		White adipose	1—2 (25°C)	746
	Newborn	Brown adipose	3—4 (25°C)	746
		Adrenal gland	3—5 (25°C)	746
		Ovary	0.5—1 (25°C)	746
(Wistar)	Fed (control)	Ovary	4.8 ± 0.9 (5) (37°C)	281
	Fed (luteinizing-hormone treated[t])	Ovary[m]	5.0 ± 1.3 (5)	281
(Wistar II)	Fed (190 g)	Testis	0.25 (37°C)	118
		Testis	0.5—1 (25°C)	746
	Androgen maintained[u]	Caput epididymis	0.259 ± 0.017 (5) (25°C)	130
	Androgen deprived[v]	Caput epididymis	0.016 ± 0.001 (5)	
	Androgen maintained[u]	Cauda epididymis	0.346 ± 0.008 (5)	
	Androgen deprived[v]	Cauda epididymis	0.010 ± 0.0004 (5)	
	Normal	Epididymal spermatozoa	0.193 ± 0.011 (4)	
(Wistar)	Male (160—180 g, fed)	Mesenteric lymph node	32.8 ± 2.55 (6) (25°C)	22
	Male (160—180 g, fasted)	Mesenteric lymph node	27.7 ± 3.57 (6)	
	7 weeks old (fed)	Brain		
		cerebellum		
		vermis	13.5 ± 3.5[r] (6)	722
		hemispheres	15.7 ± 4.2[r] (6)	

Table 4 (continued)
PYRUVATE CARBOXYLASE IN VERTEBRATES

Species	Condition[a]	Tissue	Activity[b]	Ref.
		cerebrum	10.1 ± 1.1[r] (6)	
		brainstem	9.2 ± 2.5[r] (6)	
(Wistar)	Adult (55 days)	Brain	0.83 ± 0.05 (6) (37°C)	958
	Pups (1—30 days)	Brain	(0.13—0.65)	
(Wistar)	12—15 Days post-partum	Mammary gland	11[r] (37°C)	349
(Wistar)	7—14 Days post-partum	Mammary gland[m]	2.0 ± 0.1 (4) (25°C)	212
		Mammary gland	3—4 (25°C)	746
		Heart	0.2 (3) (25°C)	211
		Skeletal muscle (quadriceps)	0.2 (3)	
(Wistar)	150—200 g (fed)	Heart	0.12 (37°C)	214
	150—200 g (fed)	Skeletal muscle (hind limb)	0.03 ± 0.02[n] (5)	
(Sprague-Dawley)	Male (250—350 g, fed)	Liver[m]	29 ± 2 (11) (37°C)	12
	Male (250—350 g, fed)	Jejunal mucosa[m]	0.38 ± 0.07 (12)	
(Wistar)	200—230 g	Skin	0.30 ± 0.047[w] (6) (30°C)	645
		Muscle (vastus lateralis)	0.011 ± 0.002[w] (4)	
		Liver	5.10 ± 1.26[w] (4)	
(Sprague-Dawley)	Male (200 g, fed)	Brain[m]	0.2 (25°C)	720
		Brain[m]	0.72 ± 0.09[s] (6) (37°C)	633
(Sprague-Dawley)	Male (200 g, fed)	Brain	1.04 ± 0.07 (6) (37°C)	635
(Sprague-Dawley)	Adult (fed)	Brain		722
		cerebellum		
		vermis	16.7 ± 1.8[r] (6) (37°C)	
		hemispheres	18.3 ± 1.1[r] (6)	
		cerebrum		
		(first cortical slice)	12.0 ± 0.5[r] (3)	
		brainstem		
		(first pontine slice)	14.8 ± 0.2[r] (3)	
(Wistar)	Normal	Liver	29.9 ± 1.5[r] (6) (30°C)	366
	Tumor-bearing	Liver	71.6 ± 6.9[r] (8)	
	Tumor-bearing	Hepatoma	3.5 ± 0.7[r] (12)	
	Cell-culture	Hepatoma	1.1 ± 0.3[r] (11)	

(Buffalo)	normal	Liver[m]	4.71 ± 0.3[aa] (37°C)	173
(Sprague-Dawley)	normal	Liver[m]	4.68 ± 0.42[aa]	
(Wistar)	normal	Liver[m]	4.92 ± 0.21[aa]	
	Regenerating, 24 hr after partial hepatectomy	Liver[m]	3.35 ± 0.24	173
	Fetus, 17—19 days old	Liver[m]	0.25 ± 0.03	
	Tumor-bearing host	Liver[m]	2.81 ± 0.22	173
	Fast growth rate	Hepatoma 9A[m]	0.07 ± 0.01	
	Tumor-bearing host	Liver[m]	2.98	173
	Intermediate growth rate	Hepatoma H35[m]	0.41	
	Tumor-bearing host	Liver[m]	3.32	173
	Intermediate growth rate	Hepatoma 7800[m]	0.78	
	Tumor-bearing host	Liver[m]	2.85	173
	Slow growth rate	Hepatoma 6[m]	0.23	
	Tumor-bearing host	Liver[m]	2.79 ± 0.20	173
	Slow growth rate	Hepatoma 7787[m]	0.36 ± 0.02	
	Tumor-bearing host	Liver[m]	3.25	173
	Slow growth rate	Hepatoma 21[m]	0.49	
	Tumor-bearing host	Liver[m]	4.11 ± 0.17	173
	Slow growth rate	Hepatoma 9633[m]	0.31 ± 0.04	
	Tumor-bearing host	Liver[m]	4.30 ± 0.31	173
	Slow growth rate	Hepatoma 9618[m]	4.77 ± 0.57	
Mouse	10 Weeks old (30—40 g, fed)	Liver	13.1 ± 1.3[r] (5) (37°C)	86
(Swiss albino)	(30—40 g, fasted 24 hr)	Liver	25.9 ± 3.1[r] (4)	86
(C57 BL/6J(ob/ob)	Female (lean)	Liver	17.0 ± 1.6[r] (11) (37°C)	597
	(obese)	Liver	37.0 ± 3.0[r] (12)	597
(C57 BL/Ks)	(+/+) 2—6 months old (normal)	Liver	0.342 ± 0.017 (4) (30°C)	196
	(db/db) 2—6 months old (diabetic)	Liver	0.441 ± 0.028 (4)	
	7—14 Days post-partum	Mammary gland[m]	6.0 ± 1.5 (4) (25°C)	212
	Cell culture, day 1—12 post-confluence	3T3-L1 preadipocytes	5[r] (37°C)	518
	Cell culture, day 21 post-confluence	3T3-L1 adipocyte[m]	90[r]	518
	Cell culture	3T3-F442A preadipocytes	7[r] (37°C)	16
		3T3-F442A adipocytes	140[r]	16
		Heart	0.2 (3) (25°C)	211
Rabbit	Adult (fed)	Liver	108[r] (30°C)	439
		Kidney	183[r]	439

Table 4 (continued)
PYRUVATE CARBOXYLASE IN VERTEBRATES

Species	Condition[a]	Tissue	Activity[b]	Ref.
		Brain	10[r]	439
		Heart	16[r]	439
		Skeletal muscle	1[r]	439
		Spleen	1[r]	439
	Male (1.7—2.1 kg, fed)	Liver[m]	0.85 ± 0.1 (5) (30°C)	717
	(1.7—2.1 kg, fed)	Kidney cortex[m]	0.85 ± 0.1 (5)	717
	(1.7—2.1 kg, fed)	White adipose[m]	2.75 ± 0.7[k] (5)	717
	Lactating (18—21 days)	Mammary gland[m]	0.7 ± 0.1 (3) (25°C)	212
	Fetus (22 days)	Liver	0.92 ± 0.18 (6—9) (37°C)	148
	(25 days)	Liver	2.49 ± 0.28	148
	(28 days)	Liver	2.86 ± 0.28	148
	(30 days)	Liver	2.68 ± 0.33	148
	(31 days)	Liver	2.31 ± 0.28	148
	Newborn (1 day, fed)	Liver	3.97 ± 0.37	148
	(1 day, starved)	Liver	5.72 ± 0.28	148
	(2 days, fed)	Liver	5.54 ± 0.28	148
	(2 days, starved)	Liver[m]	5.07 ± 0.37	148
	Adult (pregnant, fed)	Liver	2.31 ± 0.37 (4)	148
	Adult (fed)	White skeletal muscle	0.03 (25°C)	211
(N.Z. White)	1.5—2.5 g (fed)	Liver[m]	7.18 ± 0.18 (8) (37°C)	305
	1.5—2.5 g (fasted 48 hr)	Liver[m]	6.13 ± 0.25 (8)	
Hamster	Cold-adapted (5°C) 3 weeks (fed)	Brown adipose	35—40[r] (37°C)	149
Guinea pig	Female (600—800 g, fed)	Liver[m]	1.40 ± 0.15 (6) (30°C)	717
	(600—800 g, fed)	Kidney cortex[m]	1.50 ± 0.12 (6)	717
	(600—800 g, fed)	White adipose[m]	19.25 ± 1.15[k] (8)	717
(Pirbright White strain)	Male (320—360 g, fed)	Liver	5.14 ± 0.39 (5) (30°C)	805
	(320—360 g, fasted 48 hr)	Liver	7.26 ± 0.40 (5)	805
	Lactating (4—7 days)	Mammary gland[m]	0.8 ± 0.2 (5) (25°C)	212
(Hartley)	Fetus (35—40 days)	Liver	0.1 ± 0.05 (4) (37°C)	666
	(45—50 days)	Liver	1.8 ± 0.1 (5)	
	(55—60 days)	Liver	1.8 ± 0.1 (5)	
	(64—72 days, term)	Liver	2.5 ± 0.1 (5)	(433, 685,
	(2 days)	Liver	1.4 ± 0.1 (5)	700)

	(6 days)	Liver	1.5 ± 0.2 (5)	
	(8 days)	Liver	5.1 ± 0.1 (5)	
	(12 days)	Liver	3.3 ± 0.2 (5)	
	(20 days)	Liver	2.3 ± 0.2 (5)	
	(28 days)	Liver	1.8 ± 0.2 (5)	
	Adult	Liver	1.5 ± 0.1 (5)	
	Fetus (35—40 days)	Kidney	0.09 ± 0.01 (5) (37°C)	667
	(45—50 days)	Kidney	0.12 ± 0.01 (5)	
	(55—60 days)	Kidney	0.25 ± 0.02 (5)	
	Post-natal (0 hr)	Kidney	0.25 ± 0.03 (5)	
	(12 hr)	Kidney	0.25 ± 0.05 (5)	
	(24 hr)	Kidney	0.42 ± 0.12 (5)	
	(2 days)	Kidney	0.41 ± 0.05 (5)	
	(7 days)	Kidney	0.61 ± 0.05 (5)	
	(14 days)	Kidney	0.98 ± 0.03 (5)	
	(21 days)	Kidney	1.12 ± 0.06 (5)	
	(28 days)	Kidney	1.51 ± 0.17 (5)	
	Adult	Kidney	0.72 ± 0.07 (5)	
	Male (300—400 g, fed)	Liver[m]	7.90 ± 0.6 (10) (37°C)	13
	Male (300—400 g, fed)	Jejunal mucosa[m]	0.97 ± 0.07 (10)	
	Male (300—400 g, starved 72 hr)	Jejunal mucosa	1.37 ± 0.09 (8)	
Sheep	Fed[z]	Liver	3.86 ± 0.40[r] (12) (30°C)	278
(Merino)	Fasted 6 days[z]	Liver	14.78 ± 1.95[r] (7)	
	Cortisone[z]	Liver	1.88 ± 0.28[r] (5)	
	Dexamethasone[z]	Liver	1.69 ± 0.50[r] (5)	
	Phlorhizin (fed)[z]	Liver	3.68 ± 0.29[r] (2)	
	Phlorhizin (fasted 5 days)[z]	Liver	13.32 ± 0.99[r] (2)	
	Diabetes (alloxan)[z]	Liver	18.37 ± 1.72[r] (4)	
	Diabetes (depancreatized)[z]	Liver	14.06 ± 3.41[r] (3)	
	Pregnant (fasted 5—7 days)[z]	Liver	9.97 ± 2.79[r] (4)	
	Fed[z]	Kidney	7.10 ± 0.55[r] (9) (30°C)	278
	Fasted[z]	Kidney	10.76 ± 3.24[r] (5)	
	Cortisone[z]	Kidney	3.98 ± 0.16[r] (4)	
	Dexamethasone[z]	Kidney	3.38 ± 0.10[r] (2)	
	Phlorhizin (fed)[z]	Kidney	6.78 ± 0.29[r] (2)	
	Phlorhizin (fasted 5 days)[z]	Kidney	8.98 ± 0.99[r] (2)	
	Diabetes (alloxan)[z]	Kidney	5.76 ± 0.43[r] (4)	
	diabetes (depancreatized)[z]	Kidney	5.35 ± 0.24[r] (3)	

Table 4 (continued)
PYRUVATE CARBOXYLASE IN VERTEBRATES

Species	Condition[a]	Tissue	Activity[b]	Ref.
	Pregnant (fasted 5—7 days)[x]	Kidney	10.05 ± 1.03[r] (3)	
	Wethers (4—6 months, pasture fed)	White adipose	0.009 ± 0.006 (4) (37°C)	50
	Wethers (4—6 months, carbohydrate fed)	White adipose	0.025 ± 0.012 (4)	
	Wethers (5 months, abomasal glucose infusion)	White adipose	0.041 ± 0.026 (3)	
	Wethers (6 months, i.v. glucose infusion)	White adipose	0.012 ± 0.005 (4)	
	Lactating (8 weeks)	Mammary gland[m]	0.15 ± 0.08 (3) (25°C)	212
	Fetus (55 days)	Liver	1.3 ± 0.12 (6—12) (37°C)	936
	(85 days)	Liver	1.0 ± 0.1	
	(115 days)	Liver	1.9 ± 0.18	
	(135 days)	Liver	3.0 ± 0.15	
	(148 days)	Liver	2.9 ± 0.15	
	Newborn (5 days)	Liver	5.06 ± 0.2	
	(20 days)	Liver	5.16 ± 0.1	
	Adult	Liver	3.74 ± 0.19	
(Dorset Horn × Suffolk)	Adult	Liver	0.176 ± 0.004 (4) (37°C)	236
	Fetus (120 days)[y]	Liver	0.046 ± 0.005 (4)	
	(120 days)[y]	Placental cotyledon	0.034 ± 0.015 (4)	
	Pregnant (55—85 days)[y]	Placental cotyledon	0.034 ± 0.013 (5) (37°C)	257
	(86—116 days)[y]	Placental cotyledon	0.049 (1)	
	(117-147 days)[y]	Placental cotyledon	0.131 ± 0.017 (7)	
Cow	Nonlactating (fed)	Liver[n]	5.62 ± 0.42 (5) (37°C)	56
	Lactating (fed, normal)	Liver[n]	10.90 ± 2.74 (5)	
	Lactating (starved 4 days)	Liver[n]	18.82 ± 2.69 (4)	
	Lactating (fed, spontaneously ketotic)	Liver[n]	9.41 ± 2.20 (6)	
	Nonlactating (fed)	Liver	1.30 (2) (25°C)	45
	Lactating (normal)	Liver	0.76 ± 0.37 (6)	
	Lactating (ketotic)	Liver	1.34 ± 0.46 (6)	
	Pregnant (80—280 days)	Liver	0.47—0.7 ± 0.07 (58)	663
	Fetus (80—120 days)	Liver	(37°C)	
	(160 days)	Liver	0.05 ± 0.01	

	(180 days)	Liver	0.17 ± 0.05	
	(240 days)	Liver	0.25 ± 0.07	
			0.48 ± 0.05	
		Liver	10 (37°C)	275
		Brain	1.13	
	Lactating	Mammary gland[m]	0.18 ± 0.03 (3) (25°C)	212
(Angus × Hereford)	Steers	White adipose	0.042 ± 0.005 (4) (37°C)	798
Pig	Fetus (104 days)	Liver	1.6 ± 0.2 (8) (25°C)	566
	(110 days)	Liver	1.0 ± 0.1 (15)	
	Piglet (at birth)	Liver	1.5 ± 0.3 (19)	(849)
	2 days	Liver	2.0 ± 0.25 (10)	
	6 days	Liver	2.5 ± 0.6 (10)	
	14 days	Liver	1.1 ± 0.2 (8)	
	21 days	Liver	0.65 ± 0.1 (6)	
	60 days	Liver	2.2 ± 0.4 (6)	
	Adult	Liver[m]	2.5 (30°C)	937
(Yorkshire × Hampshire)	Piglet (1 day old, fed)	Liver	53.6 ± 11.0[r] (9) (35°C)	391
	(1 day old, fasted 24 hr)	Liver	82.2 ± 16.1[r] (7)	
	(5 days old, fasted 24 hr)	Liver	125.5 ± 13.3[r] (7)	
	(24 days old, fasted 24 hr)	Liver	178.0 ± 23.7[r] (7)	
(Large white × Landrace)	wheat/soybean diet (4 weeks)	Blood	16.2 ± 3.0[g] (4) (38°C)	955
	wheat/soybean diet (6 weeks)	Blood	8.4 ± 3.8[g] (4)	
	wheat/soybean diet (8 weeks)	Blood	5.4 ± 1.1[g] (4)	
	wheat/soybean diet (12 weeks)	Blood	4.0 ± 1.5[g] (4)	
	Lactating (biopsy)	Mammary gland	0.3 ± 0.3 (3) (25°C)	212
Dog	Adult (male, fed)	Liver[m]	0.90	309
	Adult (male, fed)	Kidney[m]	0.59	
	Fetus	Liver	1.2 ± 0.1 (30°C)	595
	Newborn (12 hr)	Liver	1.12 ± 0.1	
	(60 hr)	Liver	3.6 ± 0.3	
	(120 hr)	Liver	6.8 ± 0.4	
	(210 hr)	Liver	11.3 ± 1.1	
	Adult	Liver	15.8 ± 1.3	
(Beagle)	Fetus (60 days)	Liver	0.36 ± 0.06 (4) (30°C)	184
	Newborn (3 hr premature)	Liver	1.97 ± 0.21 (5)	
	(1 day)	Liver	4.53 ± 0.37 (11)	
	(3 days)	Liver	4.18 ± 0.31 (8)	
	(5 days)	Liver	3.98 ± 0.25 (10)	

Table 4 (continued)
PYRUVATE CARBOXYLASE IN VERTEBRATES

Species	Condition[a]	Tissue	Activity[b]	Ref.
Rhesus monkey	2 months old	Liver	0.53 ± 0.015 (2) (37°C)	34
	Adult	Liver	0.695 ± 0.02	
Human	Children (2 months—10 years)	Liver	(0.9—12.2)[m] (25°C)	(690, 767, 855)
	Various ages (14 days to adult)	Liver	(0.032—6.2) (30°C)	(118, 134, 230, 416
	Newborn (premature newborn to full term, 10 days)	Liver	(4.6—8.0) (37°C)	531
	Children (3 months—4 years)	Liver	6.2 ± 1.0 (37°C)	531
	Adult	Liver	(9.5—11.3) (37°C)	(34, 589)
	Adult	Liver	17.5[r] (30°C)	366
	Tumor-bearing (adult)	Liver	21.5 ± 4.8[r] (8)	
	Tumor-bearing (adult)	Hepatoma	4.5 ± 0.8[r] (12)	(886)
	Fetus	Liver	3.4 ± 0.6[r] (11)	
	Tumor-bearing (adult)	Liver	7.2 ± 2.5 (4)(37°C)	994
	Primary hepatocellular carcinoma	Tumor	0.2 ± 0.3 (5)	994
	16-year-old (normal) accidental death	Kidney	5.05 ± 0.24(37°C)	34
	Autopsy sample (5 hr after death)	Kidney	0.99 (37°C)	345
	Autopsy sample (tissue frozen 16 days)	Kidney	0.37	
	Autopsy sample (tissue frozen 3 months)	Kidney	0.16	(886)
	Autopsy sample (24 hr after death)	Kidney	1.08 (37°C)	
	5 month old (autopsy sample 15 hr after death)	Kidney	3.4 (37°C)	
	5 month old (autopsy sample, tissue frozen 7 days)	Kidney	2.7	
	Male (4 year, with Menkes kinky hair disease[ff])	Brain		
	cerebellum		0.24[r] (37°C)	886
	cerebrum		0.78[r]	
	Adult (myocardial infarction[ff])	Brain		
	cerebellum		0.11[r] (37°C)	886
	cerebrum		0.34[r]	
	Adult (brain hemorrhage[ff]) cerebrum	Brain	0.39[r] (37°C)	886

14 year old (autopsy)	Brain	0.44 ± 0.03 (37°C)	34
9 month old (autopsy 4 hr after accidental death)	Brain	0.21 (37°C)	635
Children (3.5—12 years) with degenerative central nervous system disease; biopsy	Brain	0.25 ± 0.05 (4) (37°C)	635
	Fibroblasts	1.52 ± 0.12[r] (22) (37°C)	529
	Fibroblasts	(0.52—1.35[r]) (37°C)	(32, 35, 273, 368, 371, 529, 585, 624, 668, 707, 728, 854)
Children	Mixed peripheral leukocytes	0.089 ± 0.005[r] (2) (37°C)	
Adults	Mixed peripheral leukocytes	0.095 ± 0.014[r] (3) (37°C)	35
Adults	Circulating lymphocytes	0.154 ± 0.028[r] (5) (37°C)	35
9 *am* (14—15 hr, postprandial)	Lymphocyte	0.104 ± 0.022[r] (7) (37°C)	32
3.30 *pm* (3—4 hr, postprandial)	Lymphocyte	0.150 ± 0.017[r] (8)	32
3.30—4.00 *pm* (8 hr postprandial)	Lymphocyte	0.149 ± 0.025[r] (4)	32
3.30—4.00 *pm* (20—21 hr postprandial)	Lymphocyte	0.167 ± 0.019[r] (7)	32
	Lymphocytes	0.0073 ± 0.0026[r] (7) (37°C)	868
	Lymphoblasts	0.259 ± 0.034[r] (7) (37°C)	854
	Leukocytes	0.038 ± 0.009[r] (5) (37°C)	854
	Amniotic fluid cells	(0.46—0.97[r]) (37°C)	(585, 623, 273, 368, 529)
	Platelets	0.0016[r] (4) (30°C)	739

[a] Refers to the stage of development, sex, nutritional, or enocrine status, or other factors which might influence the activity of pyruvate carboxylase.

[b] Unless otherwise indicated by superscript, activity is expressed as micromoles per minute per gram fresh weight of tissue at the indicated temperature; values are the means ± standard error of the mean for the number of animals indicated in parentheses. In the interests of brevity, values in some cases are given as a range (in parentheses). Additional data for the same tissue of the same species (though not necessarily the same strain) are to be found in the references given in parentheses. In an earlier survey of the reported activities of some enzymes of carbohydrate metabolism in selected tissues from various vertebrate species,[762] the following factors were used to convert the data presented in the literature to a common basis for comparison (viz., μmol/min/g wet weight at 37°C):

Table 4 (continued)
PYRUVATE CARBOXYLASE IN VERTEBRATES

1. Activity/g wet wt. = Activity/g dry wt. × 0.28
   = Activity/mg protein × 200 (for liver kidney, heart, skeletal muscle)
   = Activity/mg protein × 120 (brain)
   = Activity/mg protein × 12 (adipose tissue)
2. Activity at 37°C = Activity at 37°C × 1.5
   = Activity at 25°C × 2

[c] Activity expressed as μmol/min/10^4 oocytes or embryo.

[d] Activity expressed as nmol/min/mg protein; N.B. assays based on NADH oxidation using only crude mitochondrial fraction.

[e] Activity, expressed as nmol $^{14}CO_2$ fixed/min/mg protein at 30°C, was 96% dependent on acetyl-CoA, and was inhibited (90%) by avidin.

[f] Activity undetectable, but no evidence found for the presence of an inhibitor.

[g] Activity expressed as nmol $^{14}CO_2$ fixed/g hemoglobin/min.

[h] Activity expressed as nanomoles per minute per 100 μg of DNA.

[j] Abbreviation for adrenalectomized.

[k] Cortisol 5 mg/100 g body wt. by stomach tube; values of activity corrected for water retained by tissue under influence of the hormone.

[m] Indicates activity demonstrated to be exclusively or very largely mitochondrial.

[n] Indicates the value is the sum of the mitochondrial and cytoplasmic activities given separately by the authors; see Section V.B.

[o] The letter "o" was skipped to avoid confusion with other symbols.

[p] Rats starved for varying periods (16 to 72 hr) but no further increment in activity observed after 16 hr.

[q] Diabetes induced by injection of alloxan (60 mg/kg body wt) 4 days prior to killing for assay.

[r] Values are nmoles $^{14}CO_2$ fixed/min/mg protein, i.e., mU/mg protein; see footnote "b" for approximate conversion factors.

[s] Equivalent to 58 nmoles $^{14}CO_2$ fixed/min/mg mitochondrial protein; similar activity (66 nmoles/min/mg mitochondrial protein at 37°C reported by Patel and Tilghman.[636]

[t] Luteinizing hormones (20 μg/100 g body wt) administered i.p. 2 hr before death; corresponding specific activities are 30.3 ± 5.0 mU/mg protein (control) and 40.3 ± 9.7 (luteinizing - hormone-treated).

[u] Rats castrated as in 'v' (below); hormone replacement therapy carried out by daily s.c. injections of testosterone propionate in maize mℓ (1 mg/mℓ) at a dose of 1 mg/kg body wt.

[v] Either the epididymis was rendered free of spermatozoa by bilateral ligation of the efferent ducts or bilateral castration was carried out and animals kept for at least 6 days before daily s.c. injections of maize oil only.

[w] For comparison, values in mU/mg protein are 13.1 (skin), 46.4 (liver), and 0.28 (skeletal muscle).

[x] Activity expressed as nmol/min/mg mitochondrial protein.

[y] Days after conception.

[z] Ewes, 2 to 3 years old; cortisone acetate (2.5 mg/kg body wt) administered i.m. for 3 to 5 days; dexamethasone (0.25 mg/kg body wt.) likewise; phlorhizin administered s.c. in oil, 0.14 g (day 1); 0.07 g (days 2 to 6); alloxan diabetes induced by i.v. injection of 60 mg alloxan/kg body wt; pregnant ewes at 110 to 130 days gestation.

[aa] Means ± S.E.M. for 3 to 8 animals.

[bb] Triamcinolone diacetate (5 mg) administered i.p. three times at 12-hr intervals and sacrificed 3 hr after last injection.

[cc] Diabetic rats sacrificed 4 to 10 days after a single dose of alloxan (18 mg/100 g body wt).

[dd] Diabetic rats with blood glucose levels over 350 mg/100 mℓ were treated with 8 U/day of protamine zinc insulin for 2 to 3 days to reduce their blood glucose level to below 100 mg/100 mℓ.

[ee] Based on mean pyruvate carboxylase activity of 3.4 ± 1.9 (5) U/mg mitochondrial protein, and 10 mg mitochondrial protein per gram of skeletal muscle.

zymic activity under the "nonstandard" conditions. Although conversion factors are given in the footnotes to Table 4, it is clear that these should be used with caution when making comparisons between species or even between animals of the same species in different physiological states (e.g., see Henning et al.[397] and Moon and Johnston[574]). Furthermore, without detailed characterization of the enzyme from different sources, it cannot be assumed that even the response to the assay temperature will be identical for any two species.[46,808,929] An additional reason for caution in making interspecies comparisons is the considerable variation within one species (e.g., rat) between different workers' values for liver pyruvate carboxylase activity (viz., 2 to 15 U/g wet weight at 37°C). It seems unlikely that these discrepancies are merely attributable to strain differences, since Chang and Morris[173] found almost identical activities in the livers of Wistar, Sprague-Dawley, and Buffalo rats when treated and assayed under the same conditions.

Inconsistencies between the values for enzymic activity of a particular tissue measured by different investigators could arise at a number of stages during the determination, from the age or size of the animal and details of its nutrition or handling (often unspecified), through the method of tissue sampling and extraction to specific aspects of the assay procedure.

b. Factors Affecting the Physiological State of the Tissues

It is readily apparent from the data summarized in Table 4 that, depending on the particular species and tissue in question, a number of factors affecting the physiological state of the tissue can also influence the level of pyruvate carboxylase activity expressed. These include the animal's age, sex, reproductive status (pregnant, nonpregnant, or lactating), and dietary or hormonal status (fed, fasted, diabetic, corticosteroid-treated, etc.). A detailed discussion of the influences of these factors on the regulation of the total amount of enzyme protein found in liver, kidney, and adipose tissue is presented in Chapter 5, Section 1.B.

Several groups[32,668] have reported also that in certain strains of fibroblasts in cell culture the activity of pyruvate carboxylase can vary considerably depending on the time of harvest with respect to the period post-confluence and the time since the last medium change.

Similarly, Atkin[32] has drawn attention to a possible diurnal variation in the pyruvate carboxylasc activity of human leukocytes (see Table 4).

c. Tissue Sampling and Preparation

Many authors[34,531,574,577,650,730] have commented on the importance of the state of the tissue at the time of sampling and following its storage in the frozen state, when immediate processing to a final determination is impractical under the circumstances. Where such is the case, collection into liquid nitrogen and subsequent powdering of the tissue, as described by Wallace and Newsholme,[929] yields complete recovery of activity, though it will preclude investigation of the intracellular distribution of the enzyme. Another effective procedure is described by Atkin et al.[32] Lyophilized mitochondria or tissue homogenates retain their full activity of pyruvate carboxylase for at least 12 months if stored dessicated at −10 to −20°C.

Although pyruvate carboxylase has only an average rate of turnover in vivo,[170,944] it is highly susceptible to proteolysis,[75] and loses activity readily (3 to 5 hr) if held in a crude extract in vitro at either 23 or 0°C.[57,118,130,397] Loss of activity due to "cold-lability" is an additional complication encountered when dealing with some forms of pyruvate carboxylase.[51,281,397,426] Samples should, therefore, be processed rapidly from the fresh, frozen, or lyophilized state through to the assay, taking steps to minimize loss of activity due to proteolysis.

Latency of the mitochondrial pyruvate carboxylase activity to its substrates is readily overcome by disruption of the inner mitochondrial membrane by freeze-drying,[57] by freezing and thawing,[366] by sonication, or with detergents. Both non-ionic detergents (viz., 0.5% NP-40;[16] 0.06 to 0.5% Triton X-100;[518,700,746] Lubrol WX [0.5 to 1 mg/mg protein][173]) and deoxycholate (0.1 to 1.0%)[212,246,808] have been used successfully, but there is species variation in the optimum concentration to be used.[805]

A convenient procedure which has consistently yielded a full and reproducible measure of pyruvate carboxylase activity has been described by Ballard et al.[57]

d. Assay Procedures

Most values in Table 4 have been determined by the fixation of radioactivity from $H^{14}CO_3^-$ into oxaloacetate (or citrate, malate, or aspartate derived from it) as described originally by Utter and Keech,[899] and subsequently adapted by others.[52,743] Though fairly time-consuming, the procedure is relatively straightforward and can be employed at a high degree of sensitivity as illustrated by Utter and associates.[34]

For precise quantitation of the enzymic activity, the specific radioactivity of the $H^{14}CO_3^-$ in the assay mix, including the tissue extract, should be determined using the same counting procedures for the radioactivity as for the acid-stable radioactivity incorporated by the tissue extract.

It is essential, also, that appropriate control incubations be included to assess the extent of isotope incorporation that is not dependent on the simultaneous presence of MgATP, pyruvate, and acetyl-CoA, all of which are essential for the activity of pyruvate carboxylase in vertebrate tissues, except under special conditions (see Chapter 4, Section III.B).

Avidin, the egg-white protein with an extraordinarily high affinity for biotin,[337] can be used to inhibit pyruvate carboxylase and thus is an effective tool in correctly assigning pyruvate-dependent CO_2-fixation activity to this enzyme in a crude tissue extract.[51,275,279,439]

Henning and Seubert[396] have described in detail an alternative assay based on the incorporation of 2-[^{14}C]-pyruvate into citrate, while Berndt et al.[98] have more recently described a spectrophotometric variant of that procedure.

Under some circumstances, where interfering enzyme activities (e.g., lactate dehydrogenase,[211] NADH oxidase, or acetyl-CoA hydrolase) are insignificant (e.g., after partial purification), more convenient coupled spectrophotometric assays have been used. However, since it is extremely rare for pyruvate carboxylase activity to be present in excess over lactate dehydrogenase or NADH oxidase in tissue homogenates, most workers have settled for the radiochemical assay in preference to the spectrophotometric assay employing malate dehydrogenase to couple NADH oxidation to oxaloacetate production.[899]

An alternative spectrophotometric assay, in which pyruvate carboxylation is coupled to the citrate synthase reaction, has been described by Martin and Denton[532] and has been applied successfully to a wide range of tissues.[211,212,532,597,746,859] In this procedure oxaloacetate formation is measured by monitoring the CoA-SH released in the citrate synthase reaction using 5,5′-dithiobis-(2-nitrobenzoic acid).

Whichever assay procedure is used, it is essential that the conditions be optimized for maximal activity of the particular pyruvate carboxylase in question, since it is clear that while they are generally similar, not all forms of the enzyme are exactly the same.

Furthermore, in view of the value of the comparative approach in understanding metabolic regulation,[211,650,746] it would be of lasting significance if individual workers would attempt to relate the absolute values of their observations on specific tissues to those of others by the inclusion of additional information demonstrating the efficacy of their tissue extraction and assay procedures on some relatively "standard sample" such as liver from a Wistar rat of specified age, sex, and nutritional status.

2. Interspecies Comparisons

Notwithstanding the considerable variation between the results of different workers studying the same species, as noted above, some broad generalizations can be made in comparing the levels of pyruvate carboxylase activity in the livers of the vertebrate species summarized in Table 4.

The values reported for fish, amphibia, and reptiles are mostly much lower than those found in the avian and mammalian livers, even after making allowance for the differences in assay temperatures. However, too few fish species have been examined yet to reveal any clear pattern within the group.

The apparent absence of pyruvate carboxylase activity from toad liver, though surprising, is consistent with the lack of an active gluconeogenic pathway from lactate in the amphibia.[650] Conversely, the relatively similar activity of pyruvate carboxylase in toad gastric mucosa[279] and rat[57] intestinal mucosa should be viewed in the light of a significant gluconeogenic capacity in the toad gastric mucosa.[279] The relatively high level of pyruvate carboxylase activity in frog gastric mucosa is, however, apparently related to the dependence of that tissue on aerobic oxidative metabolism for acid secretion.[164]

Amphibian brain tissues have a considerable capacity for CO_2 fixation. Although no direct measurements of pyruvate carboxylase in frog or toad brain seem to have been made in vitro, Tachiki and Baxter[850] concluded from their detailed studies of metabolite labeling patterns in vivo in the toad, *Bufo boreas*, that 40+% of ^{14}C atoms in aspartate of brain tissue appeared to have been incorporated by this pathway. This is comparable with the situation in isolated lobster nerve,[174] but about four times higher than that observed in mammalian brain.

All avian species examined so far have exhibited substantial liver pyruvate carboxylase activity. This is consistent with the very high rate of gluconeogenesis from lactate by isolated hepatocytes from pigeon in comparison with those of rat and guinea pig.[804]

Among the mammalian species, the highest levels of liver pyruvate carboxylase activity in fed, nonpregnant/nonlactating animals are found in rat and mouse, followed by rabbit, guinea pig, and dog, ahead of man and the larger domestic species (e.g., sheep, pig, and cow). However, these differences do not merely reflect the variation in basal metabolic rate expected from the range of sizes among these species, but also are influenced by their differing patterns of metabolism, particularly the contrasts between ruminants and nonruminants. The peculiarities of ruminant nutrition and intermediary metabolism in liver and adipose tissue throughout development have been discussed by Ballard et al.[55] in comparison with the situation in monogastric species. Further evidence of this distinction is to be seen in the different responses to pregnancy and lactation of liver pyruvate carboxylase in the sheep[278] and cow[56] compared with the rat.[797]

While there are many gaps in any catalogue of species for the reported occurrence of pyruvate carboxylase, it is clear that further reports of this widespread enzyme are best appreciated in the context of the rates of gluconeogenesis or other causes of anaplerotic CO_2 fixation, and alongside the measured activities of the other key enzymes of the relevant pathways.

3. Comparison Between Tissues

Perhaps the most meaningful comparisons between the relative levels of pyruvate activities in various tissues of the same species are those where all the measurements have been made at one time by the same investigator(s).[57,118,393,746] In the chicken, liver pyruvate carboxylase activity is greater than in the kidney,[393,776,777] and very much higher than in other tissues tested so far.[393]

Among the mammalian species the activities in liver and kidney are similar, and significantly greater than those of lactating mammary gland, adrenal gland, brown or white adipose tissue, and brain, which are in turn much higher than those in testis, epididymis, heart, skeletal muscle, intestine, spleen, and blood cells. Some of these comparisons are seen clearly in Figure 5 in terms of activity per gram wet weight and per milligram of protein. The relative significance of pyruvate carboxylase in white adipose tissue[51] is seen more readily when the activity is expressed on a protein basis.

Although the potential flux through pyruvate carboxylase in epididymis, skeletal muscle, or heart is very much less than in liver or kidney, the anaplerotic function of this enzyme may be no less important to the normal functions of these tissues, as recent results suggest.[214,647,811] Similarly, the highly androgen-dependent character of epididymal pyruvate carboxylase[130,131] suggests that there are large differences in the flux of metabolites into anabolic pathways in this tissue during different endocrine states.

Further discussion of the roles or, in some cases, proposed functions of pyruvate carboxylase in the various tissues of vertebrates is present in Chapter 5, Section II.A.

B. Intracellular Localization of Pyruvate Carboxylase in Vertebrate Tissues

1. Background

From the earliest report on the properties of this enzyme,[439] it appeared that it was located in the mitochondria of chicken liver. This observation has since been confirmed and extended to include a number of other species and tissues. In addition to the many studies indicated in Table 4 (see footnote 'm'), there have been others specifically on the intracellular localization of pyruvate carboxylase.[57,118,244-246,484,527,532,644,717,859,862,931-933,946]

In the overwhelming majority of all these investigations, it has been concluded with differing degrees of justification that this enzyme is mitochondrial. An exclusively mitochondrial location for pyruvate carboxylase has important implications for the mechanisms and regulation of the flow of carbon between the mitochondria and the cytosol during periods of gluconeogenesis from pyruvate and its precursors, or during lipogenesis from glucose.[385,488,489,527,532,780,935]

Although its concentration in liver is very low,[959] oxaloacetate is the product and/or substrate of many enzymes.[893] Together with PEP and pyruvate, oxaloacetate is a key component in the interlocking of glycolysis, gluconeogenesis, the Krebs tricarboxylic acid cycle, and several related processes. The localization, then, of one of the enzymes linking two members of this "triangle" has been of considerable interest in understanding the regulation of this metabolic crossroads, particularly in view of the fact that the relative proportions of PEP carboxykinase in the mitochondria vs. the cytosol vary so markedly from species to species (for a review, see Söling and Kleineke[804]).

2. Experimental Approaches

Several different experimental approaches have been employed and will be discussed below.

a. Differential Centrifugation

Based on the classical procedures described by Hogeboom,[410] the tissue homogenate is separated by differential centrifugation into nuclear, mitochondrial (sometimes heavy mitochondrial and light mitochondrial),[217] microsomal, and cytosolic fractions. As has been shown by DeDuve and Baudhuin[216] and Leighton et al.,[495] the fraction designated mitochondrial is in fact a mixture of several kinds of subcellular organelles. This mixture can be resolved by a further fractionation using sucrose-density gradient centrifugation.

Using rate-zonal centrifugation, Seubert and associates[946] conclusively eliminated peroxisomes, lysosomes, and microsomes as alternative locations for pyruvate carboxylase activity, and also demonstrated, as had others previously (for a review, see Greville[341]) that rat liver mitochondria are heterogeneous with respect to their enzyme

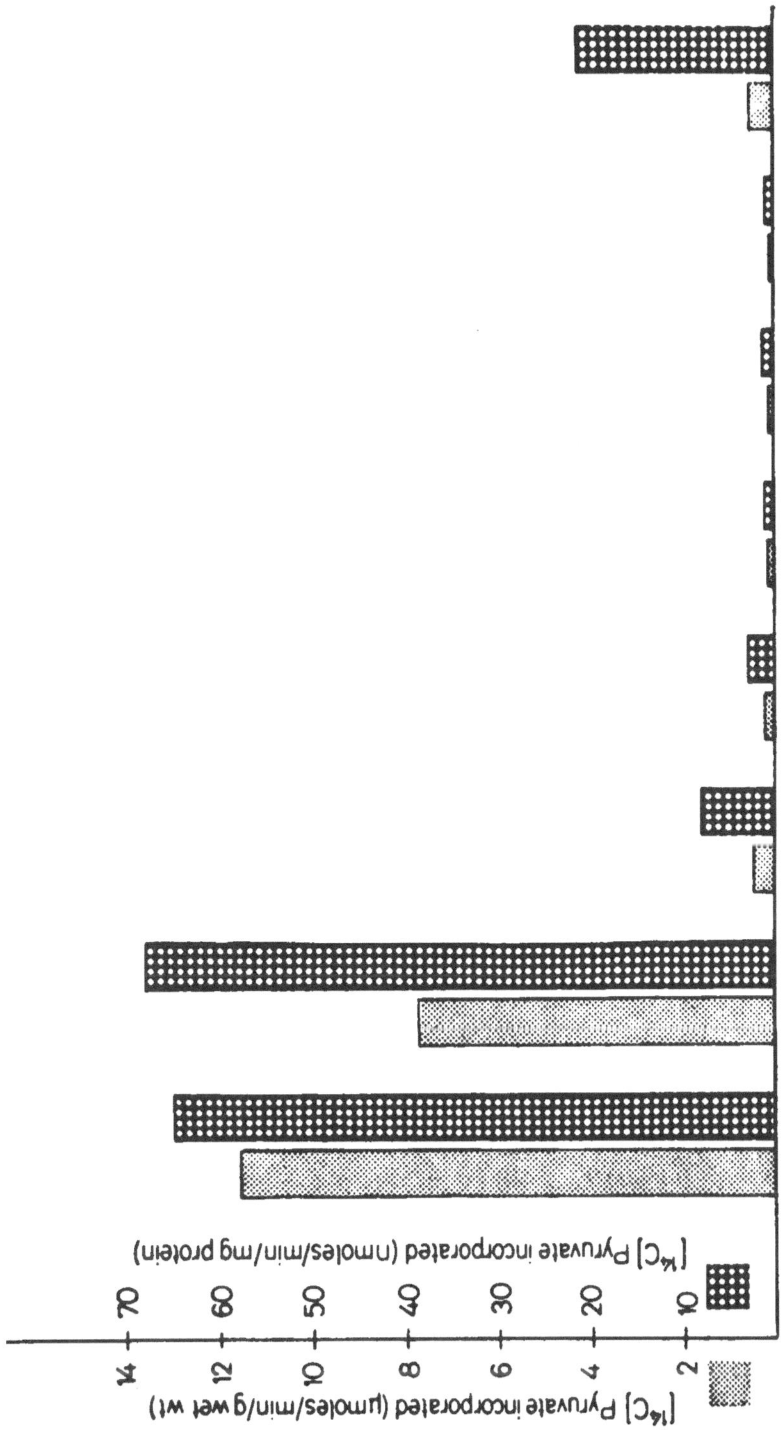

FIGURE 5. The distribution of pyruvate carboxylase in various rat tissues. (From Bottger, I., Wieland, O., Brdiczka, D., and Pette, D., *Eur. J. Biochem.*, 8, 113, 1969. With permission.)

content. Pyruvate carboxylase, along with malate dehydrogenase and aspartate aminotransferase, is found in the faster-sedimenting mitochondria (cf., ornithine aminotransferase or glutamate dehydrogenase). Further fractionation of mitochondria by treatment with digitonin[125,735] to yield the component fractions of outer membrane/intermembrane space, inner membrane, and matrix has consistently shown that pyruvate carboxylase is to be found in the matrix fraction along with glutamate dehydrogenase, the soluble enzymes of the citric acid cycle (viz., citrate synthase, malate dehydrogenase, isocitrate dehydrogenase, and fumarase), and the mitochondrial isoenzyme of PEP carboxykinase (e.g., Figure 6).[125,527,862,946]

Depending on the nature of the tissue,[349] the physiological state of the animal,[527,862] and the composition of the homogenizing medium,[946] there will be a varying amount of leakage of the soluble contents of the mitochondria into the cytosol. Measurements of pyruvate carboxylase in subcellular fractions isolated by differential centrifugation usually show some activity in the particle-free supernatant.[57,309] However, where appropriate marker enzymes (i.e., from the same compartment of the mitochondria) have also been determined, the apparently extramitochondrial occurrence of pyruvate carboxylase can be completely accounted for by mitochondrial breakage.[118,527,717,777,859,862,946] Thus, e.g., the 10-fold increase in liver pyruvate carboxylase observed in diabetic sheep was to a large extent attributable to a massive 37-fold increase in an apparently cytosolic form of this enzyme. The data clearly show, however, that the percentages of the two exclusively mitochondrial matrix enzymes (viz., citrate synthase and glutamate dehydrogenase) found in the cytosol also increase in parallel with the increased proportion of pyruvate carboxylase activity in the cytosol. This suggests that the mitochondria of diabetic sheep are remarkably fragile, as subsequent electron microscope studies confirmed (see Section V.B.2.d).

b. Differential (or Fractional) Extraction

This procedure, described by Pette,[644] is based on subjecting a tissue mince or homogenate to progressively more effective extraction procedures, so that enzymes of the cytosol are extracted first and mitochondrial enzymes later. Employing this approach, Böttger et al.[118] found that the pyruvate carboxylase activity of rat liver followed exactly the same distribution as the intramitochondrial enzymes citrate synthase, 3-hydroxyacyl-dehydrogenase and glutamate dehydrogenase (see Figure 7). They therefore concluded that pyruvate carboxylase has an exclusively mitochondrial location in rat liver. However, Dugal,[244-246] using basically this same approach as well as several other fractional extraction procedures,[246] found that while 20 to 50% of the pyruvate carboxylase activity of rat, pig, bovine and chicken liver was present in the initial supernatants, there were only small amounts of glutamate dehydrogenase activity extracted until 0.1% deoxycholic acid was included in the medium. On the basis of these results and some earlier reports of some portion of the pyruvate carboxylase being in the particle-free supernatant, Dugal concluded that 40% of this enzyme is in "an extramitochondrial, noncytoplasmic location, possibly in a distinct mitochondrial population containing little or no glutamate dehydrogenase, or in a new subcellular compartment".

However, as shown by Walter,[931-933] Lardy,[319] and Seubert[946] and colleagues, glutamate dehydrogenase binds readily in certain media to subcellular fractions such as microsomes and nuclei, whereas this is not the case for pyruvate carboxylase. Thus, even when equal percentages of glutamate dehydrogenase and pyruvate carboxylase are released from the mitochondria during homogenization, a higher proportion of the pyruvate carboxylase will be detected in the particle-free supernatant because a significant fraction (60% in sucrose-triethanolamine-HCL; 95% in sucrose-sodium acetate medium[931]) of the glutamate dehydrogenase will have been sedimented with the parti-

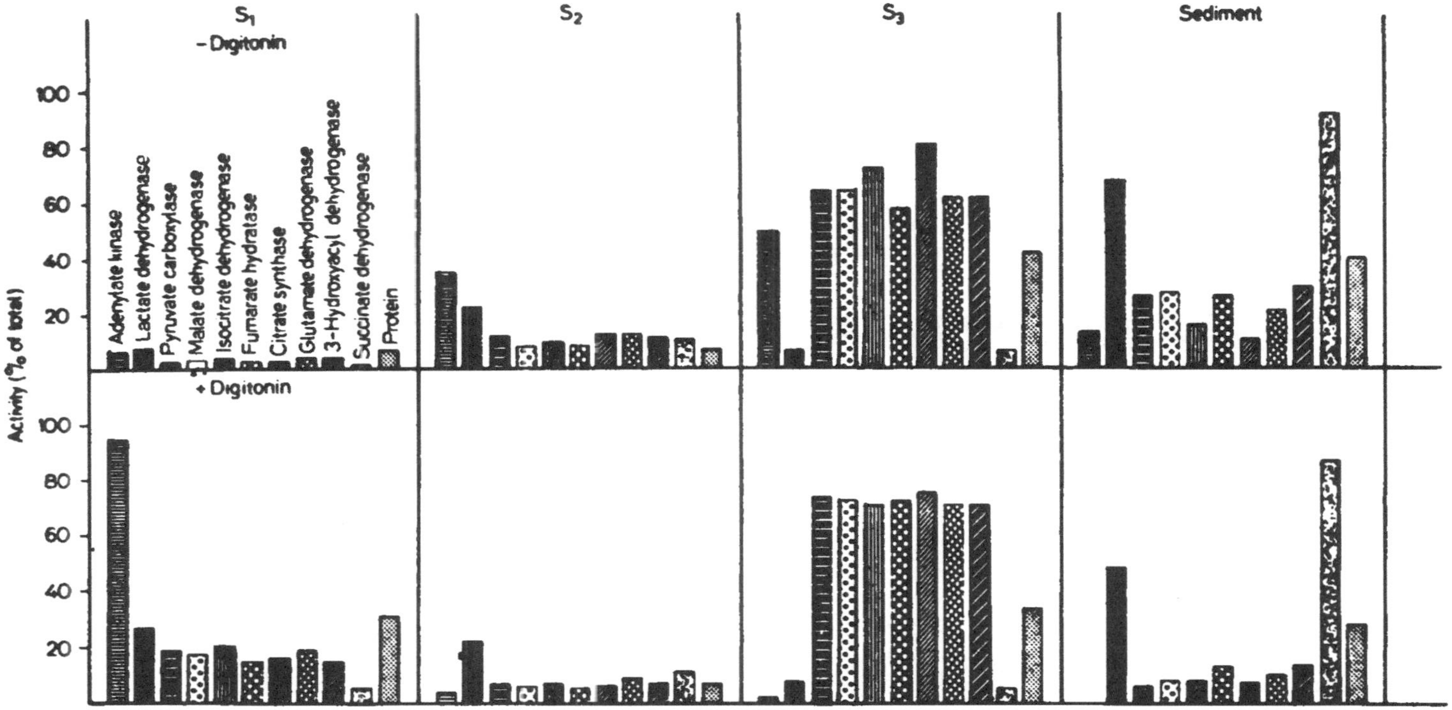

FIGURE 6. Fractional extraction of isolated rat liver mitochondria. Washed mitochondria are suspended in sucrose medium (0.44 *M* sucrose, 10 m*M* triethanolamine-HCl, 2 m*M* EDTA, pH 7.4) to a concentration of 20 to 30 mg mitochondrial protein per milliliter of suspension. Added to 10 mg of mitochondrial protein are 2.1 mg of digitonin and the suspension is incubated for 20 min at +4°C. After 10 min centrifugation at 9500 × g the supernatant S1 is obtained. The sediment is resuspended in the sucrose medium and centrifuged again for 10 min at 9500 × g. The supernatant is S2. The sediment is suspended in 0.1 *M* phosphate buffer, pH 7.2, and exposed for 4 × 10 sec to sonication in a Branson Sonifier. After 15 min centrifugation at 144,000 × g supernatant 3 (S3) is obtained. The sediment of this centrifugation is resuspended. For each fraction, the enzyme activities are calculated as percentage of the total activities. Control: mitochondria, identical treatment but without digitonin. Mean values of three experiments. (From Böttger, I., Wieland, O., Brdiczka, D., and Pette, D., *Eur. J. Biochem.*, 8, 113, 1969. With permission.)

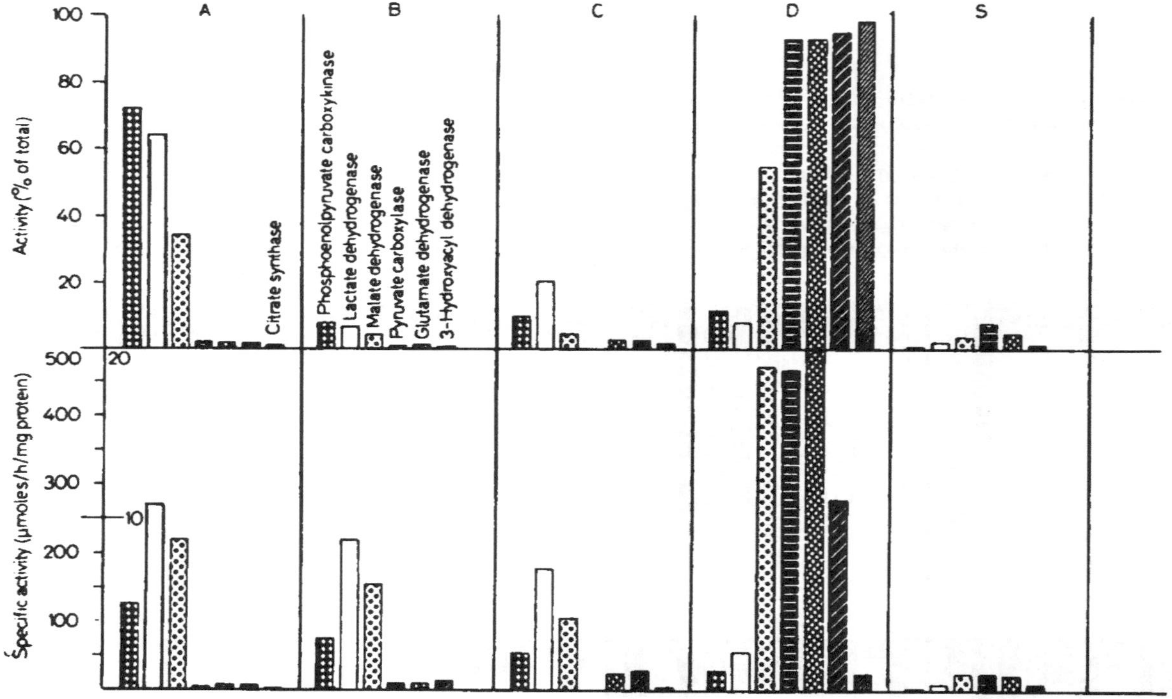

FIGURE 7. Fractional extraction of rat liver. The tissue was passed through a special meat grinder and was suspended in a 19-fold volume of sucrose medium (0.3 *M* sucrose, 10 m*M* triethanolamine-HCl, 2 m*M* EDTA, pH 7.2). After 15 min stirring the suspension was centrifuged for 15 min at 144,000 × g. The supernatant is fraction A. Fraction B is obtained by repeating the extraction for 15 min in fresh sucrose medium, and fraction C results from a third extraction which is performed in 0.1 *M* phosphate buffer, pH 7.2. Fraction D is the supernatant which is obtained after disintegrating the sediment of Fraction C in 0.1 *M* phosphate buffer by 2-min treatment with the Ultra-Turrax (Janke und Kunkel K.G., Staufen i. Br., Germany). Fraction S is the resuspended final sediment. Enzyme activities were measured in all fractions and for each fraction the percentage of the total activity and the specific activity were calculated. Because of the great discrepancies of the specific activities there are two different scales. The outer scale (0 to 500) gives the values for the specific activities of lactate dehydrogenase, malate dehydrogenase, glutamate dehydrogenase, and 3-hydroxyacyl dehydrogenase, while the inner scale (0 to 20) is for pyruvate carboxylase, PEP carboxykinase, and citrate synthase. Mean values of four experiments, except PEP carboxykinase and the specific activity of pyruvate carboxylase, which were determined in one experiment. (From Böttger, I., Wieland, O., Brdiczka, D., and Pette, D., *Eur. J. Biochem.*, 8, 113, 1969. With permission.)

culate or membranous fraction. With a phosphate-containing[931] or glycylglycine-containing[946] homogenization medium in which glutamate dehydrogenase does not bind to the organelles, equal percentages of both enzymes are found extramitochondrially.

It seems reasonable to conclude, therefore, that liver pyruvate carboxylase is localized exclusively in the mitochondria,[931] albeit in a more rapidly sedimenting subpopulation of these organelles.[946] Any lingering doubts about this conclusion would best be dispelled by resorting to an alternative experimental approach.

c. Immunochemical Techniques

When an enzymatic activity is detected in two (or more) intracellular compartments, the question arises, as discussed in Section III.A.2, as to whether:

1. There exist two (or more) discrete isoenzymic forms of this enzyme
2. The same enzyme actually does occur in vivo in several locations within the cell, perhaps in some equilibrium distribution which is influenced by the physiological state of the cell
3. The observed distribution of activity is an artifactual result created by the methods employed in vitro

Immunological techniques have proved to be an effective means of resolving such questions; the mitochondrial and cytoplasmic activities of malate dehydrogenase,[342] aspartate aminotransferase,[602] and PEP carboxykinase[53] have been clearly distinguished as different molecular forms by their lack of immunological cross-reaction. Genetic studies[779] have corroborated this conclusion for the malate dehydrogenase, and protein sequence data[76] have confirmed the separate identities of the aspartate aminotransferases. The two isoenzymes of PEP carboxykinase are also readily distinguishable, at least in the rat, by their different responses to hormonal stimuli.[605,781]

Antibodies, produced by rabbits immunized with pyruvate carboxylase purified to homogeneity from both rat liver[57] and sheep liver[862] mitochondria, have been shown to react identically in qualitative (Ouchterlony)[862] and quantitative (inhibition of activity and immunoprecipitation) tests with both the mitochondrial and cytoplasmic activities of this enzyme.[57,862] Given the discriminating nature of immunological techniques, it seems highly likely, though not as yet proven, that these two activities are the same molecular species, and therefore probably have the same intracellular origin. Furthermore, these results are consistent with the conclusions of the marker enzyme studies that there is only a single pyruvate carboxylase present in the liver in vivo, and that this is located in the mitochondria.

d. Electron Microscopy

Although electron microscopy was employed hand-in-hand with enzymatic assays in the development of tissue fractionation procedures,[495,629,957] it has received surprisingly little attention in the resolution of the controversy over the intracellular localization of pyruvate carboxylase. However, as illustrated in Figures 8 and 9, Taylor et al.[862] produced clear electron microscopic evidence to substantiate their contention, based on the distribution of marker enzymes and immunochemical studies, that the very substantial increase (over 30-fold) in pyruvate carboxylase activity in the cytosol was attributable to the release of this enzyme from the matrix subcompartment of the mitochondria during tissue homogenization. Comparison of sections taken from the livers of normal and diabetic sheep (Figure 8) reveals that, as well as there being more fat droplets and fewer glycogen granules evident in the diabetic liver, the most striking difference is the two- to threefold increase in the axial dimensions of many of the

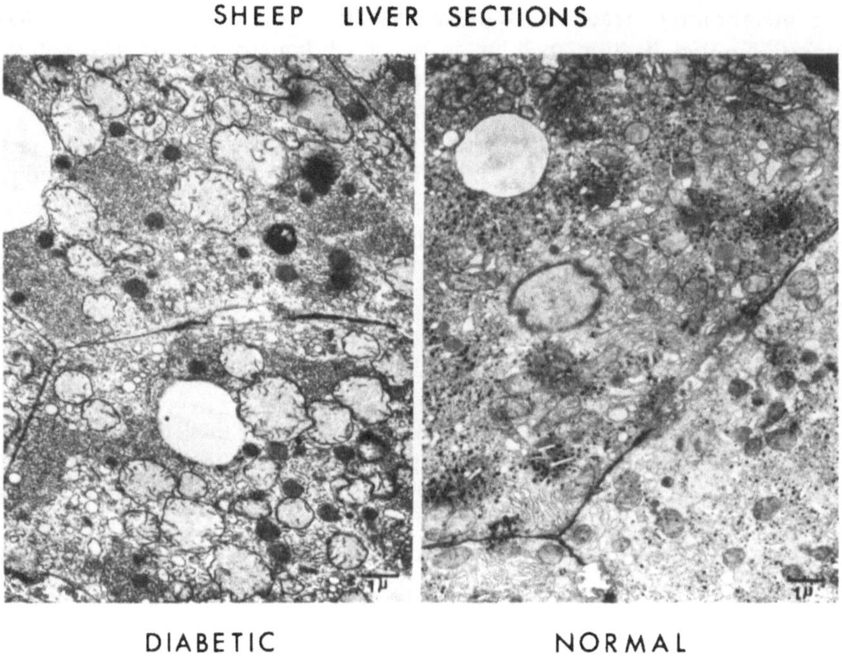

FIGURE 8. Electron micrographs from normal and diabetic sheep. The samples were fixed for 30 min with 1% (w/v) osmium tetroxide, dehydrated in acetone, and embedded in araldite. Sections were cut with glass knives, stained with uranyl acetate and lead acetate, and viewed in a Siemens Elmiskop I with an 80-kV and 50-μm objective aperture. The bar indicates 1 μm. (From Taylor, P. H., Wallace, J. C., and Keech, D. B., *Biochim. Biophys. Acta*, 237, 179, 1971. With permission.)

mitochondria in the diabetic sheep liver. It would seem likely that these swollen, lightly staining mitochondria with abnormal cristae would be more fragile than those of normal tissue.

Indeed, comparison in the electron microscope of the mitochondrial fraction isolated from normal and diabetic sheep liver (Figure 9) shows that none of the larger mitochondria has been isolated intact, but that there is a much reduced number of mitochondria of normal appearance accompanied by a large amount of membrane fragments and vesicles. It would be expected that the matrix enzymes of these disrupted giant mitochondria would be isolated in the cytosol fraction.

Abnormalities in the size (an increase of 60%), appearance, and fragility of liver mitochondria have also been observed for diabetic rats,[364,374,536] but the effects are not as extreme as in the sheep.

Electron microscopy was also used by Weiss et al.[946] to demonstrate that the various subfractions of mitochondria, which they isolated by rate-zonal centrifugation on sucrose density gradients, had retained their outer membrane and had a condensed conformation.

Similarly, Landriscina et al.[484] used electron microscopy to monitor the effectiveness of their fractionation of pigeon liver mitochondria. In agreement with similar studies in other species, they showed unequivocally that pyruvate carboxylase, along with $NADP^+$-isocitrate dehydrogenase, NAD^+-isocitrate dehydrogenase, glutamate dehydrogenase, and aspartate aminotransferase, were localized in a fraction which was practically free of both the inner and outer membranes, and which they identified as the mitochondrial matrix.

ISOLATED LIVER MITOCHONDRIA

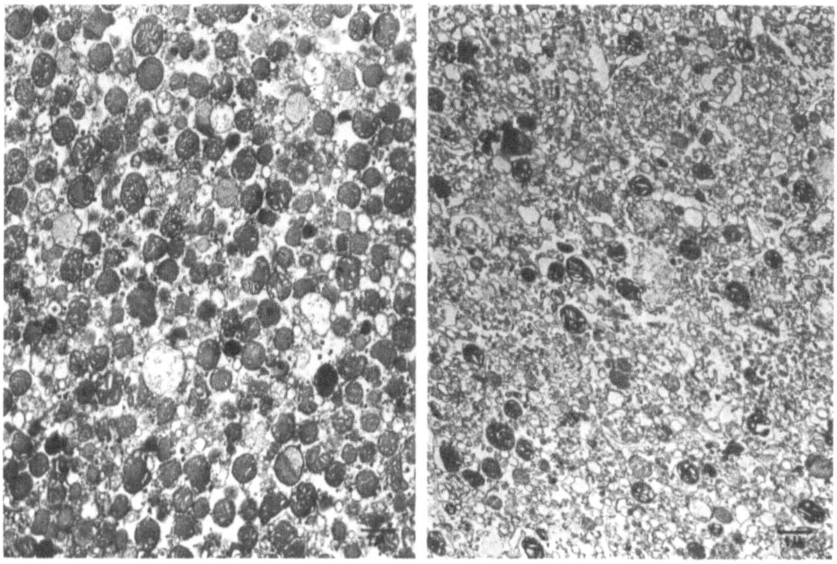

NORMAL DIABETIC

FIGURE 9. Electron micrographs of isolated liver mitochondria from normal and diabetic sheep. The mitochondria were isolated as described by Keech and Utter.[439] The mitochondrial pellet was fixed with 2% (w/v) glutaraldehyde for 1 hr followed by treatment with 1% (w/v) osmium tetroxide for 30 min. Samples were then dehydrated in acetone and embedded in araldite. Sections were then cut and viewed as described in Figure 8. (From Taylor, P. H., Wallace, J. C., and Keech, D. B., *Biochim. Biophys. Acta*, 237, 179, 1971. With permission.)

Histochemical studies in the electron microscope have demonstrated that the mitochondria of the liver and other tissues are enzymatically heterogeneous.[341] The application of histochemical techniques employing specific antibodies linked to peroxidase[41] should complement the direct histochemical assays and extend the range of enzymes that can be localized by direct visualization in the electron microscope. While not with out its own technical difficulties and limitations, this approach avoids many of the problems inherent in enzyme localizations based on disruptive procedures such as homogenization, and is particularly applicable to tissues having a more complex ultrastructure than does liver.

Similarly, fluorescent-labeled antibodies are an effective means of localizing intracellular structures and have been used successfully by Gratzner et al.[332] to confirm the mitochondrial location of pyruvate carboxylase in 3T3 fibroblasts undergoing conversion to adipocytes.

VI. CONCLUSIONS

Summarized in this chapter is the evidence which clearly demonstrates that pyruvate carboxylase is one of the major anaplerotic enzymes. Its widespread occurrence in various species from all major life forms except plants is significant and could be of value in tracing evolutionary relationships.

Although as noted above, in some organisms the functions of pyruvate carboxylase are either wholly or partly taken over by other anaplerotic enzymes, there are nevertheless many different cell types in which this enzyme performs essential functions. These are discussed further in Chapters 5 and 6.

The properties of pyruvate carboxylase, particularly with regard to its regulation by acyl-CoA compounds and by aspartate or α-ketoglutarate, vary considerably. These aspects, as well as details of the structure of the purified forms of this enzyme, are considered further in Chapters 3 and 5.

Despite the many hundreds of investigations involving pyruvate carboxylase in the past 2 decades, it is evident from the gaps in Tables 1 to 4 (some obvious, other implicit) that much remains to be clarified about the occurrence of this enzyme and the many factors affecting its activity. This is particularly true for some microbial species, for the invertebrates, and among the vertebrate species not commonly found in laboratories. When viewed in the context of the levels of activity and the regulatory properties of other key enzymes of intermediary metabolism, such comparisons of the properties of the different forms of an enzyme like pyruvate carboxylase as it occurs in widely differing organisms can be most revealing.

ACKNOWLEDGMENTS

I should like to thank Ms. J. Burton for the care and patience involved in typing the manuscript, and Ms. J. Brazier for preparing the diagrams for publication.

It is a pleasure to acknowledge also the valuable assistance of Ms. J. Lloyd and Ms. E. Randva of the Medical Library, University of Adelaide, with the bibliographic searches conducted using data bases available on the Australian Medline Network.

Work from the author's laboratory included in this review was supported by the Australian Research Grants Scheme and the University of Adelaide.

Chapter 3

THE STRUCTURE OF PYRUVATE CARBOXYLASE

John C. Wallace and Simon B. Easterbrook-Smith

TABLE OF CONTENTS

I. INTRODUCTION

Despite the widespread occurrence of pyruvate carboxylase in many varied life forms (see Chapter 2), the purification of this enzyme to homogeneity has so far only been reported for some 12 sources (viz., the livers of chicken,[749,755] turkey,[751] rat,[545,773] sheep[328] and ox,[751] sheep kidney,[29,46] *Saccharomyces cereviseae,*[157,836,996] *Aspergillus nidulans,*[616] *Bacillus stearothermophilus,*[163,503] *Arthrobacter globiformis,*[350] *Pseudomonas citronellolis,*[193,856] and *Ps. fluorescens.*[570] Partial purifications have also been reported for the pyruvate carboxylases from human[767] and pig[937] livers, rat brain,[525] locust flight muscle,[708] *Amphidinium carterae,*[17] *Aspergillus niger,*[269] *Azotobacter vinelandii,*[758] and *Ps. aeruginosa.*[671]

In view of the considerable confusion over the physical properties of pyruvate carboxylase created in the past two decades by studies of impure samples of this enzyme (see Sections III.A.1 and III.B.1), the need for homogeneous preparations of fully active enzyme as starting material for structural studies can hardly be overstressed.

II. PURIFICATION OF PYRUVATE CARBOXYLASE

A. Animal Sources

1. The Starting Material and its Extraction

As discussed in Chapter 5 (Section I.B.2.a), in most species investigated there is a two- to threefold increase in total liver pyruvate carboxylase activity after a period of starvation. Most investigators have, therefore, prepared liver mitochondria from fasted animals to obtain the highest specific activity at the outset. However, as noted in Chapter 2 (Section V.B.2.d), there could be a point reached in the period of fasting in some species where increasing mitochondrial fragility could limit or even reduce the recovery of these organelles intact from the tissue homogenate and hence also the yield of this mitochondrial matrix enzyme.

Lyophilized EDTA-washed mitochondria are the usual starting material. Given the problem that proteolysis has been with several biotin carboxylases,[75,983] it would also seem prudent to include protease inhibitors (e.g., 2 to 100 μM phenylmethylsulfonyl fluoride,[193,749] Trasylol® (200 Kallikrein inactivator units/mℓ)[519] in the extraction medium as several investigators have recommended.

2. Fractionation with Ammonium Sulfate

Pyruvate carboxylases from vertebrate sources are precipitated between 25 and 33% saturation with respect to ammonium sulfate (equivalent at 25°C to 1.03 to 1.37 *M*).

This precipitation is carried out at room temperature because it appears all vertebrate pyruvate carboxylases are inactivated by the combination of such high ionic strength and low temperature.

Several investigators[545,755,773] have achieved a significant purification by successive "back-extractions" of the above pellet with a series of ammonium sulfate-EDTA solutions of decreasing concentration.

However, this fractionation procedure does not generally yield homogeneous enzyme and has been superceded (or, in some cases, abbreviated and succeeded) by anion exchange chromatography on various DEAE-substituted matrixes.

3. Anion Exchange Chromatography on DEAE-Matrixes

After the enzyme has been precipitated with ammonium sulfate, it must be desalted in order for it to bind to anion exchange material. This can be effected by gel filtration on Sephadex G-25 or its equivalent, but for large-scale preparations this desalting can be achieved more conveniently by precipitation of the enzyme with polyethylene glycol.[328] In some cases this step can be used selectively to effect a purification of two- to threefold in its own right.[525]

Although various chromatographic purifications had been attempted previously (with only rather disappointing results), the key to the success of the DEAE-Sephadex-A50 procedure introduced by Scrutton and Fung[749] was their unconventional use of a linear gradient of increasing ammonium sulfate concentration in a phosphate-buffered medium. Providing a reducing agent (e.g., dithioerythritol [0.1 m*M*]) is present in all buffers, this step regularly achieves a 10 to 20-fold purification with a 60 to 70% recovery of enzymic activity. In those situations where there is a particularly high ratio of glutamate dehydrogenase to pyruvate carboxylase activity (e.g., in ruminant liver), an improved separation of these two enzymes can be achieved during this anion-exchange chromatography by combining a pH gradient along with the ionic strength gradient, as described by Goss et al.[328]

4. Gel Permeation Chromatography

Largely because of the unusually high molecular weight of pyruvate carboxylase (see Section III.A.1), many investigators have found gel filtration on Sephadex G-200[29] or Sepharose 6B[329] to be an effective purification step.

Recent advances in high-performance liquid chromatography (HPLC) technology have enabled rapid (~8.5-min) separations, albeit in small quantities (~10 to 15 mg) of pyruvate carboxylase from its subunits or from other proteins differing by more than ~25% from it in molecular weight using gel permeation chromatogrphy on TSK G3000SW columns as described by Khew et al.[442]

5. Gradient Centrifugation

Another purification procedure highly effective because of both the high molecular weight of pyruvate carboxylase and the protective effect of sucrose and glycerol (see Section V.B), but relatively limited in its scale of operation, is density-gradient centrifugation in either sucrose[545] or glycerol.[937]

6. Affinity Chromatography on Avidin-Sepharose

The specific affinity of avidin for biotin[336] suggests this readily available egg-white protein would represent a powerful basis for affinity chromatography of biotin-containing carboxylases.

However, as early workers discovered,[95,335,346] the extreme tightness of this noncovalent binding presented problems with the elution of these enzymes from the avidin-Sepharose columns. Following the publication by Henrikson et al.[398] of a modified procedure for preparing avidin monomer affinity columns, this method has been applied successfully, with some alterations, to several mammalian forms of pyruvate carboxylase.[442]

B. Microbial Sources

1. Growth of the Organisms

As discussed in Chapter 2 (Sections II.B.1 and III.A.1.a), the level of pyruvate carboxylase activity in some organisms is strongly influenced by the carbon source in the growth medium. Accordingly, lactate is the substrate most commonly used for the large-scale culture of the organisms from which this enzyme has been purified.[193,350,758,996] However, pyruvate,[509,758] succinate,[504,758] acetate,[163,509,618] and low concentrations of sucrose,[269] glucose,[570,615] and methanol[509] have also been used, as their effects on the expression of pyruvate carboxylase vary considerably from species to species (see also Chapter 2, Sections II.A.1 and III.A.1).

2. Preparation of Cell-Free Extracts

The microbial sources for the purification of pyruvate carboxylase have, so far, either been prokaryotes or have had an extramitochondrial location for this enzyme where they were eukaryotes.[358,618] Hence, the significant purification achieved in isolating mitochondria as the first step in purifying vertebrate pyruvate carboxylases is not an available option.

Instead, the microbial cell wall, a formidable barrier in some organisms (e.g., yeast) must be breached and the cellular contents freed of the copious quantities of nucleic acids they contain so that protein separation procedures may be initiated. Cell lysis has been effected by treatment with lysozyme,[163,503,570] sonication,[17,671] the French Pressure Cell,[193,856] N_2 cavitation in the Cell Disruption "Bomb" (Parr Instrument Co.),[615,758] or by grinding with glass beads (~100-μm diameter) in a colloid mill (Eppenbach)[996] or continuous Dyno-Mill homogenizer (W. A. Bachofen, Switzerland).[350]

Most extraction media have contained inhibitors of proteases (e.g., phenylmethylsulfonylfluoride [PMSF] [5 to 100 μ*M*]), reducing agents (e.g., dithioerythritol [DTE] [0.1 to 1.0 m*M*]), Mg^{2+} (1 to 5 m*M*), and EDTA (0.1 to 5 m*M*), and have been used at low temperatures to minimize proteolytic cleavage during the isolation of the enzyme.

After removal of cell debris by centrifugation, the extract in most cases has been treated with protamine sulfate (0.04 to 0.2% w/v final concentration, ~0.02 to 0.2 mg/mg protein) or lysozyme[193] to remove the nucleic acids by precipitation.

3. Fractionation with Ammonium Sulfate

Initial precipitations with ammonium sulfate for all microbial pyruvate carboxylases purified so far appear to have used higher concentrations (at least 44% saturation [≡ 1.8 *M*][157] and some as high as 80% saturation [≡ 3.28 *M*]).[671] than have been employed with the vertebrate forms of this enzyme.

Similarly, where "back-extractions" with progressively decreasing concentrations of ammonium sulfate have been used,[157,269,350,615,996] these too have been higher than those employed with the pyruvate carboxylases of animal origins.

Typically, the enzyme has a specific activity of 1 to 3 μmol oxaloacetate formed per minute per milligram protein after this stage in the purification, with this being a 10- to 100-fold increase over the initial extract, depending on the species concerned.

4. Purification by Chromatography

a. Gel Permeation

Some investigators have combined a necessary desalting step after ammonium sulfate fractionation with the opportunity to select for pyruvate carboxylase on the basis of its high molecular weight by molecular size exclusion chromatography. Using such media as Sephadex G200,[350,570] Sepharose 6B, or Bio-Gel A-1.5m,[17] they have typically achieved a 5- to 10-fold purification with this procedure. Other purification schedules[157,163,193,503,758] have included gel permeation chromatography as a step after hydroxylapatite or anion exchange chromatography.

b. Cation-Exchange Chromatography

Hydroxylapatite — Successful purifications have been obtained by chromatography on columns of hydroxylapatite using step-wise elution with increasing concentrations of potassium phosphate buffer.[157,163,350,570] Typically, this has resulted in a two- to threefold increase in specific activity over the Sephadex G-200 purified starting material.

Cellulose phosphate — Cellulose phosphate has also been used to purify microbial pyruvate carboxylases,[503,996] although the enzyme in one case was not adsorbed by the column.[996]

c. Anion Exchange Chromatography

Several microbial pyruvate carboxylases have also been purified substantially on DEAE-Sephadex[193,503,758] or DEAE-cellulose.[17]

d. Affinity Chromatography on Avidin-Sepharose

The first pyruvate carboxylase purified by affinity chromatography on avidin-Sepharose was that from *B. stearothermophilus.*[506] Scrutton and colleagues[616] have recently shown how readily a variation of this approach not involving urea can yield native enzyme of greater than 98% purity when applied after only minimal fractionation of crude extracts of *Aspergillus nidulans* with protamine and ammonium sulfates.

III. PHYSICOCHEMICAL PROPERTIES OF THE ENZYME AND ITS SUBUNITS

A. Molecular Weight

1. Native Enzyme

The values determined by various means for the molecular weights of the catalytically active forms of pyruvate carboxylase from a number of species are summarized in Table 1. In general, it would seem that the most common form of the active enzyme is a tetramer with an aggregate molecular weight in the range of 440,000 to 600,000. Earlier estimates as high as 660 kDa for the predominant native form of pyruvate carboxylase in chicken liver[759] and yeast[768] have been revised by Utter and colleagues.[896] These overestimates may have arisen because of: (1) the presence of contaminating proteins (e.g., an octameric species of 724 kDa (see Section III.B.1), (2) the formation of inactive higher-molecular-weight aggregates (see Section V.C), (3) difficulties in applying the standard Archibald approach-to-equilibrium method to such a large protein,[887] or (4) a combination of these problems.

All have now been eliminated and the values obtained since are consistent with those expected for an obvious tetrameric species (see Section III.B.1) with protomers of approximately 120 kDa.

Table 1
THE MOLECULAR WEIGHTS AND SEDIMENTATION COEFFICIENTS OF THE CATALYTICALLY ACTIVE FORMS AND SUBUNITS OF PYRUVATE CARBOXYLASE ISOLATED FROM VARIOUS SOURCES

Source	Molecular species	Catalytic activity	Molecular weight[a]	Sedimentation coefficient	Method[b]	Ref.
Chicken liver	Octamer	+[c]	—	22.7S	A	857
	Tetramer	+[c]	—	16.4S	A	857
	Monomer	—	—	7.5S	B	857
	Tetramer	+[c]	500	—	C	896
	Monomer	—	120	—	D	75
	Monomer	—	125	—	E	75
	Monomer	—	112	—	E	124
Turkey liver	Tetramer	+[c]	440	15.1S	C/A	896/857
	Monomer	—	110	—	E	75
Rat liver	Tetramer	+[d]	—	16.2S	A	857
	Dimer	+[d]	—	12—13S	A	857
	Monomer	+[e]	—	5.1—6.1S	A	857
	Tetramer	+	—	13.2S	A	592
	Dimer	+	—	10.7S	A	592
	Monomer	+	—	7.5S	A	592
	Tetramer	+[c]	—	15.1S	B	543
	Monomer	—	130	—	E	545
	Monomer	—	130	—	E	75
	Monomer	—	116	—	E	124
Pig liver	Tetramer	+	520	—	F/G	937
	Dimer	+	260	—	F/G	937
	Monomer	+	130	—	F	937
	Monomer	—	130	—	E	75, 937
Pig brain	Tetramer	+[c]	—	14.8S	H	525
Human liver	Monomer	—	130	—	E	767
Calf liver	Tetramer	+[c]	480	16.1	C/A	896/857
	Monomer	—	120	—	E	75
Sheep liver	Tetramer	+[c]	480	—	G/I/J	328/442/326
	Monomer	—	117	—	E	124
	Monomer	—	113	—	L	326
Sheep kidney	Tetramer	+[c]	480	—	C	46
	Tetramer	+[c]	450	15.1 ± 0.2S	J/B	46
	Monomer	—	115	—	D	46
	Monomer	—	110	—	K	46
	Monomer	—	120	—	L	46
	Monomer	—	118	—	E	124
Yeast	Tetramer	+[d]	—	16.3S[g]/15.8S[h]	A	857
(S. cerevisiae)	Tetramer	+[d]	475	—	M	75
	Monomer	—	120	—	E	75, 194
Aspergillus	Tetramer	+[d]	(4 × 133)	—	R	616
nidulans	Monomer	—	133	—	E	616
Amphidinum carterae	—	+	340/≥400	—	G/N	17
Pseudomonas citronellolis	Dimer ($\alpha_2\beta_2$)	+[f]	—	12.96S/12.28S	A/B	856
	Dimer ($\alpha_2\beta_2$)	+[f]	256	—	C	856
	Subunit-α	—	65	—	E	75
	Subunit-β	—	54	—	E	75
	Tetramer ($\alpha_4\beta_4$)	+[f]	454/510	—	M/G	193
	Tetramer ($\alpha_4\beta_4$)	+[f]	530	—	O/P	193
	Dimer ($\alpha_2\beta_2$)	+[f]	228	—	P	193
Ps. fluorescens	?	+[f]	300	—	G	570

Table 1 (continued)
THE MOLECULAR WEIGHTS AND SEDIMENTATION COEFFICIENTS OF THE CATALYTICALLY ACTIVE FORMS AND SUBUNITS OF PYRUVATE CARBOXYLASE ISOLATED FROM VARIOUS SOURCES

Source	Molecular species	Catalytic activity	Molecular weight[a]	Sedimentation coefficient	Method[b]	Ref.
Ps. aeruginosa	?	+[f]	220	—	G,H	671
Ps. oleovorans	?	+[f]	300	—	G	509
Bacillus stearothermophillus	Tetramer	+[c]	600	—	G	505
	Monomer	—	150	—	E	505
Thermophilic *Bacillus*	Tetramer	+[c]	600	—	G	504
	Tetramer	+[c]	558	—	J	504
	Monomer	—	140—155	—	E	504
	Monomer	—	140	—	E	140
Azotobacter vinelandii	—	+[f]	296	12.78S	J,A	758
Arthrobacter globiformis	Tetramer	+[c]	(580)	15.0S/14.2S	B/A	350
	Dimer	—	290	9.2S—10.7S	Q	350
	Monomer	—	130	—	E	350

[a] Molecular weight in kilodaltons.
[b] A denotes the reaction enzyme sedimentation technique;[856] B denotes velocity sedimentation with Schlieren optics;[732] C denotes the equilibrium sedimentation technique;[887] D denotes the equilibrium sedimentation in 6*M* guanidine-HCl by the meniscus-depletion method;[998] E denotes an SDS-PAGE method; F denotes glycerol gradient centrifugation; G denotes gel filtration on Sephadex G200 or equivalent; H denotes sucrose density gradient centrifugation; I denotes high-performance gel filtration on TSK 3000 SW; J denotes the molecular weight calculated from values for $S°_{20,w}$ and $D°_{20,w}$;[732] K denotes sedimentation equilibrium in SDS; L denotes gel filtration in SDS on Sephadex G200 or 150; M denotes the meniscus-depletion sedimentation equilibrium method of Yphantis;[998] N denotes gradient gel electrophoresis; O denotes scanning transmission electron microscopy; P denotes disc gel electrophoresis; Q denotes the sedimentation equilibrium (Archibald) method;[732] R denotes structure assessed by electron microscopy with negative staining.
[c] Activity dependent on the presence of acetyl-CoA.
[d] Activity evident in the absence of acetyl-CoA, but also activated by acetyl-CoA.
[e] Active in the absence of acetyl-CoA if high (0.29 *M*) concentration of bicarbonate present.
[f] Activity unaffected by the presence of acetyl-CoA.
[g] Sedimentation in the presence of acetyl-CoA.
[h] Sedimentation in the absence of acetyl-CoA.

There are, however, other catalytically active forms of this enzyme:

1. When concentrated solutions of chicken liver pyruvate carboxylase are dialyzed against its substrates, approximately 20% of it associates and remains catalytically active as a 22.7S species which has been tentatively identified as an octamer.[857]
2. Several groups[592,857] have shown by reacting enzyme sedimentation studies that rat liver pyruvate carboxylase exists as an associating/dissociating mixture of tetramers (16S), dimers (12 to 13S), and monomers (6 to 7S). The relative proportions of these three species and the level of their expressed activity is strongly influenced by the concentration and nature of specific ions present in the solution, by the temperature of the system, and by the presence or absence of acetyl-CoA (see also Section V.B). At low concentrations of enzyme and salt, the major active form in the presence of acetyl-CoA is the tetramer, whereas in the absence of this allosteric effector it is the dimer. High concentrations of $KHCO_3$ (0.29 *M*) promote the dissociation of rat liver pyruvate carboxylase to dimers and subsequently to monomers even in the presence of acetyl-CoA. Both species are active

in the absence of acetyl-CoA, though the dimers at least are still activated by this effector.[857] Nakashima et al.[592] have shown that incubation of rat liver pyruvate carboxylase at 0°C in the presence of ammonium chloride (30 to 100 m*M*) reversibly dissociates the tetramer to enzymically active dimers and monomers which exhibit the same responsiveness to acetyl-CoA as does the tetramer.

3. Enzymically active tetramers, dimers, and monomers have been reported also for pig liver pyruvate carboxylase subjected to glycerol gradient centrifugation, but upon gel filtration on Sephadex G200, activity was evident in only the tetramer and dimer fractions.[937] However, neither of these "discontinuous" techniques is as definitive or unequivocal as the reacting enzyme sedimentation approach since it is possible for active tetramers to form by reassociation of the separated monomers or dimers after the separation process by centrifugation or gel permeation has been completed.
4. Taylor et al.[856] showed, by reacting enzyme sedimentation studies, that pyruvate carboxylase from *Ps. citronellolis* had a $S^\circ_{20,w}$ of 12.3S, which was consistent with the molecular weight of 256,000 they determined for it by sedimentation equilibrium. Subsequently, however, Cohen et al.[193] showed that while they could confirm by disc electrophoresis the existence under some circumstances of this apparently dimeric form of the enzyme, the predominant form of the native enzyme has a molecular weight estimated to be between 454,000 and 530,000, depending on the technique employed.
5. From reacting enzyme sedimentation and gel filtration studies, Scrutton and Taylor[758] concluded that the molecular weight of the catalytically active form of pyruvate carboxylase in *Azotobacter vinelandii* OP was only 296,000.
6. Gel filtration and sucrose density gradient studies have indicated that the pyruvate carboxylases of *Ps. fluorescens*[570] *Ps. oleovorans,*[509] and *Ps. aeruginosa*[671] are 300,000, 300,000, and 220,000 to 230,000, respectively.
7. Similarly, the pyruvate carboxylase from the dinoflagellate *Amphidinium carterae* appears by gel filtration to be only 340 kDa, but may be as much as 400 kDa by gradient gel electrophoresis.[17] Further studies of the molecular weight and subunit composition of these last-mentioned five microbial species are necessary to establish whether the native forms of this enzyme are genuinely smaller than the other, more fully characterized examples in other species.

2. Subunits

It is clear from the data in Table 1 that, with the exception of *Ps. citronellolis,* all pyruvate carboxylases studied so far have subunits in the molecular weight range from 110,000 to 150,000. Where the determination has been made using a technique with sufficient resolving power, such as SDS-PAGE, in all cases except *Ps. citronellolis,* the authors have reported all the subunits of a particular enzyme to be of identical size. Some discrepancies between different workers' estimates for various species, even some quite closely related, may be more apparent than real. One source of error is the widespread acceptance[447,943] of the molecular weight of the β-galactosidase protomer as 130,000, whereas sequencing studies[284] have since shown the correct value is 116,248. As there have been, until recently, few other commercially available proteins with high-molecular-weight subunits, β-galactosidase has been widely used as a standard in SDS-PAGE determinations of subunit molecular weights. Nonetheless, there does appear to be a small but distinct difference in subunit molecular weight between avian and mammalian species, as noted by Barden et al.[75] The protomers of the microbial pyruvate carboxylases, except *Ps. citronellolis,* also seem to be slightly larger than those of this enzyme from vertebrate livers (see Table 1).

Earlier reports that the approximately 130-kDa subunits of rat[545] and pig[937] liver pyruvate carboxylases were composed of at least three different polypeptides have been discounted by Barden et al.[75] on the grounds that when adequate precautions are taken to prevent proteolysis during dissociation of the enzyme there is no evidence for the existence of smaller components. Similarly, the suggestion[904] that the 2.3 to 2.6S species, evident upon ultracentrifugal analysis of chicken liver pyruvate carboxylase treated with 1% SDS,[759] represented dissimilar subunits of molecular weight in the range 40,000 to 50,000, has been refuted by the results of Bais et al.[47] These authors showed that, whereas for most globular proteins the relationship between sedimentation coefficient (S) and molecular weight (M) is given by:[365]

$$\log S = \log k_1 + 0.67 \log M \qquad (1)$$

where k_1 is a constant; this relationship for SDS-protein complexes is given by:[47]

$$\log S = \log k_2 + 0.27 \log M \qquad (2)$$

Thus, although a sedimentation coefficient of approximately 2.5S for a pyruvate carboxylase subunit of 120 kDa is consistent with those observed for the protomers of eight other well-characterized proteins, such measurements are an insensitive and rather imprecise guide as to the molecular weight of a polypeptide.

However, even when precautions are taken to prevent proteolysis, it is clear that the pyruvate carboxylase of *Ps. citronellolis* is composed of two dissimilar types of polypeptides, the α of 65 kDa and the β of 54 kDa, present in equimolar amounts.[75,193]

B. Three-Dimensional Structure

1. Electron Microscopic Studies of Subunit Arrangement

Soon after the first physicochemical study of pyruvate carboxylase[759] revealed that it had a large molecular weight, preparations of this enzyme were examined in the electron microscope. Unfortunately, a highly photogenic impurity exhibiting a square planar image and having some of the characteristics expected of pyruvate carboxylase was mistakenly reported to be this enzyme.[907] Like pyruvate carboxylase, this apparently square planar tetrameric molecule was "cold labile", and it had dimensions that were broadly consistent with the molecular weight of that enzyme. Despite indications that highly purified yeast pyruvate carboxylase had a rhomboidal appearance in the electron microscope,[906] it was widely accepted that vertebrate pyruvate carboxylases had a square-planar tetrameric structure.[381,768,896] It was rationalized that this structural difference was consistent with the quite different catalytic properties of the microbial enzyme, particularly in relation to its activation by acyl-CoA compounds.

However, independent studies by Goss et al.[328] and Cohen et al.[192] clearly showed that the distinctive square-planar tetrameric image belonged to a contaminant devoid of biotin and probably composed in fact of eight subunits, each of 88 kDa. A cubic octomeric structure would be consistent with its elution from Sepharose 4B gel filtration columns ahead of pyruvate carboxylase, with its unusually vivid appearance in the electron microscope, and with the occurrence of some images appearing to be composed of two parallel cylindrical structures most probably representing a side view of such a molecule.[328] Highly purified preparations of both chicken[192,328] and sheep[328] liver pyruvate carboxylases were seen to be composed of molecules similar to the yeast enzyme and having a "splayed tetrahedral" structure[328] or rhombic appearance.[192] There were no square planar tetrameric structures evident in the pyruvate carboxylase-containing fractions in either DEAE-Sephadex or Sepharose 4B column eluates.

There are fundamental differences between a splayed tetrahedral structure and a rhombus, and so further electron microscopic studies of chicken, sheep, and rat liver pyruvate carboxylases were undertaken by Mayer et al.[538] Cross-linking agents (e.g., glutaraldehyde) or protective ligands are often used to stabilize the native structure of

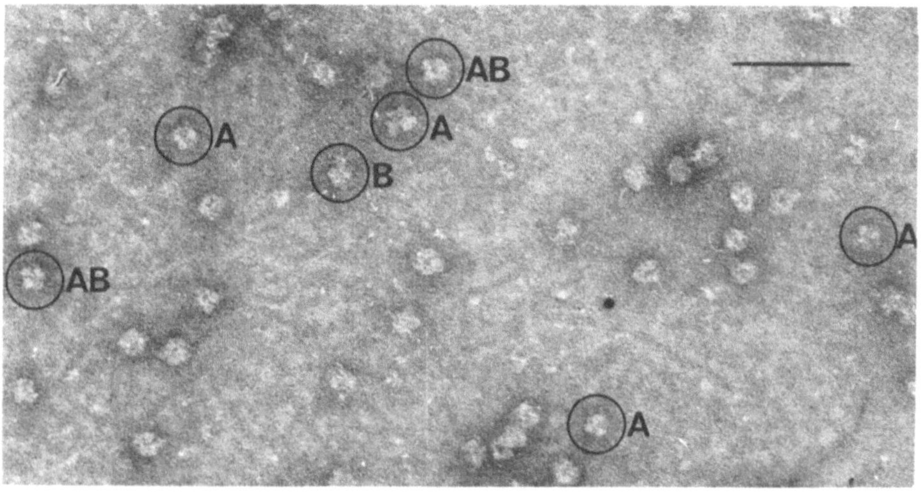

FIGURE 1. Pyruvate carboxylase isolated from chicken liver mitochondria mounted in the presence of acetyl-CoA. The sample was prepared from a protein solution (0.05 mg/m*l*) containing 0.25 m*M* acetyl-coenzyme A. Besides particle projections of diffuse appearance, there are present: (A) projections with three intensity maxima arranged at the corners of an equilateral triangle; (B) projections with four intensity maxima positioned at the corners of a rhomboid; and (AB) projections with three clear-cut intensity maxima and a barely visible fourth intensity maximum. The bar represents 80 nm. (From Mayer, F., Wallace, J. C., and Keech, D. B., *Eur. J. Biochem.*, 112, 265, 1980. With permission.)

proteins during their dilution, mounting, and staining for electron microscopy. Acetyl-CoA is known to protect pyruvate carboxylase completely against cold inactivation,[426,759,761] dilution inactivation,[29] and even low concentrations of SDS and urea,[761] as well as to improve the frequency of tetramers in electron micrographs of this enzyme isolated from a thermophilic *Bacillus*.[504] Mayer et al.,[538] therefore, investigated the effect of this allosteric activator of pyruvate carboxylase on the preservation of this enzyme during its preparation for electron microscopy. As is evident in Figure 1, in the presence of acetyl-CoA the enzyme appears as three types of images: (1) the majority as three readily visible intensity maxima in a compact triangular layout (type A) as would be expected of a tetrahedral structure, (2) a number showing three clear-cut intensity maxima with a fourth just visible (type AB), and (3) a much smaller percentage with four intensity maxima arranged at the corners of a rhomboid (type B). In the absence of acetyl-CoA the predominant forms are rhomboidal images in which two strong-intensity maxima are positioned on the short axis and two less prominent intensity maxima are evident on the long axis (see Figure 2). These and the type B images (Figure 1) appear to represent views of tetrahedral structures similar to those shown as type A (Figure 1) that have been deformed by the mounting and staining procedures, and have adopted a "splayed tetrahedron" structure. Further evidence in support of this view has been obtained subsequently[442] in the course of studies on dilution inactivation (see Section V.C). However, no tetramers in which all four subunits occupied equivalent positions and were equally well resolved (as would be expected of a rhombus) were observed under the staining conditions used by Mayer et al.[538] As all four subunits of chicken liver pyruvate carboxylase are of identical molecular weight,[75,328] the apparently smaller subunits seen in the type B (Figure 1 and 2) images probably represent either different depths of negative stain surrounding the subunits due to their being in different planes or a different orientation of the subunits.

Additional evidence favoring the conclusion that the type A and type AB images of Figure 1 are merely different views of the same tetrahedral-like structure was obtained

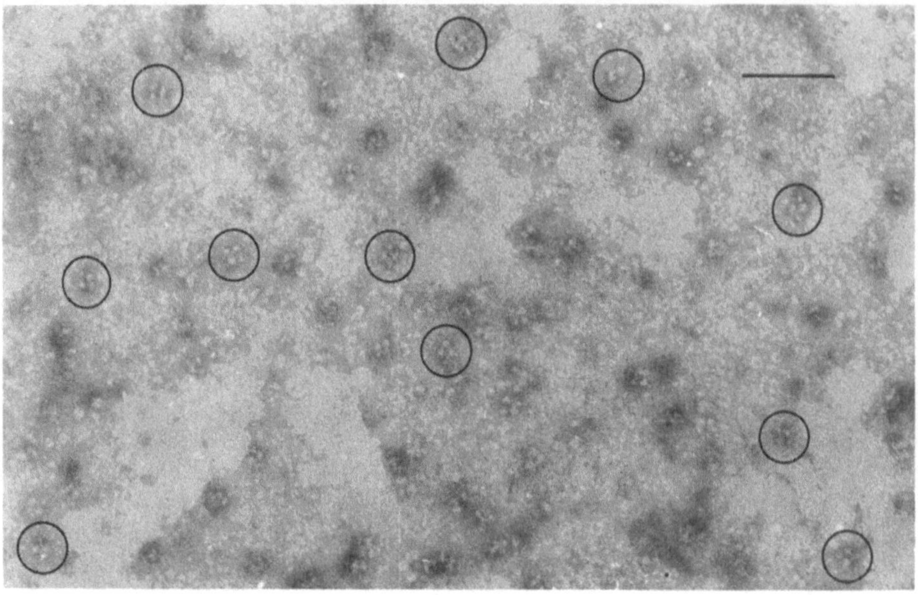

FIGURE 2. Survey micrograph of chicken liver pyruvate carboxylase in the absence of acetyl-CoA. The enzyme was mounted for electron microscopy immediately after dilution to 0.05 mg/m*l* in the absence of acetyl-CoA. Projections of particle with distinct appearance are circled. The predominant forms are rhomboidal images in which two strong-intensity maxima are positioned on the short axis and two less prominent intensity maxima are evident on the long axis. The bar represents 80 nm. (The data are from Mayer, F., Wallace, J. C., and Keech, D. B., *Eur. J. Biochem.*, 112, 265, 1980. With permission.)

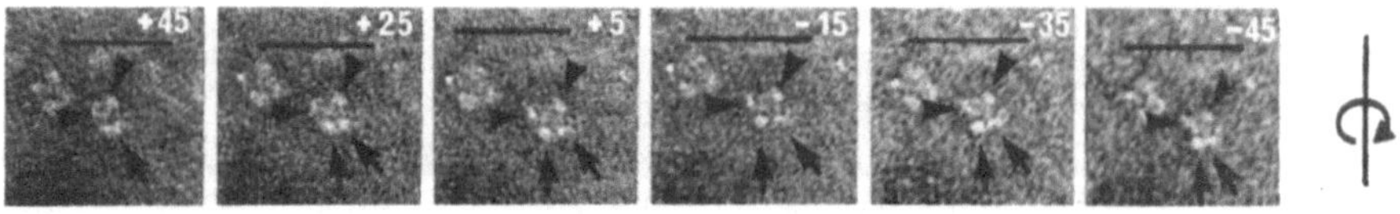

FIGURE 3. Chicken liver pyruvate carboxylase prepared for electron microscopy in the absence of acetyl-CoA. The samples were prepared as in Figure 2. The tilting angles and tilting axis used are indicated. The bar represents 40 nm. A particle projection showing three intensity maxima (projection 1, see arrowheads) is transformed, by appropriate tilting, into projections exhibiting four intensity maxima (projections 2 to 6). The data are from Mayer, F., Wallace, J. C., and Keech, D. B., *Eur. J. Biochem.*, 112, 265, 1980. With permission.)

by examining samples while tilting the stage of the electron microscope through +45 to −45° (see Figure 3).

Further detailed examination of individual particle projections (see Figure 4) revealed a "cleft" 1.2 to 2.0 nm wide running along the longitudinal midline of all subunits, apparently dividing each into at least two domains. This "cleft" was not evident, or barely so, in molecules mounted in the presence of acetyl-CoA.

A model consistent with all the features seen in images of the isolated pyruvate carboxylase molecule was proposed by Mayer et al.[538] (see Figures 5 and 6). It is a tetrahedron-like structure composed of two pairs of subunits in different planes orthogonal to each other, with the distal tips of the subunits within each pair tilted up relative to the plane in which the proximal ends abut, and with the opposing pairs contacting each other on their convex surfaces. The midline cleft is depicted in the

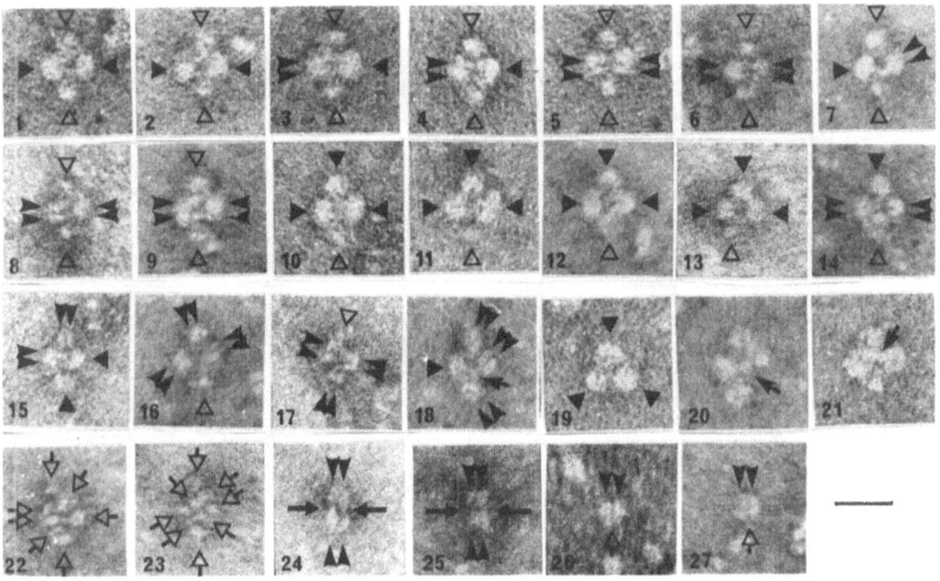

FIGURE 4. Gallery of particle projections from a sample prepared for electron microscopy in the absence of acetyl-CoA. Intensity maxima with low brightness and/or small diameter are indicated by (△), strong intensity maxima by (▲), strong intensity maxima subdivided into two submaxima by a "cleft" or line of negative stain running along the longitudinal midline of the subunit by (▲▲), and intensity maxima in particles with nonsymmetrical outlines by (⇧). The bar represents 20 nm. (Projections 1 to 9) Two strong intensity maxima are positioned on the short axis (15 to 17 nm), and two intensity maxima of less brightness and apparently smaller diameter are situated on the long axis (18 to 21 nm). (Projections 10 to 13) The images are composed of three intensity maxima of similar size and brightness, with a fourth region of less brightness and apparently smaller diameter. (Projection 19) An image of three intensity maxima of similar diameter and brightness. (Projections 3 to 9) Projections showing one or two fine white lines running parallel to the short axis of the particle and connecting the two bright intensity maxima on this axis. (Projection 18, 20, 21) Arrows indicate possible presence of protein in the center of pyruvate carboxylase molecule. (Projections 5, 8, 14 to 18) Projections showing a line of negative stain along the longitudinal midline of one or more of the major intensity maxima, suggesting the subunits may consist of two distinctly visible domains separated by a "cleft" 1.2 to 2.0 nm wide. (Projections 24, 25) Images composed of only two intensity submaxima with a contact region between them suggesting these are dimers, with each subunit exhibiting a "cleft". (Projections 26, 27) Images composed of only a pair of intensity submaxima indicating the presence of monomers with the "cleft" still evident. (From Mayer, F., Wallace, J. C., and Keech, D. B., *Eur. J. Biochem.*, 112, 265, 1980. With permission.)

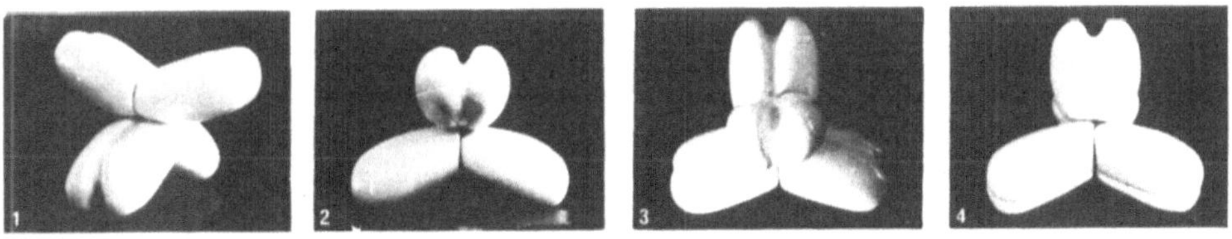

FIGURE 5. Painted clay models of chicken and sheep pyruvate carboxylase molecules. The models are as derived from electron microscopic investigations of negatively stained intact and flattened molecules. In views 1 to 4, the model is viewed from different directions. In views 3 and 4 the model is tilted so that three of the four subunits are in contact with the support while the fourth subunit appears to stand almost vertical to the support (viewed from above in 3, viewed from below the support plane in 4). Such a configuration accounts for the many projections showing three intensity maxima and also those with a fourth intensity maximum partially visible. (From Mayer, F., Wallace, J. C., and Keech, D. B., *Eur. J. Biochem.*, 112, 265, 1980. With permission.)

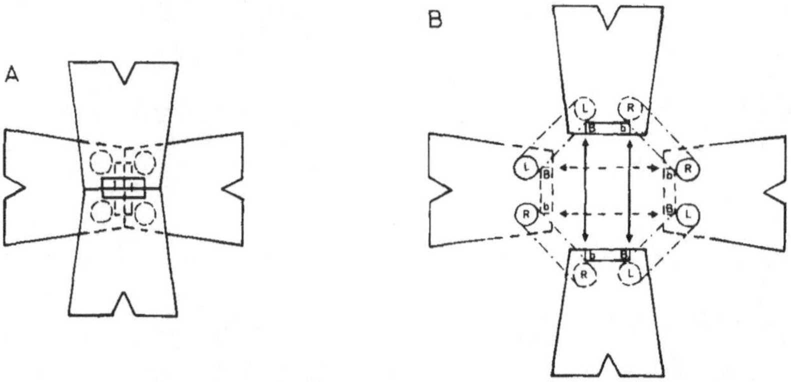

FIGURE 6. Diagrammatic view of chicken, sheep, or rat pyruvate carboxylase. (A) A face view of the enzyme; (B) An exploded view of the enzyme to illustrate the possible intersubunit contact regions. (From Mayer, F. Wallace, J. C., and Keech, D. B., *Eur. J. Biochem.*, 112, 265, 1980. With permission.)

model as being more prominent on the "external" faces of the subunits, but this is conjectural at this stage and requires further investigation.

The individual subunits have dimensions of length 8.5 nm, breadth 7.0 to 7.5 nm, and thickness 3.5 to 3.75 nm, giving each a volume, as measured by immersion of a scale model, of 160 nm^3. Taking this volume, together with the partial specific volume calculated from the amino acid composition to be 0.73 mℓ g^{-1}, one can compute the subunit molecular weight to be 116,000, in remarkably good agreement with values obtained by other physicochemical techniques (see Section III.A.2).

This model is similar to one proposed for the pyruvate carboxylase isolated from a thermophilic *Bacillus*,[504] except that the opposing pairs of subunits in the bacterial enzyme were viewed as interacting on their concave surfaces. However, the ease with which the chicken enzyme became flattened during the mounting procedures led Mayer et al.[538] to favor the "convex" arrangement of dimers within each molecule. This view has since been substantiated by the observations of Johannssen et al.[429] on complexes of avidin and chicken liver pyruvate carboxylase (see Section III.B.2).

Although no model has yet been proposed for the pyruvate carboxylase isolated from *Ps. citronellolis*, the images evident in the few micrographs published[193] (viz., mainly trimeric but with a proportion also revealing a fourth subunit) are nevertheless consistent with a tetrahedron-like structure.

Recent studies[616] of the pyruvate carboxylase purified from *Aspergillus nidulans* have clearly shown that despite the general similarity in regulatory properties of this enzyme to that from *Saccharomyces cereviseae*, its molecular structure as revealed by electron microscopy is very similar to that described for the pyruvate carboxylases from vertebrate liver mitochondria. The *A. nidulans* enzyme appears under the electron microscope as either rhomboid or triangular projections, and in areas of very deep stain also as a third type of image consisting of two intensity maxima. As a tilt series clearly shows (see Figure 7), these different projections are interconvertible, and are consistent with those originating from a tetrameric enzyme molecule in which the subunits are located at the corners of a tetrahedron.

The similarity in structure of this microbial pyruvate carboxylase to the enzymes from vertebrates also appears to extend to the molecular organization within the subunit, since there is evident here also a cleft running along the longitudinal axis of the subunits. Although significantly better preservation of molecular structure is observed if *A. nidulans* pyruvate carboxylase is prepared for electron microscopy in the presence of acetyl-CoA, the diameter of this cleft (1.0 ± 0.2 nm) is the same in the presence or absence of acetyl-CoA.

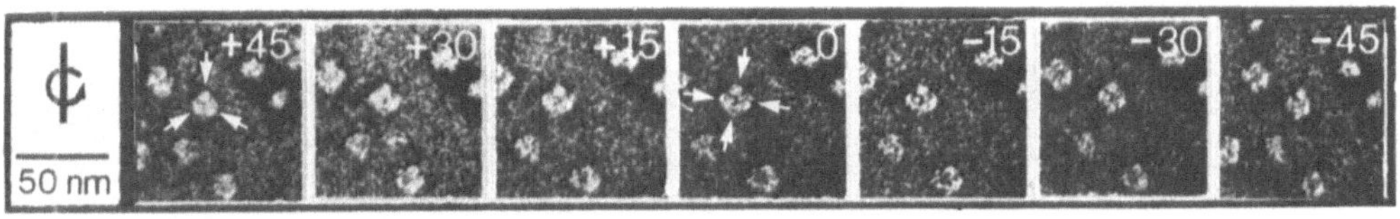

FIGURE 7. Tilt series of electron micrographs of a sample of *Aspergillus nidulans* pyruvate carboxylase. The tilt angles and tilt axis are indicated on the figure. The projections of the subunits are indicated by the arrows. The bar represents 50 nm. A triangular projection is converted to a rhomboid projection by tilting the same tetrameric enzyme molecule. (From Osmani, S. A., Mayer, F., Marston, F. A. O., Selmes, I. P., and Scrutton, M. C., *Eur. J. Biochem.*, 139, 509, 1984. With permission.)

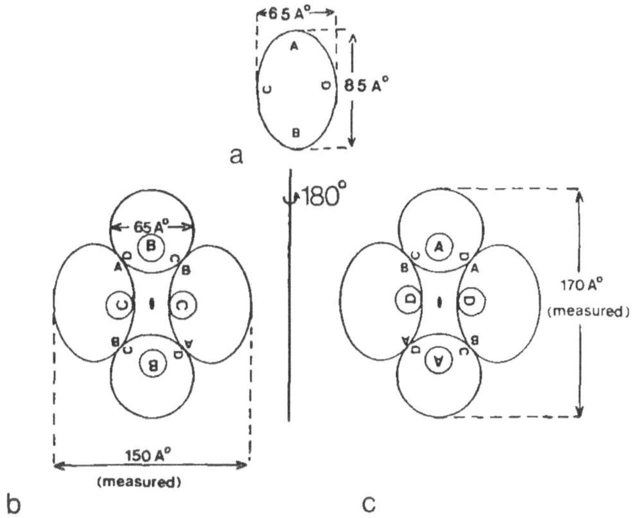

FIGURE 8. Diagrammatic model for the quaternary structure of yeast pyruvate carboxylase illustrating its binding sites. (a) The subunits of the enzyme are prolate ellipsoids approximately 8.5 nm long and 6.5 nm wide, with each having four separate binding sites, A, B, C, and D. For convenience the protomer is shown with all four binding sites on the same face. However, the authors state the model does not require that binding domains A and B lie on the same face of the protomer as C and D. (b,c) These two figures represent two separate views of the same molecule as it is rotated 180° about the vertical axis. The two subunits on the left and right sides of each drawing lie with their long axes in the plane of the paper. The two subunits at the top and bottom of each tetramer lie with their long axes perpendicular to the plane of the paper. The twofold axis of symmetry of this model is reflected by the presence of two sets of bonding pairs: A-D and B-C. The circled letters show the open binding sites on each face of the tetramer. (From Cohen N. D., Utter, M. F., Wrigley, N. G., and Barrett, A. N., *Biochemistry*, 18, 2197, 1979. With permission.)

However, the model proposed by Cohen et al.[194] for yeast pyruvate carboxylase is fundamentally different. After considering a number of computer-generated alternatives, these authors chose a model in which ellipsoidal subunits are arranged with their centers located in the same plane at the corners of a rhombus, but with the diagonally opposite pairs of subunits aligned in orthogonal planes (see Figure 8). This configuration implies either that the subunits are of two different types or that each of the apparently identical subunits has four different bonding domains (see Figure 9), two of which are unoccupied. The presence of functional unoccupied binding sites could

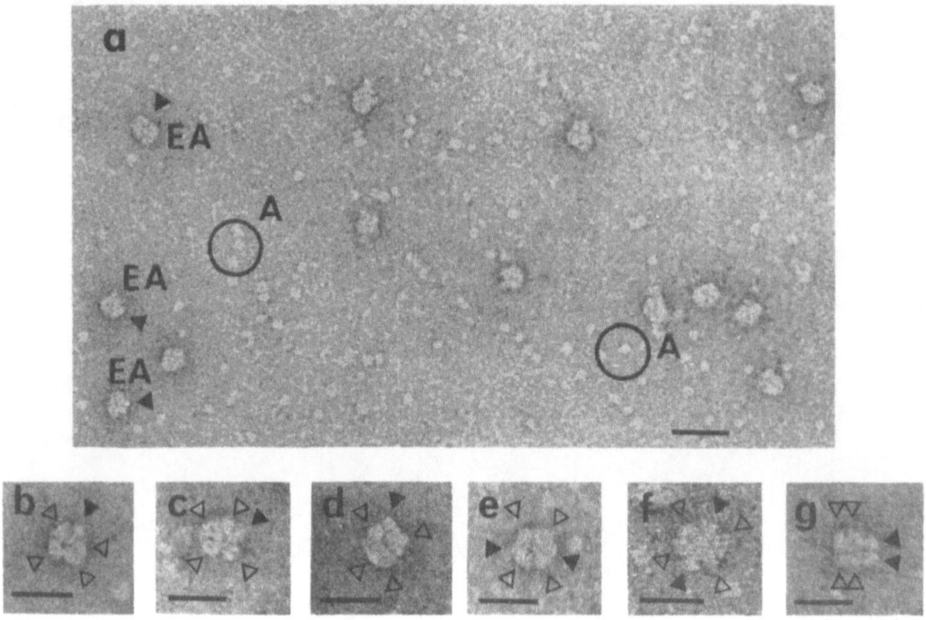

FIGURE 9. Electron micrographs of purified chicken liver pyruvate carboxylase and complexes of the enzyme with avidin after negative staining. In the low-magnification micrographs the bars represent 40 nm; in the high-magnification micrographs the bars represent 20 nm. Sample prepared from a solution containing avidin-pyruvate carboxylase in a ratio of 9:1. (a) Single enzyme-avidin complexes are visible (EA), and also many free avidin molecules (A). Bound avidin molecules (▲) are also indicated. (b to g) Selected single enzyme tetramers shown at higher magnification complexed with one avidin molecule (b to d) or two avidin molecules (e to g); the enzyme subunits (Δ) and the avidin molecules (▲) are indicated. (e,f) Two bound avidin molecules are located on opposite sides of pyruvate carboxylase tetramer; the enzyme (e) has a rhomboid-like shape, whereas in (f) the enzyme seems to be slightly flattened, resulting in a rectangular shape. (g) An enzyme tetramer with two bound avidin molecules on one side. (From Johannssen, W., Attwood, P. V., Wallace, J. C., and Keech, D. B., *Eur. J. Biochem.*, 133, 201, 1983. With permission.)

facilitate polymerization of the tetramer to larger-molecular-weight forms, but there is no evidence for such an occurrence. On the other hand, the subunits of yeast pyruvate carboxylase are indistinguishable on the basis of size or overall charge, all contain biotin, and all have blocked N-termini.[194] Though it remains to be proven that these protomers are identical, they clearly are very similar (see also Section III.H).

While at present one cannot exclude the possibility that the yeast pyruvate carboxylase has a unique structure, unlike that of any other regulated pyruvate carboxylase studied so far, it also seems possible that the rhomboidal images resulted from an artifactual distortion of the enzyme molecules due to a lack of support for their structure by the surrounding negative stain. Thus, further investigation of this molecule is required under different conditions of staining and with use of the tilting stage to examine the same molecule at several different orientations.

Certainly the rhomboidal model (Figure 8) cannot account for images in which only three mass centers are evident (see type A, Figure 1 and Figure 4, Projection 19) nor those with protein clearly evident in the center of the molecule (see Projections 3 to 10 of Figure 4). There are some examples of this latter situation also in published micrographs of yeast pyruvate carboxylase (see Figures 4B and 8 of Reference 194 and Figure 4 of Reference 896).

2. Localization of the Active Site

Crystals of pyruvate carboxylase have been obtained,[287,801] but not in sufficient quantity nor of adequate quality so far to permit a detailed structural analysis to be undertaken by X-ray diffraction techniques. This gap in our knowledge will become an increasingly urgent priority to be filled once the complete sequence of this polypeptide is established (see Section III.G.3).

With biotin-dependent enzymes having this prosthetic group covalently attached at each active site, they offer excellent prospects for the electron microscopic localization of their catalytic centers by use of the biotin-binding protein, avidin. This 68-kDa tetramer binds 4 mol of biotin with a high degree of specificity and remarkable affinity (Kd = 10^{-15} *M*).[336]

Wood and co-workers[338] first used this biotin-specific probe to demonstrate by electron microscopy that the biotin-containing subunits of transcarboxylase (methylmalonyl-CoA:pyruvate carboxytransferase [EC 2.1.3.1]) are at the interface between the 5S keto-acid binding subunits and the 12S actyl-CoA binding subunits of that enzyme.

When Duggleby et al.[248] investigated the kinetics of inhibition of pyruvate carboxylase by avidin they noted that at low ratios of the concentrations of avidin to enzyme (<8:1), a stoichiometry of binding of one biotin-binding site to one biotin occurred, i.e., on average one tetramer of avidin to one tetramer of pyruvate carboxylase. It is known from the elegant work of Green et al.[337] and Safer et al.[716] that the biotin-binding sites of avidin are arranged in pairs 2.0 nm apart, with each pair located on opposite sides of the tetrameric avidin molecule and thus separated by 4.1 to 4.5 nm. Noting particularly the considerable disparity in size between pyruvate carboxylase (440 to 480 kDa) and avidin (68 kDa), Johannssen et al.[429] sought to understand how the above configuration of biotin-binding sites in avidin could be accommodated in a 1:1 stoichiometry with the tetrahedronlike arrangement of subunits in pyruvate carboxylase (see Figure 5). These authors therefore examined a series of samples in which the ratio of avidin to pyruvate carboxylase ranged from 9:1 tol:4, with acetyl-CoA (0.25 m*M*) being present in all cases for maximum preservation of the native structure of the enzyme. As can be seen in Figure 9, when the ratio of avidin to enzyme was 9:1 there were only single enzyme-tetramer/avidin complexes and free avidin molecules evident. The enzyme-bound avidin molecules were difficult to see, with many complexes appearing to have only one avidin molecule present (see Figure 9b to d), and those with two having the avidin molecules on opposite sides of the enzyme tetramer (see Figure 9e to f), except the very rare case shown in Figure 9g with two avidin molecules bound to the same side of the enzyme. As the ratio of avidin to enzyme decreased to 1:2, polymerization into linear unbranching chain-like structures became increasingly evident (see Figures 10 and 11). However, no polymers were seen when the ratio of avidin to enzyme equalled 1:4.

Within the chains all the enzyme molecules were the same size and shape, indicating that all had the same orientation to the electron beam (see Figure 11b). Furthermore, the dimensions of the pyruvate carboxylase molecules, both in the chains and as isolated molecules complexed with avidin, were the same as those measured by Mayer et al.[538] in the absence of avidin, indicating that there had not been any significant distortion of the enzyme during formation of the complex with avidin. In the well-preserved chains the avidin molecules were not clearly evident, but in a few cases where the polymer structure appeared loose there were avidin molecules visible between the enzyme tetramers and these were localized at the centers of the long axis of the enzyme molecules (see Figure 10b and c). To account for their observations, Johannssen et al.[429] proposed the model shown in Figures 12 and 13, from which it can be seen that the biotin moieties and hence the active sites are localized in the midline cleft on the external surface of each subunit, within approximately 3 nm of the intersubunit junction. The location of an active site in a cleft region between two domains is now recognized to be a general phenomenon[740] and had been postulated for pyruvate carboxylase by Mayer et al.[538] Such a location would also be entirely consistent with current proposals on the evolution and the dual subsite structure of the catalytic centers of

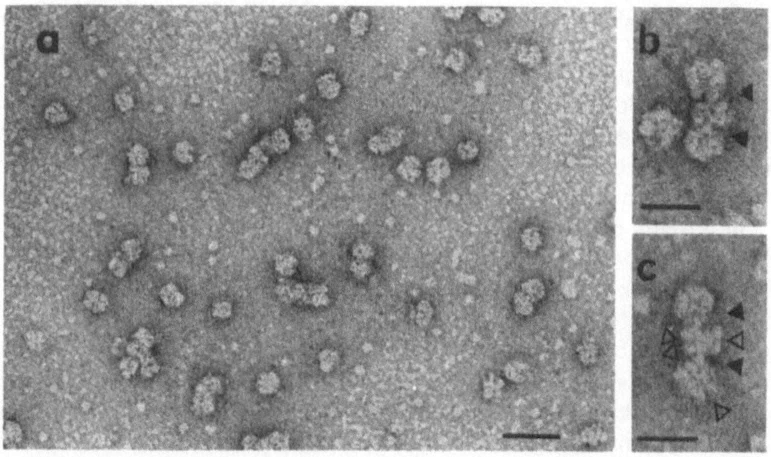

FIGURE 10. Electron micrographs of purified chicken liver pyruvate carboxylase and complexes of the enzyme with avidin after negative staining. In the low-magnification micrographs the bars represent 40 nm; in the high-magnification micrographs the bars represent 20 nm. Sample prepared from a solution containing avidin-pyruvate carboxylase in the ratio 2:1. (a) Single enzyme-avidin complexes and polymers of those that are visible as short chains. (b,c) Chain-like polymers shown at higher magnification; the avidin molecules between the enzyme tetramers are indicated by (▲). (c) At two of the three enzyme tetramers the protein intensity maxima are divided by a line of negative stain (Δ) running parallel to the long axis of the enzyme tetramer. (From Johannsen, W., Attwood, P. V., Wallace, J. C., and Keech, D. B., *Eur. J. Biochem.*, 133, 201, 1983. With permission.)

biotin enzymes (see Section IV.A; also Chapter 4, Section I) and with the effects of acetyl-CoA on the conformation and function of pyruvate carboxylase in particular (see Section V.C; and Chapter 4, Section III).

Alternative locations for the prosthetic group (and hence the orientations of the bound avidin molecules) were eliminated after considering the effects they would have on the configuration of the avidin-enzyme complexes in terms of the distortion of the dimensions of the enzyme molecules, the orientation of the chain with respect to the electron beam, or the occurrence of branching or other nonlinear complexes. Similar conclusions have also been reached recently regarding the location of biotin in pyruvate carboxylase from *A. nidulans*.[616]

It is also obvious from Figure 12c why, in the samples where avidin is present in excess (viz. Figure 9a), it is nevertheless difficult to discern in most cases more than one avidin molecule bound per enzyme tetramer. It can be seen that with a well-preserved tetrahedral-like enzyme molecule in its most stable orientation with respect to the support (viz., resting on three of its subunits) the image of only one avidin molecule is clearly visible outside the perimeter of the image of the enzyme as the mass center of the other avidin is superimposed on one of the subunits.

Several recent studies[37,616] have shown that when pyruvate carboxylase is complexed with an excess of avidin (in the presence of acetyl-CoA in the case of the chicken liver enzyme), most of the single molecules exhibit a triangular projection and are well preserved, even when diluted and mounted for electron microscopy in the absence of acetyl-CoA. This suggests, as can be inferred also from Figure 13a, that combination of pyruvate carboxylase with avidin stabilizes this tetrameric enzyme in its tetrahedral-like configuration. Thus, by the pair of biotin-binding sites on the same face of an avidin molecule binding both the biotin moieties from a pair of adjacent enzyme monomers, avidin can act as a bridge between the two subunits of a dimer within the tetrameric molecule. It is conceivable that this bridging action could stabilize the mon-

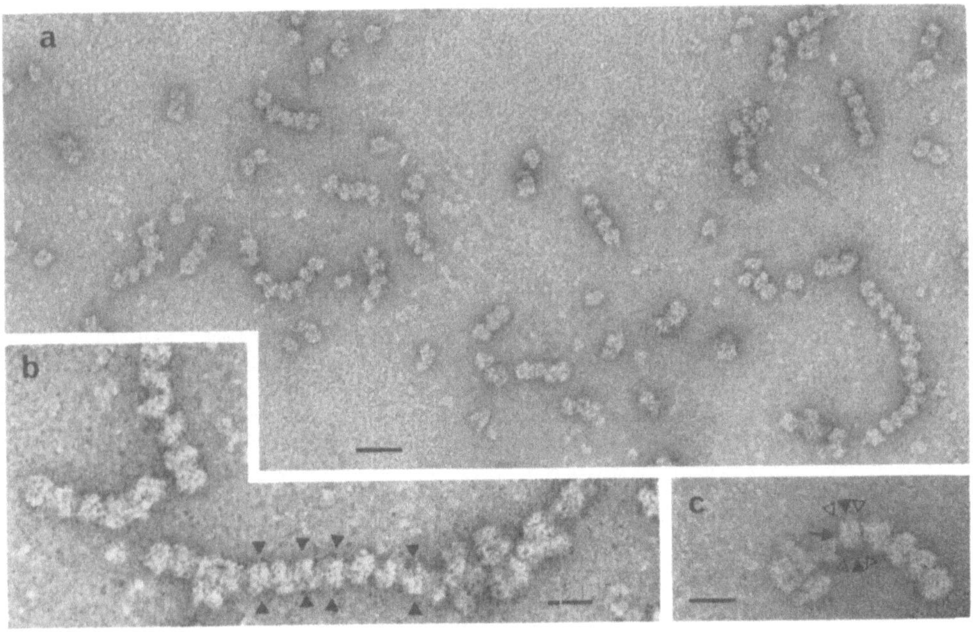

FIGURE 11. Electron micrographs of purified chicken liver pyruvate carboxylase and complexes of the enzyme with avidin after negative staining. In the low-magnification micrographs the bars represent 40 nm; in the high-magnification micrographs the bars represent 20 nm. Sample prepared from a solution containing avidin-pyruvate carboxylase in the ratio 1:2. (a) The chain-like structure of the polymer is obvious. (b,c) Selected chains of avidin-enzyme complexes at higher magnification. Most of the enzyme particles have a rectangular shape. Sometimes protein intensity maxima of the enzyme tetramers are divided by a line of negative staining material across the longitudinal axis of the chain (▲). (c) An additional line of negative-staining material can be observed orthogonal to the long axis of the enzyme tetramer (♠). The positions of the enzyme subunits are indicated by (△). (From Johannssen, W., Attwood, P. V., Wallace, J. C., and Keech, D. B., *Eur. J. Biochem.*, 133, 201, 1983. With permission.)

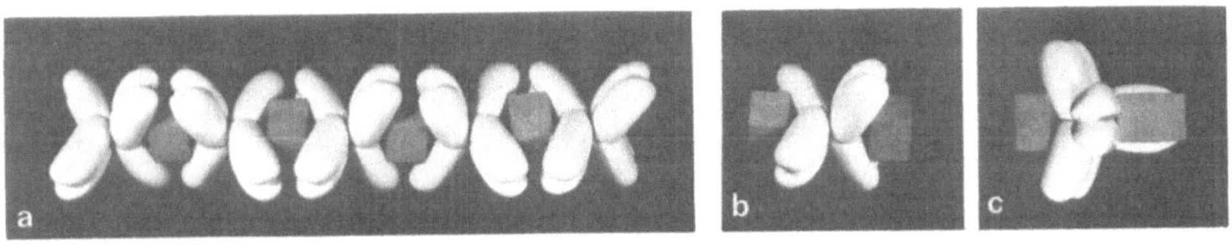

FIGURE 12. Models of the complexes between avidin and chicken liver pyruvate carboxylase using the model of the enzyme tetramer proposed by Mayer et al.[538] (a) Chain-like structure of the polymer. In this orientation of the enzyme tetramers two of the four subunits are in contact with the support. This projection shows a rectangular shape of the enzyme with its long axis orthogonal to the long axis of the polymer. The enzyme tetramers are connected to each other by avidin molecules positioned on opposite sides of an enzyme tetramer. (b) Single avidin-enzyme complex at the same orientation as described in (a). The protein masses of both bound avidin molecules are separated from the protein masses of the enzyme subunits and are visible. (c) Single avidin-enzyme complex at an orientation in which three of the four subunits are in contact with the support. This projection shows a triangular profile of the enzyme and only one protein mass of the two bound avidin molecules (at the left side) is clearly visible. On the right side the protein masses of the avidin and of one enzyme subunit are superimposed. (From Johannssen, W., Attwood, P. V., Wallace, J. C., and Keech, D. B., *Eur. J. Biochem.*, 133, 201, 1983. With permission.)

omer/monomer interaction within the enzyme dimer, and that this could in turn result in stronger dimer/dimer interactions, thereby stabilizing the overall tetrameric molecule. It was evident with the chicken liver enzyme, at a ratio of avidin to enzyme of

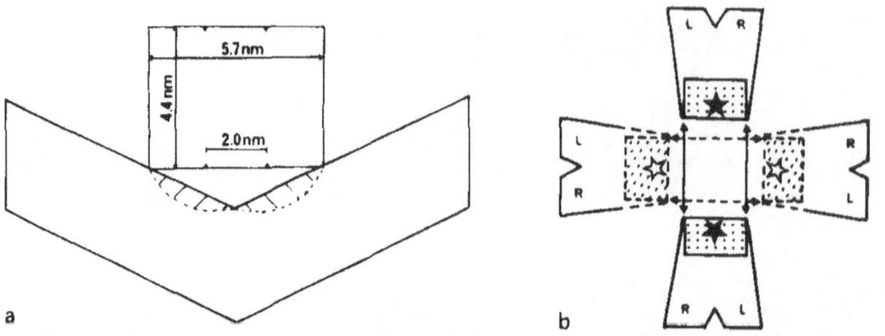

FIGURE 13. Diagrammatic views of a complex between avidin and chicken liver pyruvate carboxylase. (a) Side view of half of an enzyme tetramer with a bound avidin molecule in the same configuration as in Figure 12. The areas of the active sites on the enzyme subunits (shaded) and the biotin-binding sites on the avidin are indicated (▲). (b) Exploded face-view of the enzyme tetramer (after Mayer et al.[538]) with an indication of the bound avidin molecule (shaded) with the sites of the biotin-binding areas indicated (★) on the avidin bound to the upper pair of pyruvate carboxylase subunits; (☆) on the avidin bound to the lower pair of pyruvate carboxylase subunits). (From Johannssen, W., Attwood, P. V., Wallace, J. C., and Keech, D. B., *Eur. J. Biochem.*, 113, 201, 1983. With permission.)

1:2, that when diluted in the absence of acetyl-CoA, there was much less stabilization of the enzyme tetramers which were not part of the chains. This suggests that for this type of stabilization to occur, two avidin molecules must be bound per enzyme tetramer, i.e., both dimers in the tetramer have to be bridged by an avidin molecule.

In the case of chicken liver pyruvate carboxylase, it was found that for this stabilization by excess (10×) avidin to result or for chain formation to occur (when the ratio of avidin to enzyme was 1:2) it was essential to have acetyl-CoA present during the incubation of enzyme with avidin. This suggests that even at high enzyme concentration (16 U/mℓ) the tetrameric enzyme exists in two forms: a "loose" form and a "tight" form that are in equilibrium, with acetyl-CoA favoring formation of the latter species. As the two biotin-binding sites on the same face of an avidin molecule are only separated by 2 nm,[337,716] there is a similar restriction on the distance between the two biotins on adjacent enzyme subunits if both are to be bound by the same avidin molecule. If the dimers are more splayed in the "loose" tetramer, then a single avidin molecule may not be able to bind both biotin groups and hence not be able to stabilize the structure. Further discussion of the effects of acetyl-CoA on the structure of this enzyme is presented in Section V.

C. Biotin Content

As is evident in Table 2, whenever pyruvate carboxylase has been purified to homogeneity from normal biotin-replete sources, its biotin content has, with the exception of *Ps. citronellolis,* approached a value of 1 mol per subunit. In several cases (viz., chicken, turkey, rat, and sheep liver and kidney) it has been shown by forming complexes with avidin that every polypeptide chain in the purified enzyme has a biotin moiety covalently attached.[124] Similar conclusions have been arrived at for yeast[75] and a thermophilic *Bacillus*[604] by culturing the organisms in the presence of radioactively labeled biotin and showing a complete coincidence of radioactivity and protein when the purified enzymes were analyzed by SDS-PAGE.

Both by incorporation of [^{3}H]-biotin and by direct measurement of biotin in the isolated subunits, all the biotin in the pyruvate carboxylase of *Ps. citronellolis* has been shown to be covalently attached to only the larger 65-kDa subunit.[193]

D. Bound Metal Content

Most purified proteins contain extraneous, loosely bound metal ions which are un-

Table 2
THE BIOTIN AND BOUND-METAL CONTENTS OF PYRUVATE CARBOXYLASE ISOLATED FROM VARIOUS SOURCES

Source	Condition	Biotin content (mol/mol[a])	Metal content (g-atom/mol[a])	Ref.
Chicken liver (500 kDa)	Normal	3.8	3.3 Mn^{2+} 0.4 Mg^{2+}	751
	Mn^{2+}-Deficient[b]	3.8	0.1 Mn^{2+} 3.3 Mg^{2+}	751
Turkey liver (500 kDa)	Normal	3.45	3.30 Mn^{2+} 0.18 Mg^{2+}	751
Calf liver (500 kDa)	Normal	3.05	1.2 Mn^{2+} 1.74 Mg^{2+}	751
Rat liver (520 kDa)	Normal	3.48	1.9 Mn^{2+}[c]	545
Sheep liver (480 kDa)	Normal	4.0	1.88 Mn^{2+}[d]	46
Sheep kidney (480 kDa)	Normal	4.2	1.8Mn^{2+}[d]	46
Pig liver (520 kDa)	Normal	3.8	2.7 Zn^{2+} 1.08 Mg^{2+} 0.3 Mn^{2+}	937
Yeast (*S. cerevisiae*) (600 kDa)	Normal medium[e]	3.41	3.15 Zn^{2+}	769
	2 m*M* Co^{2+} medium	3.75	1.5 Zn^{2+} 2.0 Co^{2+}[f]	769
Pseudomonas citronellolis (467 kDa)	Normal	4.32	—[g]	193
(65-kDa subunit)	Normal	1.0	—[g]	75, 193
(54-kDa subunit)	Normal	0	—[g]	75
Thermophilic *Bacillus* (560 kDa)	Normal	3.36	3.84 Zn^{2+} 0.46 Mg^{2+} 0.1 Mn^{2+}	504

[a] The molecular weights used as the bases for calculating the biotin and bound-metal contents are given as kilodaltons in parentheses under the source.
[b] The dietary Mg was 0.28 mg/kg (cf. normal diet containing 426 mg/kg; diets of intermediate Mg content gave correspondingly intermediate values for the manganese content of pyruvate carboxylase (see Table IV of Reference 751). There was no relationship evident between specific activity and the Mn^{2+}:Mg^{2+} ratio.
[c] This enzyme was not assayed for the presence of other bound metals, e.g., Mg^{2+}.
[d] This enzyme was assayed also for Mg^{2+}, Co^{2+}, Cu^{2+}, Fe^{2+}. Ca^{2+}, and Zn^{2+}, but none was found to correspond with the pyruvate carboxylase activity.
[e] Normal medium ~60 μM Zn^{2+}: ~4 μM Co^{2+}
-supplemented medium ~60 μM Zn^{2+}: 2 m*M* Co^{2+}.
[f] 90% of Co^{2+} not stable to gel filtration in buffer containing 10 m*M* EDTA.
[g] Not reported.

related to catalytic function and which may be removed by gel filtration or dialysis in the presence of EDTA. Of those proteins which have been clearly shown to have a metal atom firmly incorporated into their structure, only a handful are manganoproteins.[895] Only two of these are enzymes: superoxide dismutase from avian liver mitochondria and certain bacteria,[291] and avian liver pyruvate carboxylase.[753,759,763]

However, as can be seen from the data summarized in Table 2, although manganese is the metal normally found in avian liver in a 1:1 stoichiometry with the biotin (i.e., 4 g-atoms per tetramer), it can be replaced there by magnesium without any decrease in specific enzymic activity when chickens are raised on a manganese-deficient diet.[751] The magnesium incorporated into the pyruvate carboxylase of manganese-deficient chickens was stable to gel filtration on Sephadex G25 in buffer containing 1 m*M* EDTA, but, unlike bound manganese, it was reduced by 40 to 50% when the enzyme was subjected to gel filtration on Sephadex G75 or G200 in buffer with 20 m*M* EDTA. When this Mg^{2+} depleted enzyme was incubated with Mn^{2+}, Co^{2+}, or Ni^{2+}, these metal ions bound effectively to form fully active hydrid metalloenzymes containing bound Mg^{2+}, Co^{2+}, or Ni^{2+} in a combined stoichiometry approximating the biotin content.

From the properties of model complexes[790] one would expect that if the metal ion is bound to pyruvate carboxylase primarily through nitrogen ligands, the affinity of the site would be two to three orders of magnitude greater for Mn^{2+} than for Mg^{2+}. On the other hand, if metal binding is affected primarily by oxygen ligands, the affinity for Mn^{2+} would be less than tenfold greater than for Mg^{2+}. The relative lack of stability of the Mg^{2+} enzyme, then, would be consistent with a metal binding site composed of nitrogen ligands. However, this and other possible suggestions to explain the natural preferential incorporation into this enzyme of Mn^{2+}, when it is present at only ~1% of the Mg^{2+} concentration in liver,[867] are some of the numerous unsolved questions regarding the bound metals of pyruvate carboxylase.

In the mammalian species studied so far, manganese has been shown to co-chromatograph with pyruvate carboxylase activity in elution profiles from both anion exchange and gel filtration columns, but it is present only to the extent of about 2 g-atoms per mole. In the case of calf liver pyruvate carboxylase, magnesium has also been shown to be the only other metal to co-chromatograph with enzymic activity, and it occurred at a concentration which, when combined with the manganese content, was equivalent to the biotin present.[751] On the basis of the data presently available it is not possible to distinguish between the possibility of there being a true hybrid species present (i.e., two different metal atoms in the same enzyme tetramer) or there being a mixture of appropriate proportions of two distinct parent metalloenzymes.

The extent to which the Mn^{2+}:Mg^{2+} ratio in pyruvate carboxylase can be varied by dietary manipulation in herbivorous species is unknown at present.

In view of the other data available regarding the apparently identical nature of the subunits of eukaryotic and some bacterial pyruvate carboxylases (see Section III.H), the alternative possibility of a "hybrid" manganese/magnesium metalloenzyme should be investigated further.

Though they were analyzed for Mn^{2+}, Mg^{2+}, Ca^{2+}, Co^{2+}, Zn^{2+}, Fe^{2+}, and Cu^{2+}, neither sheep liver nor sheep kidney pyruvate carboxylase appeared to contain any bound metal other than Mn, and this only at half the concentration of the biotin.[46] There could, of course, be species differences in the tightness with which the Mg^{2+} is held, and so far its stability to gel filtration in only one concentration of EDTA (viz., 1 m*M*) has been tested.

The reported occurrence of zinc in pig liver pyruvate carboxylase of specific activity 3.2 U/mg[937] should be viewed in the light of the data presented in Figure 14 where enzyme, which had a specific activity of 18 U/mg and which, by SDS-PAGE analysis, was at least 60% pure, was clearly associated with a peak of Mn^{2+} in the eluant of a DEAE-Sephadex column. While the Mn^{2+} was present in approximately stoichiometric amounts with the biotin, there was no such consistent trend evident for either Mg^{2+} or Zn^{2+}.

However, the identification of pyruvate carboxylase from yeast as a zinc metalloenzyme is well established, with the Zn^{2+} being present, stoichiometric with the biotin, at 4 g-atom per mole (tetramer) and stable to gel filtration in 50 m*M* EDTA.[769] Of ten other metals assayed for in highly purified preparations of this enzyme, only Fe^{2+} was

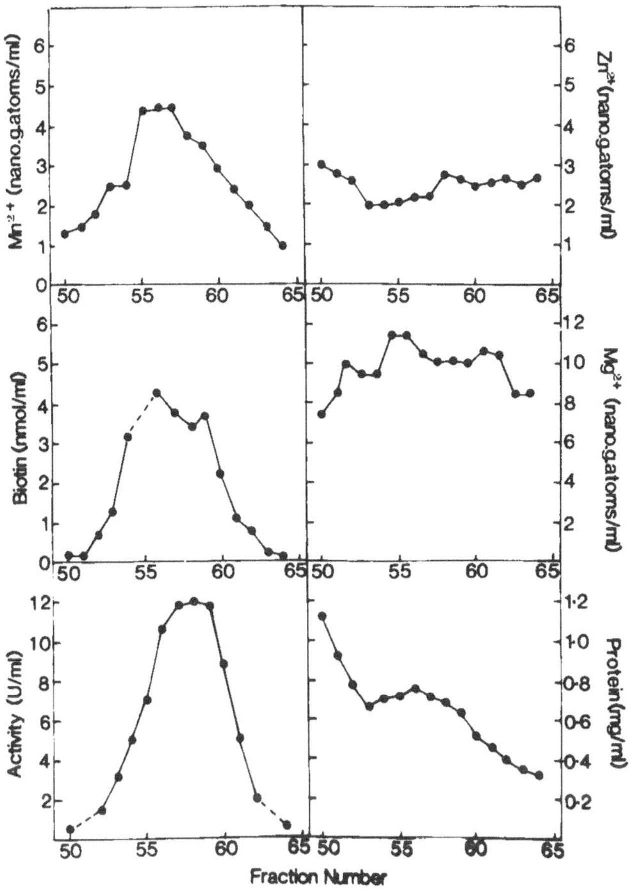

FIGURE 14. Distribution of magnesium, manganese, zinc, protein (A_{280}), biotin, and enzymic activity in elution profile when pig liver pyruvate carboxylase was subjected to chromatography on DEAE-Sephadex-A-50 (SO_4^{2-}). Pig liver pyruvate carboxylase was purified by the method described by Goss et al.[328] for sheep liver pyruvate carboxylase. The enzyme was eluted from a column (30 × 2.5 cm) of DEAE-Sephadex A-50 equilibrated in Buffer A (25 m*M* potassium phosphate buffer, pH 7.2, containing 20 m*M* ammonium sulfate, 1 m*M* EDTA, and 0.1 m*M* dithioerythritol (DTE) fusing a linear gradient (2 × 400 m*l*) from Buffer A to Buffer B (25 m*M* Tris Cl, pH 8.0, containing 200 m*M* ammonium sulfate, 1 m*M* EDTA, and 0.1 m*M* DTE). Fractions (8 m*l*) were assayed for pyruvate carboxylase activity, protein, and biotin content as described by Goss et al.[328] and for manganese, magnesium, and zinc content as described by Scrutton et al.[751]

present at a concentration greater than 1 g-atom per mole (tetramer) (compared with Zn^{2+} at 3.05 g-atoms per mole). When the yeast enzyme was passed through a column of Sephadex G200 and the eluant fractions analyzed for the distribution of catalytic activity, ^{3}H-biotin, protein, zinc, and iron, the results clearly showed a strong coincidence of Zn^{2+}, ^{3}H-biotin, and enzymic activity, but no such relationship for Fe^{2+}, indicating that iron does not behave as a constitutive component of this enzyme.

Growth of yeast in cobalt-supplemented media (Co^{2+} concentration 2 m*M*; ratio of Co^{2+}:Zn^{2+} equal to 30 in comparison with 0.06 in unsupplemented medium) resulted in a partial substitution of Co^{2+} for Zn^{2+} in the pyruvate carboxylase of these organisms, though the combined amounts (of Co^{2+} and Zn^{2+}) were still approximately equal to the biotin content. However, unlike the Zn^{2+}, the Co^{2+} was not stable to gel filtration in the presence of 10 m*M* EDTA, and concomitant with its removal there was a comparable decrease in specific activity of the enzyme.[769] While these results might suggest a catalytic role for Zn^{2+} in yeast pyruvate carboxylase, there is no direct proof of this yet. Furthermore, in view of the more recent work on the location of the bound manganese in relation to the active site of chicken liver pyruvate carboxylase (see Section IV.D.1), the role of the bound metal ion in this enzyme is very much an open question.

Analysis of the pyruvate carboxylase purified from a thermophilic *Bacillus* revealed that it too has one atom of Zn^{2+} per subunit, whereas Mg^{2+} and Mn^{2+} were present in negligible amounts and Co^{2+} was absent.[504]

E. Amino Acid Composition

The available data on the amino acid composition of pyruvate carboxylase are summarized in Table 3 (residues per mole) and Table 4 (mol %). Of these, the first published figures for the chicken liver enzyme[759] must be erroneous in their absolute values since they were based on a molecular weight of 655,000, now recognized to be much too high (see Section III.A.1). Some contamination with the square planar molecule (see Section III.B.1) may account for the differences between that early analysis and later work by Bais[46] on homogeneous samples, even when they are compared on a mol % basis (see Table 4). There is, however, a reasonable degree of overall similarity between the results for different mammalian species determined in different laboratories.

When compared with the Reeck[674] set of 207 mutually nonhomologous proteins by the methods described by Reisner and Westwood,[679] chicken liver pyruvate carboxylase is relatively low in its content of aspartic, threonine, tyrosine, tryptophan, and lysine, but has a substantially higher-than-average level of arginine, proline, glycine, alanine, and, to a lesser extent, leucine. The mammalian pyruvate carboxylases are less exceptional, being relatively low in only aspartic and to some extent tryptophan, and having somewhat higher-than-average contents of arginine, glutamate, and proline. The *Pseudomonas* enzyme is somewhat lower than average in its lysine, aspartic, threonine, and proline contents and exceptionally high in only glycine. Comparisons of the compositions of the separated subunits of the *Pseudomonas* enzyme by the method of Cornish-Bowden[206,207] suggested little or no homology between the two polypeptides.[327]

Protein-bound carbohydrate has been investigated in the *Pseudomonas*[327] and chicken liver enzymes,[124] but no evidence for its presence was found.

F. Spectral Properties

Ultraviolet absorption — The reported values for the $A^{1\%}_{280}$ of various purified pyruvate carboxylases range from 0.5 to 0.83. They are lower than average, consistent with the lower-than-average content of tryptophan in the enzyme and to a lesser extent, tyrosine.

Fluorescence — The intrinsic fluorescence of chicken liver pyruvate carboxylase is unaffected during the inactivation and dissociation of the enzyme at 2°C.[904]

Circular dichroism — Only limited use has been made so far of circular dichroism to study the structure and conformational changes of this enzyme. The data available to date for chicken liver pyruvate carboxylase[554] suggest a helix content of approximately 27% and β-pleated sheet of ~7%. The addition of acetyl-CoA and MgATP caused small changes in the near-UV region, but no detectable effects on the far-UV spectrum.

G. Primary Structure

1. N-Terminal Sequence

All attempts so far[121,194,802] to determine the sequence of amino acids in the polypeptide chain of pyruvate carboxylase using the Edman degradation procedure have been thwarted by the presence of a blocking group on the N-terminal residue. In the case of chicken and sheep liver pyruvate carboxylases, this blocking group has been identified

Table 3
AMINO ACID COMPOSITION OF PYRUVATE CARBOXYLASE ISOLATED FROM VARIOUS SOURCES

Source	Mol wt.[b]	ASX[c]	THR[d]	SER[d]	GLX[e]	PRO	GLY	ALA	VAL[d]	CYS	MET	ILE[d]	LEU	TYR	PHE	LYS	HIS	ARG	TRP[f]	Ref.
Chicken l.[g]	655	447	207	208	420	112	503	630	451	55[h]	73	200	499	203	290	165	124	322	50	759
Chicken l.	480	302	165	210	461	257	421	439	317	44[i]	70	168	392	91	156	145	105	258	32	46
Pigeon l.	480	396	171	253	509	269	423	437	276	—[j]	61	172	340	97	141	155	70	278	—	46
Pig l.	520	376	220	260	428	232	352	328	264	52[i]	64	228	280	116	168	252	116	216	56	937
Rat l.	480	336	201	223	458	219	346	347	276	—	102	206	344	111	166	196	118	209	—	46
Guinea pig l.	480	341	213	245	542	197	390	319	245	—	64	191	296	98	165	202	98	192	—	46
Sheep l.	480	353	234	232	446	215	349	345	281	46[i]	83	209	329	118	162	201	114	217	36	46
Sheep k.[g]	480	332	208	248	468	222	390	347	290	44[i]	80	208	352	106	158	194	113	198	—	46
Ps. citronellolis	476	292	196	160	404	116	780	392	316	52[h]	80	224	316	92	128	188	84	168	24	327
Ps. citronellolis	65	44	27	34	51	13	116	54	43	—	8	26	47	13	15	30	10	25	—	327
α-Subunit	65	44	27	34	51	13	116	54	43	—	8	26	47	13	15	30	10	25	—	327
Ps. citronellolis β-Subunit	54	36	20	35	55	10	66	45	30	—	7	27	36	12	13	23	8	27	—	327

[a] For the molecular weight as indicated in column headed "Mol wt.".
[b] Molecular weight in kilodaltons.
[c] Aspartic acid plus asparagine.
[d] Corrected for destruction or incomplete release during hydrolysis.
[e] Glutamic acid plus glutamine.
[f] Determined spectrophotometrically, as described by the authors referred to under "Ref."
[g] "l" Denotes liver, "k" denotes kidney.
[h] Determined as cysteic acid.
[i] Determined as S-carboxymethylcysteine.
[j] " — " denotes "not determined."

Table 4
AMINO ACID COMPOSITION OF PYRUVATE CARBOXYLASE ISOLATED FROM VARIOUS SOURCES

Moles %[a]

Source	ASX[b]	THR	SER	GLX[c]	PRO	GLY	ALA	VAL	CYS	MET	ILE	LEU	TYR	PHE	LYS	HIS	ARG	Ref.
Chicken l.[d]	9.1	4.2	4.2	8.6	2.3	10.2	12.8	9.2	1.1	1.5	4.1	10.2	4.1	5.9	3.4	2.5	6.6	759
Chicken l.	7.5	4.1	5.2	11.5	6.4	10.5	11.0	7.9	1.1	1.7	4.2	9.8	2.3	3.9	3.6	2.6	6.4	46
Pigeon l.	8.7	4.2	6.2	12.4	6.6	10.4	10.7	6.7	1.1[e]	1.5	4.2	8.3	2.3	3.9	3.9	1.7	6.8	46
Pig l.	9.5	5.6	6.6	10.8	5.9	8.9	8.3	6.7	1.3	1.6	5.8	7.1	2.9	4.3	6.4	2.9	5.5	937
Rat l.	8.7	5.2	5.8	11.9	5.7	9.0	9.0	7.2	1.1[e]	2.6	5.3	8.9	2.9	4.3	5.1	3.1	5.4	46
Guinea pig l.	8.9	5.5	6.4	14.1	5.1	10.2	8.3	6.4	1.1[e]	1.7	5.0	7.7	2.6	4.3	5.3	2.6	5.0	46
Sheep l.	9.0	5.9	5.9	11.3	5.5	8.9	8.8	7.1	1.2	2.1	5.3	8.4	3.0	4.1	5.1	2.9	5.5	46
Sheep k.[d]	8.4	5.3	6.3	11.8	5.6	9.9	8.8	7.3	1.1	2.0	5.3	8.9	2.7	4.0	4.9	2.9	5.0	46
Ps. citronellolis	7.3	4.9	4.0	10.1	2.9	19.6	9.8	7.9	1.3	2.0	5.6	7.9	2.3	3.2	4.7	2.1	4.2	327
Ps. citronellolis α-Subunit	7.8	4.8	6.0	9.1	2.3	20.6	9.6	7.6	1.3[f]	1.4	4.6	8.3	2.3	2.7	5.3	1.8	4.4	327
Ps. citronellolis β-Subunit	7.9	4.4	7.7	12.1	2.2	14.5	9.9	6.6	1.3[f]	1.5	5.9	7.9	2.6	2.9	5.0	1.8	5.9	327

[a] Calculated as the "residues per 100 residues total" (not including tryptophan) from the data in Table 3.
[b] Aspartic acid plus asparagine.
[c] Glutamic acid plus glutamine.
[d] "l" Denotes liver; "k" denotes kidney.
[e] Calculated assuming 44 cysteines per tetramer (480 kDa).
[f] Both subunits assumed to be equivalent in cysteine content.

by the hydrazinolysis procedure of Schmer and Kreil,[734] as modified by Doolittle,[243] to be an acetyl group.[268] No evidence was obtained for the presence of either a formyl or a pyroglutamyl group as the N-blocking entity even though "positive control" samples indicated the methods were effective.[268]

2. Biotin Attachment Site

From the arginine plus lysine content of vertebrate liver pyruvate carboxylases (Table 3), one would expect that there would be, assuming the subunits are identical, approximately 100 peptide fragments in tryptic digests of these enzymes. The extraordinarily high affinity of avidin for biotin[336] provided a basis for specific and relatively facile procedures for both detecting and isolating the biotinyl tryptic peptide.[712] When the amino acid sequences of the biotin-containing tryptic peptides of pyruvate carboxylases isolated from sheep, chicken, and turkey livers were determined, the 19-residue fragments from the two avian species were found to be identical (Table 5). Furthermore, with the exception of residue 19, and two quite conservative substitutions at residues 2 and 17, the same sequence occurs in the sheep liver pyruvate carboxylases as well.

More remarkable, though, is the clear evidence of homology in this region of their primary structures between these three pyruvate carboxylases and two different biotin-containing enzymes of bacterial origins, viz., transcarboxylase [E.C. 2.1.3.1] from *Propionibacterium shermanii*,[526] and acetyl-CoA carboxylase [E.C. 6.1.4.2] from *E. coli*.[843]

The occurrence of the identical sequence, ALA-MET-BIOCYTIN-MET in all five enzymes (where biocytin is the ε-*N*-biotinyl lysine), together with the absence of any bulky aromatic residues, implies this structure may be related to the requirements of the holoenzyme synthetases [EC 6.3.4.1] that attach biotin to this one specific lysine in the apoenzyme. This suggestion would be consistent with the results of McAllister and Coon,[541] who showed that almost regardless of their origin (viz., rabbit liver, yeast, and *Propionibacterium shermanii*) the holoenzyme synthetases were, with one exception, mutually cross-reactive, forming active enzymes with apo-propionyl-CoA carboxylase [E.C.6.4.1.3] from rat liver, apo-β-methylcrotonyl-CoA carboxylase [E.C.6.4.1.4] from *Achromobacter*, and apo-transcarboxylase from *P. shermanii*. In view of the lack of such cross-reaction between the *S. cerevisiae* and *B. stearothermophilus* systems,[839] it will be interesting to see what differences in the structure of the biotin attachment site there are between these species.

3. cDNA Sequencing Studies

The advent and application of recombinant DNA technology,[793] together with the development of facile methods for sequencing DNA,[725] have opened a new era of scientific discovery. One immediate consequence of these techniques is that it is now possible, without being dependent on the availability of substantial quantities of highly purified protein, to deduce the entire amino acid sequence of even a very large polypeptide chain by determining the sequence of the coding region of the gene which encodes that protein.

An essential first step in this approach is the selection of recombinant clones containing sequences coding for the protein in question. While there are various indirect methods of identifying such a clone, there is no doubt that the selection of a suitable recombinant clone is greatly facilitated by the use of a synthetic cDNA primer, as exemplified by the studies of Houghton et al.,[422] Hudson et al.,[423] and Krieg et al.[479]

Although very little of the sequence of pyruvate carboxylase is known, there is, among the 19-residue sequence flanking the biotin attachment site,[712] a sequence of five amino acids from which may be inferred a 14-base cDNA sequence complementary to the biotin carboxylase mRNA and containing only two possible mismatches. Thus,

```
                                   Biotin
                                     |
Protein   NH2- ... ALA - MET - LYS - MET - GLU - THR ... COOH

                    U                              U
                    C           G           A      C
mRNA      5' ...   GC  - AUG - AA  - AUG - GA  -  AC   ...  3'
                    G           A           G      A
                    A                              G

                    A                              A
                    G           C           T      G
cDNA      3' ...   CG  - TAC - TT  - TAC - CT  -  TG   ...  5'
                    C           T           C      T
                    T                              C
```

From the nucleotide sequences of chicken α[683] and β-[682] globins and histones H2A and H2B,[380] it would seem that in this species, AAG and GAG are the mRNA codons favored for lysine and glutamate, respectively.

Accordingly, Cassady et al.[153] have used a 14-base synthetic cDNA probe (3′-TACTTCTACCTCTG . . . 5′) to prime the synthesis of ^{32}P-labeled extended cDNA products on a poly(A+) mRNA template prepared from 17-day-old embryonic chicken liver which has been shown to synthesize elevated levels of pyruvate carboxylase.[23] Three principal cDNA products of 75, 300, and 400 bases were found and these were used to screen for pyruvate carboxylase coding sequences in a chicken genomal library incorporated in the λ vector, Charon 4A.[242] A basically similar approach has also been used by Gravel and associates[348] to isolate cDNA clones containing sequences coding for human pyruvate and propionyl-CoA carboxylases.

H. Identity of the Subunits

1. *Pseudomonas citronellolis*

On the bases of molecular weight and biotin content,[75,193] as well as amino acid composition, cyanogen bromide peptide maps, and immunochemical tests,[327] and the α (65-kDa) and β (54-kDa) subunits of pyruvate carboxylase from *Ps. citronellolis* are clearly dissimilar.

Furthermore, unlike the situation in at least one other biotin enzyme (viz., the $5S_E$ and the $12S_H$ subunits of transcarboxylase),[96] the results of the amino acid analyses and peptide mapping suggest little, if any, sequence homology between the α- and β-subunits. However, there is equally no evidence to suggest that there are more than two types of subunits comprising the intact enzyme which normally occurs as an $\alpha_4\beta_4$ structure. Using antibodies raised against the isolated subunits separated by SDS-PAGE, Goss et al.[327] showed that only the anti-α-antibodies bound to the native enzyme or inhibited enzymic activity. Identical titration curves obtained by these workers for the inhibition of total pyruvate carboxylase activity, ATP/Pi exchange activity, and pyruvate/oxaloacetate exchange activity strongly suggest that both subsites of the catalytic center (see Section IV) are located on the α-polypeptide. It would seem also, from surface labeling experiments and from the use of several proteases as surface probes, that the α-polypeptides are located on the surface of the native protein, protecting the β-polypeptides which form the stable core structure of the enzyme molecule.

Taylor et al.[858] have shown by double labeling studies that when the level of pyruvate carboxylase activity in *Ps. citronellolis* is influenced by the carbon source of the medium, it is the synthesis of these two polypeptides that is coordinately regulated rather than their assembly into protomers or the biotinylation of apoenzyme. However, it

Table 5

THE AMINO ACID SEQUENCES ADJACENT TO THE BIOCYTINS OF PYRUVATE CARBOXYLASES OF LIVER AND OF TRANSCARBOXYLASE OF *PROPIONIBACTERIUM SHERMANII* AND OF ACETYL-CoA CARBOXYLASE OF *ESCHERICHIA COLI*

Enzyme	1	2	3	4	5	6	7	8	9	10	12	12	13	14	15	16	17	18	19	Ref.
Pyruvate carboxylase																				
Sheep	Gly	Glu	Pro	Leu	Val	Leu	Ser	Ala	Met	Bct[a]	Met	Glu	Thr	Val	Val	Thr	Ser	Pro	Val	712
Chicken	Gly	Ala	Pro	Leu	Val	Leu	Ser	Ala	Met	Bct	Met	Glu	Thr	Val	Val	Thr	Ala	Pro	Arg	712
Turkey	Gly	Ala	Pro	Leu	Val	Leu	Ser	Ala	Met	Bct	Met	Glu	Thr	Val	Val	Thr	Ala	Pro	Arg	712
Transcarboxylase																				
P. shermanii	Gln	Thr	Val	Leu	Val	Leu	Glu	Ala	Met	Bct	Met	Glu	Thr	Glu	Ile	Asn	Ala	Pro	Thr	526
Acetyl-CoA carboxylase																				
E. Coli	Asn	Thr	Leu	Cys	Ile	Val	Glu	Ala	Met	Bct	Met	Met	Asn	Gln	Ile	Glu	Ala	Asp	Lys	843

[a] Bct = biocytin = biotinyl-ε-lysine.

remains to be established by genetic manipulations or by direct structural studies of the genome (see Section III.G.3) whether these α-and β-polypeptides are coded for by discrete genes or are cleaved from a single translation product.

2. Other Species

While it seems unlikely that *Ps. citronellolis* could be unique in having its pyruvate carboxylase composed of dissimilar subunits, nevertheless, in all other species for which sufficiently purified samples of this enzyme have been studied, the majority of available evidence suggests, but has not as yet proven conclusively, that the subunits are identical.

1. Within all species examined so far (see Table 1), the subunits are indistinguishable on the basis of molecular weight.
2. The subunits of pyruvate carboxylases isolated from chicken[268] and rat[400] livers, 3T3-L1 adipocytes differentiated from Swiss mouse 3T3 fibroblasts[169] and yeast[194] are each detected as a single band after isoelectric focussing in 8 *M* urea, indicating that within each species they are identical with respect to overall charge. No comparisons between species appear to have been made yet using this technique.
3. In the cases of chicken,[75] rat,[545,773] and sheep[124] liver pyruvate carboxylases as well as *A. globiformis*,[350] the biotin-binding protein, avidin, has been shown by SDS-PAGE to form tight complexes with all the subunits of those enzymes. This indicates that every polypeptide chain in these proteins has a biotin moiety attached covalently. Similar conclusions seem likely for yeast[75] and a thermophilic *Bacillus*[504] on the basis of [^{3}H]-biotin incorporation studies.
4. In the sequencing studies of the biotinyl-tryptic peptide by Rylatt et al.[712] (see Section III.F.1), there was no evidence for any other than a unique 19-residue sequence about the biotin attachment point in chicken and turkey liver pyruvate carboxylases and a unique 24-residue sequence for the sheep liver enzyme.
5. Affinity-labeling experiments with sheep liver pyruvate carboxylase indicate there are four binding sites for both ATP[252] and pyruvate[424] (in the presence of acetyl-CoA) per active enzyme tetramer (see Sections IV.B.2 and IV.D.3).
6. Similarly, modification of a single lysyl residue per biotin moiety (i.e., per catalytic center with the reagent trinitrobenzenesulfonate results in a stoichiometric loss of the acetyl-CoA-dependent catalytic activity of pyruvate carboxylases from sheep kidney,[30] rat[765] and chicken livers,[756] yeast,[765] a thermophilic *Bacillus*,[503] and *Aspergillus nidulans*[615] (see Section IV.E).
7. There are 4 mol of Mn^{2+} per tetramer in the chicken and turkey liver pyruvate carboxylases and 4 mol Zn^{2+} per tetramer for the yeast and thermophilic *Bacillus* enzymes (see Table 2). These results suggest that there is one tightly bound divalent metal ion per subunit. However, the results obtained with the mammalian species, with yeast cultured in Co^{2+}-supplemented media, and in Mn^{2+}-deficient chickens raise some as yet unanswered questions regarding the equivalence of the subunits and their ability to bind different divalent metal ions (see Section III.D).
8. Sheep kidney pyruvate carboxylase contains 44 ± 4 cysteinyl residues per tetramer (see Table 3). If the subunits are identical there would be 11 ± 1 cysteinyl residues per monomer, and thus no more than 11 ± 1 unique cysteine-containing tryptic peptides in a tryptic digest of this protein. Nonidentical subunits would result in a significantly higher number of unique peptides. Separation of the S-[2-^{14}C]-carboxymethylated tryptic peptides of sheep kidney pyruvate carboxylase by a combination of gel filtration on Sephadex G-25, and subsequent high-voltage paper electrophoresis yielded 10 approximately equally labeled peptide fractions.[49] While these preliminary results are consistent with those expected for

identical subunits, it is clear from the experience of others that only by purification and sequence analysis of all the radioactive peptides from a S-[2-^{14}C]-carboxymethylated protein digested in various ways can one be sure of determining the number of unique cysteine sequences in the enzyme subunit. Even then, it does not necessarily follow that identical sequences around the cysteinyl residues in the subunits automatically implies complete sequence homology among the subunits. With pyruvate carboxylase subunits being composed of over 1000 amino acid residues each, there is ample scope for there to be long stretches of homology interspersed among other nonidentical regions. There is, therefore, a clear need for several different approaches to this question to ensure an unequivocal conclusion.

9. As noted in Section III.G.1, all four subunits in chicken liver pyruvate carboxylase have an N-acetyl group on the amino-terminal residue(s) and are, therefore, not susceptible to sequence determination by the Edman degradation procedure. However, recent experiments by Khew et al.[441] have shown that a brief exposure of the native enzyme to low concentrations of chymotrypsin achieves a rapid but selective cleavage of each subunit. When samples of this partial digest were analyzed by SDS-PAGE it was evident that the 112-kDa polypeptide chain had been almost quantitatively converted to a 75-kDa fragment plus smaller pieces. When this 75-kDa species was isolated and subjected to Edman degradation in the gas-phase protein sequencer[403] (Applied Biosystems, Inc.), an unambiguous N-terminal sequence of 23 residues could be deduced from an analysis of the PTH-amino acid derivatives by reverse-phase HPLC. These results indicate that in all four subunits there are regions of high susceptibility to proteolysis, with cleavage in at least one of these, yielding 75-kDa fragments having identical N-terminal sequences for at least 23 residues.

Taken together, these various lines of evidence (1 to 9) are mostly consistent with the subunits of pyruvate carboxylase being identical, but this still remains to be established unequivocally.

I. Relationship of Pyruvate Carboxylase to Other Biotin-Containing Carboxylases

The similarity in mechanism of biotin-dependent enzymes (see Chapter 4, Section I) has led to the suggestion that organisms which contain multiple biotin-dependent enzymes might share common subunits.[7,312] However, at least in the cases of two organisms, this has not proved to be the case: (1) in *Ps. citronellolis*,[267] two major biotin-containing polypeptides with approximate molecular weights of 65,000 and 25,000 were shown to correspond to the biotin carboxyl carrier components of pyruvate carboxylase and acetyl-CoA carboxylase, respectively; (2) for a thermophilic *Bacillus*,[140] the biotin-containing polypeptide subunits of pyruvate carboxylase and acetyl-CoA carboxylase were shown to differ in both their molecular weights and their affinities for avidin-Sepharose. Pyruvate carboxylase activity was reconstituted from fractions containing a large biotin polypeptide of 140 kDa, whereas reconstitution of acetyl-CoA carboxylase activity was dependent on a smaller biotin polypeptide of 22 kDa. Furthermore, a mutant *Bacillus* (PC2) which is devoid of pyruvate carboxylase activity and of the 140-kDa biotin-containing subunit, nevertheless continues to produce the 22-kDa biotin subunit of acetyl-CoA carboxylase.[140]

Lynen[515,611] proposed, on the basis of the similarities in their reaction mechanisms and in their overall molecular properties, that the biotin-dependent carboxylases may represent various stages in the evolution of an enzyme "family" from common ancestral genes. Support for this attractive hypothesis is to be found in the high degree of sequence homology noted about the biotin attachment point in several different enzymes from various sources (see Section III.G.2). However, where additional sequence data are available for the two bacterial biotin enzymes,[526,843] there is little homology evident beyond the biotin attachment site. Furthermore, there has not been, until recently, any immunological data which suggested any of the different biotin-containing enzymes shared any common structural features. Thus, antibodies prepared in rabbits against purified rat liver pyruvate carboxylase were equally effective in inhibiting pyruvate carboxylases from liver, kidney, brown and white adipose tissues, and mammary gland, but were ineffective in inhibiting rat liver propionyl- or acetyl-CoA carboxylase.[57] Similarly, within the limits of sensitivity of the immunochemical techniques used, no cross-reaction was found between yeast pyruvate carboxylase and antibodies raised to yeast acetyl-CoA carboxylase, nor vice versa.[836]

However, these studies were only concerned with these carboxylases in their native states. Since (1) changes in amino acid residues due to protein evolution are tolerated more readily at protein surfaces than in their interiors[741] and (2) antibodies react primarily, if not exclusively, with the amino acid side chains of surface residues,[678] it would seem quite likely that failure to detect any immunological identity between the biotin carboxylases in their native states does not necessarily preclude the existence of common structural and functional features of these enzymes buried within the less rapidly evolving internal regions of these molecules.

Indeed, recent studies utilizing antisera raised by immunization of rabbits with denatured subunits of chicken liver pyruvate carboxylase (CLPC) and sheep liver propionyl CoA carboxylase (SLPCC) have revealed that there are some antigenic determinants common to several of the biotin enzymes.[578] These results, which have been corroborated by those of a similar study using monoclonal antibodies raised against sheep liver pyruvate carboxylase,[614] add further support to Lynen's hypothesis. The prospect of more extensive sequence data becoming available from cDNA sequencing studies[153,348,1002] encourages one to believe this issue will be resolved in the foreseeable future.

IV. CHEMISTRY OF THE CATALYTIC AND ALLOSTERIC SITES

A. Introduction

Although the details of the reaction mechanism of pyruvate carboxylase have attracted considerable attention over the last 2 decades (see Chapter 4), relatively little progress has been made in elucidating the roles of active-site amino acid residues in the catalysis and regulation of the enzymic reaction.

The two techniques best suited to yielding such structural information are X-ray diffraction[112] and high-resolution nuclear magnetic resonance (NMR);[782] to date, it has not been possible to prepare satisfactory crystals of pyruvate carboxylase and the protein is probably too large to permit NMR studies of the type which have provided so much structural information about smaller proteins, such as hen egg white lysozyme,[241] and basic pancreatic trypsin inhibitor.[920,921,987] Hence, most of the structural information concerning the ligand binding sites of pyruvate carboxylase has come from orthodox protein chemical (especially affinity labeling and chemical modification) investigations, although NMR studies have contributed to our understanding of the transcarboxylation site of the enzyme (see Section IV.D below), and electron microscopy has revealed details of the gross quaternary structure of the protein (see Section III.B.1).

Pyruvate carboxylase can be considered as being composed of four functional modules: the biotin carboxylation subsite, the biotin attachment site, the transcarboxylation subsite, and the regulatory acetyl-CoA binding site. Accordingly the available structural data are discussed below under these headings.

B. Biotin Carboxylation Subsite

1. Introduction

The biotin carboxylation subsite catalyzes the first step in the pyruvate carboxylase reaction involving the formation of new covalent bonds:

$$\text{ENZ-Biotin} + \text{MgATP}^{2-} + \text{HCO}_3^- \underset{\text{Acetyl CoA}}{\overset{\text{Mg}^{2+},\ \text{Acetyl-CoA}}{\rightleftharpoons}} \text{ENZ-Carboxybiotin} + \text{MgADP} + \text{Pi} \qquad (3)$$

Although there has been some controversy as to the position of carboxylation on biotin, the work of Guchait et al.[347] has shown that the 1′-N form of carboxybiotin is the reactive form in catalysis.

Considerable progress has been made in establishing the molecular details of the catalytic process summarized as Equation 3 (cf., Section II of Chapter 4), but little information is available concerning the structure of the biotin carboxylation subsite of the protein. The 6-amino group of ATP appears to be necessary for the magnesium chelate of this nucleotide to bind to the enzyme with high affinity.[760] The significance of this observation in terms of the amino acid residues involved in $MgATP^{2-}$ binding is unclear, but it may reflect hydrogen bonding between the 6-amino group of the purine moiety of $MgATP^{2-}$ and the sulfydryl group of a cysteinyl residue. It has been suggested that this type of interaction occurs in the $MgATP^{2-}$ binding sites of myosin[922,923] and in the Na^+/K^+ activated ATPase.[638] This hypothesis is consistent with the observation that ATP affords some protection against *N*-ethylmaleimide inactivation of chicken liver pyruvate carboxylase.[628]

2. Chemical Modification Studies

There is some evidence from affinity labeling studies for a lysyl residue in the biotin carboxylation subsite. The 2′, 3′ dialdehyde derivative of ATP (oATP) could be covalently attached to the sheep liver pyruvate carboxylase in the presence of $NaBH_4$, with concomitant inactivation of the biotin carboxylation function of the protein of the protein.[252] $MgATP^{2-}$ and acetyl-CoA protected the enzyme against this modification, while pyruvate, oxaloacetate, and bicarbonate were ineffective. Although acetyl-CoA protected the enzyme, it is unlikely that the inactivation occurred by modification of residues in the acetyl-CoA binding site, in that $MgoATP^{2-}$ in the absence of $NaBH_4$ was a linear competitive inhibitor of $MgATP^{2-}$ in both the acetyl-CoA-dependent and independent reactions, and 1.1 ± 0.1 mol of $MgoATP^{2-}$ per mole of biotin were covalently linked to the protein at 100% inactivation, regardless of whether the inactivation was carried out in the presence or absence of acetyl-CoA. Chromatography of samples of an enzymic digest of Mg−*o* $[^{14}C]ATP^{2-}$ revealed one major band of radioactivity, which co-chromatographed with the adduct of lysine and $MgoATP^{2-}$.

It has been suggested[38] that this lysyl residue may be involved in facilitating formation of the *enol* form of biotin during the process of biotin carboxylation. In the *enol* form, the 1′-N atom of the ureido ring of biotin is about 10^{10}-fold more nucleophilic than it is in the *keto* form of biotin.[390] The change in reactivity of this residue with respect to $MgoATP^{2-}$ in the presence of acetyl-CoA is consistent with this idea, in that the isotope exchange between MgATP and $[^{32}P]$-phosphate catalyzed by pyruvate carboxylase shows an absolute requirement for acetyl-CoA.[752,901] Thus, one of the consequences of binding of acetyl-CoA to the enzyme may be enhancement of the ability of the lysyl residue to promote enolization of the ureido ring of biotin, leading to effective $MgATP^{2-}$-dependent carboxylation of its 1′-N atom.

C. Biotin Attachment Site

The site of attachment of the prosthetic biotinyl moiety can be considered as the central module of pyruvate carboxylase, in that the biotin is responsible for transferring the carboxyl moiety (as 1′-*N*-carboxybiotin) from the biotin carboxylation subsite to the transcarboxylation subsite. Hence, information on the structure of this key region of the protein would be of considerable interest. To date, the only investigation of this region of the protein is that of Rylatt et al.,[712] who determined the primary structures of biotin-containing tryptic peptides isolated from the sheep, chicken, and turkey liver enzymes. The sequences of the 19-residue peptides from the avian enzymes were identical, whereas the sheep enzyme sequence differed in only three positions. In two of these positions (2; gln/ala, 17; ser/ala) the substitutions were quite conservative. Wood and Barden[983] have reported strong homology between these sequences and those of the biotin-containing tryptic peptides from two bacterial carboxylases; *E. coli* acetyl-CoA carboxylase, and *Propionibacterium shermanii* transcarboxylase. This sequence conservation may either reflect a functional requirement of this portion of the proteins, or may be required for specific enzyme-catalyzed attachment of biotin to the lysyl residue, yielding biotin.

D. Transcarboxylation Subsite

1. Introduction

The transcarboxylation subsite of pyruvate carboxylase catalyzes the reaction:

$$\text{ENZ-Carboxybiotin} + \text{Pyruvate} \rightleftharpoons \text{Oxaloacetate} + \text{ENZ-Biotin} \; \ldots\ldots \qquad (4)$$

A number of mechanisms have been suggested for this reaction, including concerted mechanisms,[301] a mechanism involving carbanion formation,[830,831] and an enolate ion relay mechanism.[38,324] Structural studies of the transcarboxylation site may help to decide which of these mechanisms is applicable; e.g., the distance between the bound Mn(II) and pyruvate is critical to one of the proposed concerted mechanisms,[301] and the enolate ion relay mechanism requires the presence of a base in the transcarboxylation site.[38,324] The structure of this site has been investigated using both magnetic resonance and chemical modification techniques.

2. Magnetic Resonance Studies

The techniques of nuclear magnetic resonance (NMR) and electron spin resonance (ESR) have been applied to pyruvate carboxylase by several investigators, with the intention of defining geometric relationships between the different functional modules of the enzyme. The use of these techniques has depended on the presence of a paramagnetic center, the bound Mn(II), in the subunits of pyruvate carboxylase. It is possible to calculate the distance between the bound Mn(II) and the nuclei of atoms of substrates or inhibitors bound to the enzyme by measuring the effects of coupling between the unpaired electron spin of the bound Mn(II) and the nuclear spins of ^{1}H and ^{13}C atoms of the substrate or inhibitor. This coupling is experimentally manifest as an alteration in the spin-lattice (T_1) or spin-spin (T_2) relaxation times of the nuclear spins of the bound ligands. However, the interpretation of such data is complex, and considerable caution should be exercised in assessing the significance of reported structural information.

In early NMR studies,[567,753,754] it was deduced that the bound Mn(II) interacts directly with substrates and inhibitors of the transcarboxylation reaction (Equation 4). A mean value of 0.35 ± 0.1 nm was calculated for the distance between the methyl group hydrogen atoms of bound pyruvate, and the bound Mn(II).[567] This value was obtained without experimental determination of the correlation time (τ_c) governing the interaction between the spins of the bound Mn(II) and the hydrogen atoms of bound pyruvate. A range of values for this τ_c [(0.18 to 29) × 10^{-12} sec] was estimated from model compound studies and the effects of the bound Mn(II) on the T_1 values of the bound pyruvate hydrogen atoms; this τ_c range allowed calculation of Mn(II)-methyl hydrogen atom distances in the range 0.25 to 0.49 nm.

However, in subsequent ^{13}C and ^{1}H NMR studies of the binding of pyruvate to the enzyme,[303,676] τ_c, determined experimentally, was found to be of the order of 5×10^{-9} sec, giving a value of 0.6 to 0.8 nm for the distance between bound pyruvate and the bound Mn(II).

In these investigations the τ_c for the pyruvate-Mn(II) interaction was calculated from the T_1:T_2 ratio of bound pyruvate hydrogen atoms[676] and from the resonance frequency dependence of the T_1 values of bound pyruvate nuclei.[303,676] The latter approach requires the assumption that the τ_c is frequency independent[249] and the fact that the measurements were made at only two resonance frequencies in both the ^{1}H and ^{13}C NMR experiments suggests that there is still some uncertainty concerning this parameter. However, as similar values of τ_c were obtained in both investigations, and given that distances calculated depend on $\tau_c^{1/6}$, it would seem that the range 0.6 to 0.8 nm is a reasonable estimate for the distance between bound pyruvate and Mn(II). A similar range (0.5 to 0.63 nm) was calculated for the distance between bound pyruvate and Co(II) in the related enzyme, transcarboxylase.[302] However, as noted by Burton et al.,[145] this distance range was calculated using the invalid assumption of an average g value for the high spin bound Co(II). Until the effects of this assumption are assessed, it would seem unwise to place undue importance on this reported distance range.

The revised pyruvate-Mn(II) distance is of some significance in various proposed models for transfer of the carboxyl group from carboxybiotin to pyruvate in the transcarboxylation reaction. These models are discussed in detail in Chapter 4. The original proposal of Mildvan et al.[568] that coordination of the bound Mn(II) to the carbonyl oxygen of pyruvate promoted abstraction of a methyl group proton is inconsistent with the revised distance range 0.6 to 0.8 nm; the Mn(II) is now too far from the pyruvate to allow such a direct interaction. However, Fung et al.[302] have noted that coordination of an inner-sphere water molecule to the Mn(II) could achieve the same effect; the water molecule could form an H bond to the carbonyl oxygen of the bound pyruvate. It should be noted that interactions of this type, which have been incorporated into concerted mechanisms for transcarboxylation,[301] may not bear directly on the mechanism of action of the enzyme, in that, as discussed in Chapter 4, a carbanion or enolate anion relay mechanism of transcarboxylation may be more likely.

It has been concluded from initial rate studies of chicken[73] and sheep[28] pyruvate carboxylase that the biotin carboxylation and transcarboxylation subsites of the enzyme are spatially distinct. This conclusion is consistent with the effects of substrates and inhibitors of the enzyme of the ESR spectrum of the bound Mn(II).[301] The ESR spectrum of the chicken liver enzyme was insensitive to components of the biotin carboxylation reaction. Also, acetyl-CoA, which is an essential activator of biotin carboxylation, had no effect on either the ESR spectrum of the enzyme[301] or on the dipolar Mn(II) water proton interaction, as measured by proton relaxation enhancement.[568] On the other hand, pyruvate and inhibitors of the transcarboxylation reaction did influence the ESR spectrum of the enzyme.

A lower limit of 1 nm for the distance between the bound Mn(II) and the biotin carboxylation site was set from the observation that there was no dipolar interaction between the paramagnetic centers of Cr(III)ATP^{-} (which is a competitive inhibitor of $MgATP^{2-}$ in initial rate studies [25]) and the bound Mn(II).[676] The sensitivity of these measurements was limited by the short electron spin relaxation time (τ_s) of Cr(III) in the ATP complex, and it was noted that an anisotropic dipolar interaction could be detected over the longer distances if a nitroxide radical, with a longer τ_s was present as one of the paramagnetic centers. The reactive sulfydryl group(s) in the transcarboxylation subsite[424] and the lysyl group in the biotin carboxylation site[252] may provide suitable attachment sites for such nitroxide radical probes.

3. *Affinity Labeling Studies*

Evidence for sulfydryl residues in the transcarboxylation site of sheep pyruvate carboxylase was obtained by Hudson et al.,[424] who showed that the pyruvate analog, bromopyruvate, inactivated both the acetyl-CoA-dependent and independent activities of the enzyme. The components of the transcarboxylation reaction, oxaloacetate and pyruvate, protected the enzyme against inactivation, while the substrates for biotin carboxylation, $MgATP^{2-}$ and bicarbonate, were ineffective. Evidence for the specificity of bromopyruvate interaction came from the observations that treatment of the enzyme with bromoacetate or iodoacetate under the same conditions resulted in an inactivation rate of less than 5% of that for bromopyruvate. This observation may be compared to the work of Palacian and Neet,[628] who showed that modification of 32 sulfydryl groups in chicken pyruvate carboxylase had no significant effect on the K_m for pyruvate obtained in initial rate kinetic studies.

Hudson and co-workers found that 1.5 amino acid residues per mole of biotin were labeled when the enzyme was completely inactivated in the absence of acetyl-CoA; a 1:1 ratio was obtained when the enzyme was inactivated in the presence of acetyl-CoA. This difference was consistent with the observation that acetyl-CoA afforded some protection against bromopyruvate inactivation of the enzyme, and raises the possibility that bromopyruvate could react with residues in the acetyl-CoA binding site, as well as those in the transcarboxylation site of the enzyme. Hudson and co-workers argued that this was unlikely, in that bromopyruvate inactivated the acetyl-CoA-dependent and independent activities of the enzyme to the same extent. This argument is inconclusive, as if the reactivity of the transcarboxylation site residue was much greater than that of the presumptive acetyl-CoA binding site residue, the kinetics of inactivation would be dominated by modification of the former residue, even if the bromopyruvate was able to modify a residue in the acetyl-CoA binding site. The observation that a plot of log (inactivation rate) vs. log (bromopyruvate) had a slope of 1.0 ± 0.24, which Hudson and co-workers interpreted as evidence that bromopyruvate was selectively modifying one residue which was essential for enzymic activity, is also inconclusive. As Horiike and McCormick[421] have noted, the interpretation of such plots is complex; the number of essential residues modified cannot always be obtained from the slope of the log-log plot. Thus, it is not clear whether the inactivation of sheep pyruvate carboxylase by bromopyruvate arises solely from modification of residues in the transcarboxylation site, or whether residues in the acetyl-CoA binding sites of the enzyme are also susceptible to modification by bromopyruvate.

Indeed, Palacian and Neet[628] have reported data which are consistent with the presence of sulfydryl residues in the acetyl-CoA binding sites of chicken pyruvate carboxylase. They showed that modification of 32 sulfydryl groups in the enzyme with *N*-ethylmaleimide was associated with a nearly threefold decrease in the affinity of the enzyme for acetyl-CoA, as judged from changes in the activation constant for acetyl-CoA in initial rate kinetic studies.

E. Acetyl-CoA Binding Site

1. *Introduction*

The extent of activation of the pyruvate carboxylase catalyzed reaction by acetyl-CoA depends upon the source of the enzyme. Pyruvate carboxylase isolated from some lower organisms[269,758,856] is unaffected by acetyl-CoA, while some prokaryote,[163,350,503] and all vertebrate[29,764] pyruvate carboxylases show at least some stimulation by acetyl-CoA.

In the vertebrate enzymes, both the biotin carboxylation[752,901] and transcarboxylation[29,30,544,764] reactions are enhanced in the presence of acetyl-CoA. The inactivation associated with dilution of vertebrate pyruvate carboxylases below 3 to 4 units/m*l*,[29] as well as the cold lability of chicken liver pyruvate carboxylase,[426,759] can be prevented by acetyl-CoA. There is evidence for the involvement of an ionizing group with a pKa of 8.6 in the acetyl-CoA activation of chicken liver pyruvate carboxylase.[761] However, with the exception of the results of chemical modification data discussed below, the nature of the amino acid residues involved in acetyl-CoA activation of pyruvate remains unknown.

2. Chemical Modification Studies

Attempts to identify the amino acid residues making up the regulatory acetyl-CoA binding site of pyruvate carboxylase have centered on characterization of the effects of modifying the protein with the lysine-selective reagents 1-fluoro-2,4-dinitrobenzene (FDNB), 2,4,6-trinitrobenzene sulfonic acid (TNBS), and pyridoxal phosphate.

Sheep kidney pyruvate carboxylase was shown to possess a lysyl residue whose integrity was essential for acetyl-CoA-dependent enzymic activity by Keech and Farrant.[438] Modification of the protein with TNBS, cyanate, or FDNB led to loss of enzymic activity. In the case of FDNB modification, one catalytically important residue per active site was modified, acetyl-CoA afforded complete protection against the inactivation, and 2,4-dinitrophenyl-ε-lysine was identified on chromatograms of the protease hydrolysate derived from [^{14}C]-FDNB-modified pyruvate carboxylase.

The importance of lysyl residue in the acetyl-CoA-mediated-activation of the enzymic reaction was confirmed by Ashman and co-workers,[30] who showed that treatment of sheep pyruvate carboxylase with TNBS inactivated the acetyl-CoA-dependent enzymic activity, while the modified protein catalyzed pyruvate carboxylation in the absence of acetyl-CoA at an increased rate compared to that of control samples. These observations provided direct evidence that the modified residues formed part of the regulatory, rather than catalytic, module of the protein. The absorption spectrum of the modified protein had maxima at 345 and 420 nm[413] and the 2,4,6-trinitrophenyl groups on the protein were insensitive to thiolysis by 2-mercaptoethanol;[774] both these observations are consistent with the modified groups being lysine side chains.

In a concurrent report, it was shown that TNBS treatment of pyruvate carboxylases from rat liver, and *Saccharomyces cerevisiae* led to desensitization of their acetyl-CoA-dependent enzymic activities.[765] In the case of the rat enzyme, the modification had little effect on the acetyl-CoA-independent enzyme activity, and the absorption spectrum[413] and thiolysis-resistance[774] of the modified enzyme were consistent with the modifiable residues being lysines. Pyruvate carboxylases isolated from a thermophilic *Bacillus*[503] and from *Aspergillus nidulans*[615] have subsequently been shown to be desensitized to acetyl-CoA activation following treatment with TNBS.

However, it is not clear whether the lysyl residues modifiable by TNBS form part of the acetyl-CoA binding sites of these enzymes, or whether they were distant from the binding site, being involved in mediating the effects of acetyl-CoA binding on the catalytic module of the enzyme. Indeed, Ashman et al.[30] suggested that the latter possibility was the case, in that the acetyl-CoA deacylase activity of sheep kidney pyruvate carboxylase was relatively insensitive to TNBS modification. The validity of this argument hinges on whether the deacylase activity is an intrinsic property of pyruvate carboxylase, and on whether there is a single acetyl-CoA binding site responsible for both activation of pyruvate carboxylation and the deacylase activity. The observation that the deacylase activity co-purifies with pyruvate carboxylase during both ion exchange and gel filtration chromatography[29,250] implies that the deacylase activity is a property of pyruvate carboxylase. Moreover, the observation of four acetyl CoA binding sites per tetramer of chicken pyruvate carboxylase by Frey and Utter[287] suggests that there is a single acetyl-CoA binding site per enzyme subunit. (Although the low acetyl-CoA concentration binding data reported by Frey and Utter are difficult to interpret, because of the deacylase activity of the protein,[250] the number of acetyl-CoA binding sites can be estimated from binding data obtained at high acetyl-CoA concentrations, where the deacylase activity is proportionately less important.)

It also seems likely that the TNBS-sensitive lysyl residues in rat and chicken liver pyruvate carboxylases do not form part of the acetyl-CoA binding sites of these proteins. This conclusion was based on two lines of evidence drawn from a study of the effects of various analogs of acetyl-CoA on the rate of TNBS inactivation of the acetyl-CoA-dependent enzymic activities of these two proteins.[756] The first piece of evidence is that there was a residual rate of TNBS-inactivation in samples of pyruvate carboxylase containing saturating levels of acetyl-CoA or its analogs. Since the magnitude of the residual inactivation rate varied for different acetyl-CoA analogs, and was independent of the purity of the enzyme, Scrutton and co-workers[756] argued that the residual inactivation could not be attributed to secondary modification of other residues, and that the modified lysyl residue did not participate directly in acetyl-CoA binding. This conclusion seems unconvincing for two reasons. First, the residual inactivation rates were very small; in the case of acetyl-CoA protection of chicken pyruvate carboxylase the residual rate was 0.002 min^{-1} (corresponding to a $t_{1/2}$ value of 346 min). Scrutton et al.[756] did not present any evidence to show that the enzyme preparations were stable over this time period in the absence of TNBS. Second, no evidence was presented to show that modification of residues other than the "essential" lysyl residue did not occur. There is no reason to suppose that the various acetyl-CoA analogs employed could not have had differential effects on the reactivities of other residues for TNBS.

However, the second piece of evidence obtained by Scrutton et al.[756] for the modified lysyl residue being distant from the acetyl-CoA binding site is more convincing. Acetyl-dephospho-CoA and acetyl-desamino-CoA both failed to afford significant protection against TNBS modification of either the rat or chicken enzymes. Both these analogs act as competitive inhibitors of acetyl-CoA in initial enzyme rate studies.[299,756] These observations, as well as the significant protection afforded by acetyl pantetheine, appear to rule out the postulated interaction between the lysyl residue and the 3′ phosphate of acetyl-CoA.[745]

The role of the TNBS-sensitive lysyl residue of sheep liver pyruvate carboxylase has been recently reexamined by Chapman-Smith.[176] Treatment of sheep pyruvate carboxylase with TNBS led to parallel loss of the acetyl-CoA-dependent ATP:Pi isotopic exchange reaction, and the overall acetyl-CoA-dependent enzymic reaction. This observation is consistent with the reported absolute dependence of the ATP:Pi exchange reaction on acetyl-CoA.[752,901]

The effects of TNBS treatment of the enzyme on the pyruvate-oxaloacetate isotopic exchange reaction were similar to those reported for the overall reaction.[30] The ability of acetyl-CoA to activate the enzyme-catalyzed exchange reaction was lost in TNBS-treated enzyme, but the modified enzyme catalyzed the exchange reaction in the absence of acetyl-CoA at an increased rate compared to that of untreated enzyme.

Chapman-Smith[176] also showed that it was not possible to reverse the desensitization of the acetyl-CoA activation of sheep pyruvate carboxylase by increasing the concentration of acetyl-CoA in the assay medium. This observation suggests that treatment of sheep pyruvate carboxylase with TNBS does not merely reduce the affinity of the enzyme for acetyl-CoA, but rather that the link between the regulatory and catalytic modules of the enzyme is broken following TNBS modification.

The effects of acetyl-CoA analogs on TNBS modification of sheep pyruvate carboxylase formed a slightly different pattern to that reported by Scrutton et al.[756] for the chicken and rat enzymes. Acetyl-dephospho-CoA afforded minimal protection against TNBS modification of sheep pyruvate carboxylase, while the modification was enhanced in the presence of acetyl phosphopantetheine. Although both 2′,5′-ADP and 3′,5′-ADP afforded some protection, the latter was more effective. In contrast, Scrutton et al.[756] found that these compounds had equivalent effects on TNBS inactivation of chicken pyruvate carboxylase.

On balance, it would seem that the TNBS-sensitive lysyl residue of pyruvate carboxylase is not directly involved in binding of acetyl-CoA to the enzyme. The evidence for this conclusion includes the retention of acetyl-CoA deacylase activity by TNBS-modified pyruvate carboxylase, and the failure of acetyl-dephospho-CoA and acetyl-desamino-CoA to afford protection against TNBS modification. The lysyl residue may form part of the link between the regulatory and catalytic sites of the enzymes. If this is the case then the differential effects of acetyl-CoA and its analogs in TNBS inactivation may arise from alterations in the reactivity of the residue as a consequence of occupancy of all or part of the regulatory site of the enzyme.

It is possible that affinity labeling studies, using protein-reactive analogs of acetyl-CoA, may provide more information about the molecular architecture of the acetyl-CoA regulatory binding site of pyruvate carboxylase. In this context the potential photoaffinity labels of benzoyl CoA developed by Myers and Utter[590] may be worthy of further investigations. These reagents (azidobenzoyl CoA and azidophenylacetyl CoA) have been shown to activate yeast pyruvate carboxylase. The highly reactive aryl nitrenes generated by photolysis of such compounds may allow successful labeling of normally unreactive amino acid residues within the acetyl-CoA binding site.

F. Summary

In conclusion, biophysical and protein chemical studies of pyruvate carboxylase over the last decade have shown the following:

1. There is a lysyl residue in the biotin carboxylation site of the sheep enzyme.[252]
2. There is a sulfydryl residue in the transcarboxylation site of the sheep enzyme;[424] there may be sulfydryl groups in the acetyl-CoA binding sites of the sheep and chicken enzymes.[424,628]
3. The pyruvate binding site is 0.6 to 0.8 nm distant from the bound Mn(II),[303,676] while the ATP binding site is at least 1 nm from this metal atom.[676]
4. The amino acid sequence surrounding the biotin attachment site is strongly conserved in the sheep and avian enzymes,[712] and is homologous to the corresponding regions of acetyl-CoA carboxylase and transcarboxylase.[983]
5. The sheep[30,438] rat,[763] yeast,[765] *Bacillus,*[503] and *Aspergillus nidulans*[615] enzymes contain a lysyl residue which is involved in activation of the enzymes by acetyl-CoA, but this residue is probably not directly involved in binding of acetyl-CoA.[30,756]

V. THE EFFECTS OF ACETYL-CoA ON THE STRUCTURE OF PYRUVATE CARBOXYLASE

A. Binding Studies

Despite the numerous studies on the effects of acyl-CoA compounds on the activity of pyruvate carboxylase (see Chapter 2, Section II.C.2; Chapter 4, Section III, and Chapter 5, Section C.III), there appears to be only one published report in which the binding of acetyl-CoA to this enzyme has been measured directly. Using a modified flow dialysis technique, Frey and Utter[287] showed chicken liver pyruvate carboxylase has four binding sites for acetyl-CoA per tetramer. The concentration of acetyl-CoA required for half-saturation of the binding sites was 13.9 μM, in good agreement with the 13.3-μM acetyl-CoA required for half-maximal catalytic activity under the same conditions. The interaction of acetyl-CoA with pyruvate carboxylase also produces a

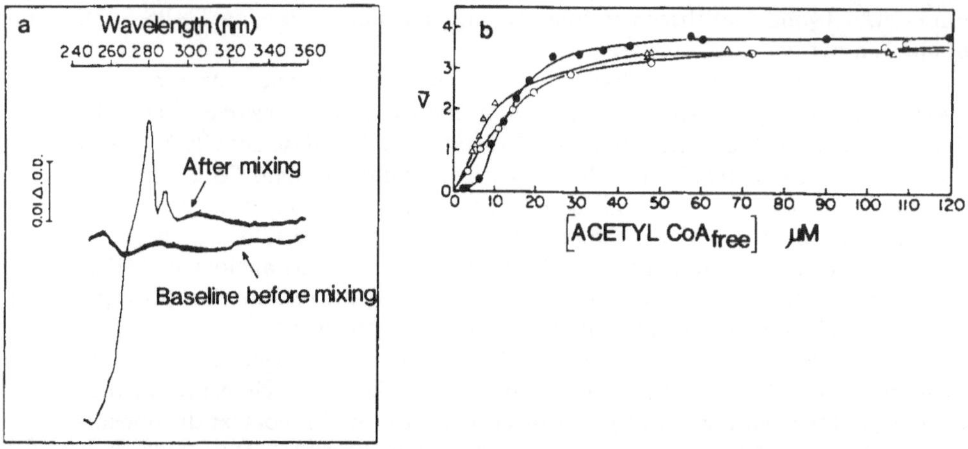

FIGURE 15. The binding of acetyl-CoA to chicken liver pyruvate carboxylase. (a) Difference spectra from the interaction of chicken liver pyruvate carboxylase (2.5 mg/ml) with acetyl-CoA (70 μM) in 10 mM potassium 4-(2-hydroxymethyl)-1-piperazinee-thanesulfonate buffer, pH 7.20, containing 0.2 M KCl and 1 mM EDTA (buffer I). The total cell path length (two sectors) was 0.88 cm. (b) Plot of acetyl-CoA binding (○), catalytic activity (●), and UV difference response (Δ) was a function of the free acetyl-CoA concentration. The curves were normalized so that $V_{max} = \bar{v} = (\Delta_{280} - \Delta_{260})$. All experiments were performed in buffer I. The free acetyl-CoA concentration for the UV difference experiment was calculated from the results of the binding experiment. (From Frey, W. H. and Utter, M. F., *J. Biol. Chem.*, 252, 51, 1977. With permission.)

small but measurable change in the UV spectrum of the enzyme, as illustrated in Figure 15a.

As is evident in Figure 15b, the concentration of acetyl-CoA required for a half-maximal response in the UV difference spectrum (viz. ~9 μM) is of a similar order to the Kd determined by the binding studies and the Ka for the activation of catalytic activity. However, it is apparent that the sigmoidicity of the plot of activation of catalysis is much greater than in the cases of either the binding of acetyl-CoA or the UV difference response. Reasons for this are discussed in Chapter 4, Section III.F. Among these is the significant rate of hydrolysis of acetyl-CoA catalyzed by even highly purified preparations of pyruvate carboxylase (~4 nmol/min/mg enzyme in the absence of the substrates,[287] and greater in their presence[761]).

Another interesting observation by Frey and Utter[287] was that the protomeric molecular form of chicken liver pyruvate carboxylase produced by cold inactivation (see Section V.B) was unable to bind acetyl-CoA even when tested over a wide range of concentrations (12 to 168 μM). This would be at least one factor contributing to the lack of catalytic activity exhibited by this 7.5S species.[857] This last result with the chicken enzyme is yet another example of the quite marked differences noted between the avian and mammalian pyruvate carboxylases. Isolated monomeric rat liver pyruvate carboxylase is catalytically competent in the absence of acetyl-CoA,[857] but is also activated by acetyl-CoA.[592]

A detailed analysis of acetyl-CoA binding to pyruvate carboxylase measured by indirect means has been virtually precluded by the very limited conformational changes induced in the enzyme by this allosteric effector. Thus, addition of saturating concentrations ($\geq$ 0.98 mM) of acetyl-CoA produced only a 10% quenching of the fluorescence emitted by the enzyme-bound probe, 8-anilinonaphthalene-1-sulfonic acid.[554] Only small changes were noted in the near-UV region and none in the far-UV circular dichroism spectra of pyruvate carboxylase upon addition of acetyl-CoA[554] The small changes observed in the UV-difference spectra[287] (see Figure 15), particularly at the low concentrations of acetyl-CoA, limited those investigators' use of this parameter.

Similarly, the effect of acetyl-CoA on the conformation of the chicken liver enzyme as revealed by ultracentrifugation was shown to be within the precision (5 to 10%) of the techniques used at the time.[761] More recent studies,[857] though, using the reacting enzyme sedimentation technique have shown that there may be a small change in the conformation of yeast pyruvate carboxylase which sediments as a tetramer in the presence (16.33 ± 0.14S) or the absence (15.8S) of acetyl-CoA.

Despite the subtlety of the change which occurs in the conformation of the enzyme upon formation of the enzyme activator complex, (of Section III.B.2) the effects of acetyl-CoA on the activity and stability of pyruvate carboxylase are profound.

Thus, the allosteric activator affords almost complete protection of chicken liver pyruvate carboxylase against inactivation by exposure to the cold,[426,759,761] 0.4 *M* urea,[426] mildly alkaline (pH 7.95) or acid (pH 6.2) conditions,[426] and even low concentrations (0.5 m*M*) of sodium dodecyl sulfate.[761]

B. Cold Inactivation

1. Chicken Liver Pyruvate Carboxylase

In common with a number of other oligomeric enzymes from a variety of sources,[113] the pyruvate carboxylases of chicken and turkey livers exhibit a loss of catalytic activity when cooled to 0 to 5°C. The effect with the chicken liver enzyme was observed first by Scrutton and Utter,[759] who found that while up to 90% of the activity was lost within 40 to 90 min of exposure to 2°C, a significant proportion (up to 90%) could be regained in approximately 20 min by rewarming the enzyme to 23°C. Loss of both the Pi-ATP and pyruvate-oxaloacetate exchange reactions at the same rate as the overall reaction[752] indicated that both subsites of the catalytic center were susceptible to inactivation at low temperature. Ultracentrifugal analysis of cold-inactivated pyruvate carboxylase indicated that loss of activity was accompanied by a transfer of protein from a 15 to a 7S species.[759]

A detailed study of the cold inactivation process was made by Irias et al.,[426] who paid particular attention to the effects of various factors (viz., reaction mix components, buffer pH, and composition, as well as other additives) on the inactivation and reactivation phenomena. They showed that acetyl-CoA, at concentrations similar to those which activate catalysis, prevented loss of both activity and tetrameric structure. It had no effect, however, in restoring either activity or structure when added to enzyme that had already been cold inactivated.

ATP, on the other hand, was effective in both reducing the loss of activity on exposure to the cold, and facilitating the regain of activity on rewarming. It even reactivated the enzyme to a limited extent without any rewarming. Ultracentrifugal analysis of the same samples showed that ATP was more effective in preserving the tetrameric structure of the enzyme than its activity, and thus indicated the existence of an inactive tetrameric species. This was also suggested by the kinetics of reactivation being first order despite there being a step in the process involving the reassembly of four monomeric units into a tetramer.

The extent of the reversibility of the cold inactivation (i.e., the reactivation process) appeared to depend on a number of variables, including the temperature, the enzyme concentration, the pH, ionic strength, and specific ionic composition of the medium, as well as the addition of specific ligands or other molecules such as methanol and various polyols.[426]

The lack of complete reversibility and the tendency to form visible precipitates during rewarming suggest the existence of a largely irreversible step which might involve denaturation of the monomers (to monomers*), rendering them incapable of reforming tetramers but susceptible to aggregation in a nonspecific manner. A minimum scheme to account for the cold inactivation of chicken liver pyruvate carboxylase would include, therefore:

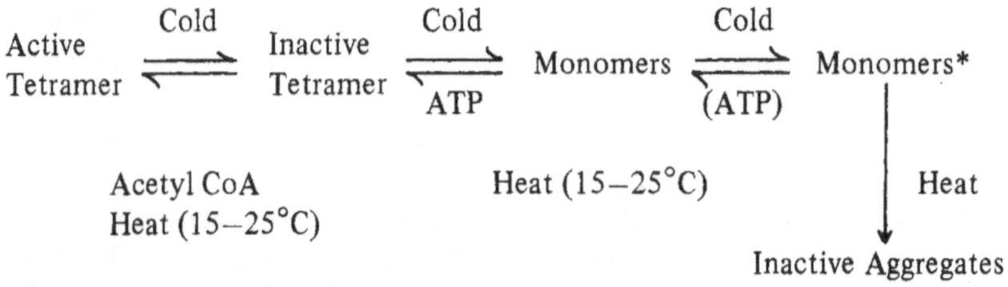

2. Other Species

Of the other non-avian forms of pyruvate carboxylase tested for cold lability, the enzyme from *A. globiformis*[350] is the only one so far which resembles the chicken liver enzyme in being stabilized by high salt concentration (100 m*M* KCl), acetyl-CoA, or ATP, and forming an inactive tetramer prior to dissociating. However, cold inactivation only induces the *A. globiformis* enzyme to dissociate to a dimer, and as yet no reactivation of this enzyme has been obtained upon rewarming it.

None of the other forms of pyruvate carboxylase tested in low ionic strength media (viz., those from calf,[930] sheep,[124] and rat liver,[545] sheep kidney,[46] rat brain,[525] yeast,[996] *Ps. fluorescens*,[570] and *B. stearothermophilus*[163]) has shown any loss of activity upon exposure to the cold for periods of several hours. However, Nakashima et al.[592] have shown that rat liver pyruvate carboxylase is inactivated reversibly and converted to protomers by incubation at 0°C in the presence of high (1 *M*) concentrations of chloride salts of monovalent cations. It is likely that it was the combination of high ionic strength (viz., high concentrations of ammonium sulfate) and low temperature which induced the loss of activity reported for the partially purified sheep kidney pyruvate carboxylase[507] that is not evident in the purified enzyme[46] in low ionic strength buffers as used to demonstrate the cold lability of the avian enzymes.

Incubation of the rat liver pyruvate carboxylase at 0°C with only 30 to 100 m*M* ammonium chloride results in dissociation of this enzyme to dimers and monomers which are enzymically active in the absence of acetyl-CoA, but retain the capacity to be activated by this allosteric effector. These results, as well as the failure of ATP or acetyl-CoA to offer any significant protection against the inactivation by 1 *M* ammonium chloride and of ATP to promote reactivation, are in stark contrast to those obtained with chicken liver pyruvate carboxylase.[426] Nevertheless, they are consistent with the rat liver pyruvate carboxylase yielding catalytically active dimers and monomers when incubated at room temperature in high (0.29 *M*) potassium bicarbonate medium,[857] whereas no catalytic activity could be demonstrated for the monomers of the chicken liver enzyme.[856,857]

As Bock and Frieden[113] have noted, cold lability of enzymes may be influenced not only by the temperature dependence of hydrophobic bond strengths, but also by the temperature dependence of specific ionizable groups which can control the dissociation or association process irrespective of which types of bond are important in maintaining quaternary structure.

C. Dilution Inactivation

Another characteristic of pyruvate carboxylase upon which acetyl-CoA can exert a major effect is the phenomenon of "dilution inactivation". This time-dependent loss of activity, which becomes apparent upon dilution of pyruvate carboxylase below ~4 U/m*l* (≃0.2 mg/m*l*) under the conditions used, was first noted by Ashman et al.[29] with the sheep kidney enzyme. The actual final level of activity after dilution was dependent on the initial activity, ranging from less than 12% of the theoretical value when diluted to 0.1 U/mℓ to 100% when diluted to 4 U/mℓ.[29]

Dilution inactivation has also been observed with the enzyme from other vertebrate sources (viz., sheep[190,191,442] and chicken[37] livers) but not with the yeast enzyme.[881] This protein-concentration-dependent effect appears to be specific for catalytically active pyruvate carboxylase protein since the addition of either bovine serum albumin or glutamate dehydrogenase failed to offer any protective effect.[29]

Acetyl-CoA, however, at similar concentrations to those affecting catalytic activity, reduced the rate of inactivation, such that at saturating concentrations it completely prevented dilution inactivation (or arrested further inactivation if added during its progress) but failed to reverse it.

Mechanistic aspects of this phenomenon were investigated by Clements et al.[190,191] and are discussed in Chapter 4, Section III.E. Recently, in a further examination of the molecular basis of this inactivation process in the sheep and chicken liver enzymes, it has been shown, using both high-performance gel permeation chromatography and electron microscopy,[37,442] that dilution inactivation is accompanied by dissociation of the tetramers to dimers and monomers which are inactive under the normal assay conditions used.

In the case of chicken pyruvate carboxylase,[39] the kinetics of both the loss of enzymic activity and dissociation are similar, with the residual activity and amount of tetrameric enzyme decaying to equilibrium levels that are dependent on the initial enzyme concentration. For both processes the decay is apparently monophasic and first order. The observed rate constants for the loss of activity show no apparent dependence on enzyme concentration and are similar to those for dissociation. Addition of acetyl-CoA to the chicken liver enzyme at various times after dilution has been observed, for the first time, to cause varying degrees of reactivation: the longer the time after dilution, the smaller the extent of reactivation. On the other hand, the reassociation of dimers and monomers induced by acetyl-CoA is apparently unaffected by the time of its addition. Also, the reactivation process appears to be significantly faster than reassociation. From this evidence, it has been concluded that the two processes are different and that reactivation is caused by activation of an inactive tetrameric form of the enzyme. Acetyl-CoA also appears to cause reassociation of dimers and/or monomers to an inactive tetramer or aggregate. From these results, then, it would seem that dilution inactivation of chicken liver pyruvate carboxylase is caused primarily by the reduction in concentration favoring dissociation of the tetrameric form of the enzyme but that the rate-limiting step is a protein conformational change which produces an enzymically inactive tetramer from the active form. This conformational change in the enzyme governs both its loss of activity and the subsequent dissociation of the tetrameric species. These changes can be summarized in a scheme thus:

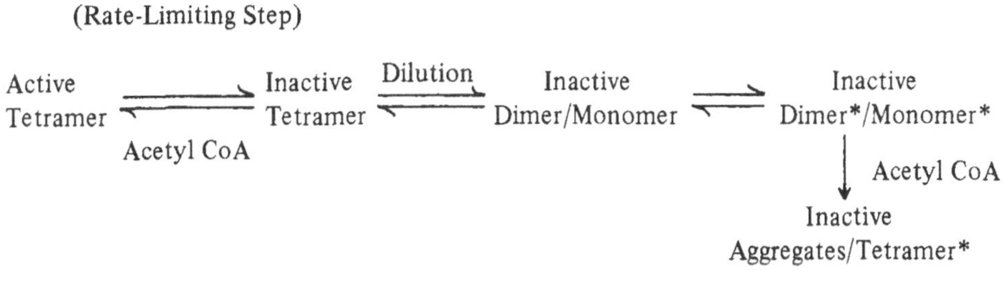

Acetyl-CoA strongly promotes reactivation of the inactive "loose" tetramer (see Section III.B.2) to the "tight" active form, but also promotes the dissociated species to re-aggregate to form inactive tetramers*.

Overall, this scheme bears a close resemblance to that described for cold inactivation of chicken liver pyruvate carboxylase (see Section V.B). In view of the enzyme concentration dependence of cold inactivation,[426] this similarity is not entirely unexpected.

The situation with the dilution inactivation of sheep liver pyruvate carboxylase is more complex: a kinetic analysis of the loss of activity on dilution to 0.5, 0.2, and 0.1 U/mℓ indicates that the process is biphasic, with an initial fast phase followed by a slower reaction. In addition, the amplitude of the fast phase appeared to increase with increasing dilution, which would seem to indicate that a concentration-dependent equilibrium is involved in this part of the overall process. With the samples at 0.2 and 0.1 U/mℓ the activity could be seen to approach zero. Thus, in the second, slower phase, an irreversible reaction or an equilibrium which strongly favors the formation of a species further from the active form would seem to be involved.

The presence of acetyl-CoA prevents inactivation from occurring and preserves the tetrameric structure as revealed by both high-performance gel filtration and electron microscopy. When added after dilution, acetyl-CoA prevents further inactivation but does not reactivate this form of the enzyme. However, acetyl-CoA does cause a rapid reassociation of the inactive monomers and dimers to form inactive tetramers. These tetramers have a larger diameter (17 to 18 nm) than active tetramers (15 to 16 nm) when viewed under the electron microscope, suggesting that they may be "flattened" or "loosely assembled".

Analysis of the process of dilution inactivation is thus another means of examining the effects which acetyl-CoA has on the structure of various pyruvate carboxylases. The differences revealed between species provide another posible route for insight into the balance of intra- and interprotomeric forces responsible for maintaining the native structure of these enzymes.

ACKNOWLEDGMENTS

One of us (J. C. Wallace) is indebted to Professor F. Mayer and Dr. W. Johannssen (Göttingen) and Dr. P. V. Attwood (Adelaide) for their thoughtful comments during the preparation of this manuscript. It is a pleasure to acknowledge the skilled technical assistance of Ms. J. Brazier both in recent experimental work and in the preparation of the figures for this manuscript. We should also like to thank Ms. L. Bradbrook for the rapid yet meticulous preparation of the typescript. The author's own experimental work reported herein has been funded by the Australian Research Grants Scheme and University of Adelaide, and this support is gratefully acknowledged.

ADDENDUM

Recently Freytag and Collier[1002] reported details of the molecular cloning of a cDNA for human pyruvate carboxylase. They showed that the biotin attachment site is 35 residues from the carboxyl-terminal end of the polypeptide as it is in two bacterial biotin carboxylases. As these authors note, this structural similarity strongly suggests that these proteins have evolved from a common ancestral gene (see also Section III.I). In human, baboon, and rat liver, the pyruvate carboxylase mRNA was shown to be 4.2 kilobases in length. By analysis of genomic DNA isolated from human-Chinese hamster somatic cell hybrids, Freytag and Collier localized the pyruvate carboxylase gene on the long arm of human chromosome 11.

ABBREVIATIONS

EDTA	—	Ethylene diamine tetraacetic acid
DEAE	—	Diethylaminoethyl
SDS-PAGE	—	Sodium dodecyl sulfate-polyacrylamide gel electrophoresis
kDa	—	Kilodaltons
U	—	Unit of enzymic activity equal to formation of 1 μmole product per minute.
PTH	—	Phenylthiohydantoin

Chapter 4

THE REACTION MECHANISM

D. B. Keech and P. V. Attwood

TABLE OF CONTENTS

I. INTRODUCTION

Probably one of the most important steps forward in understanding the mechanism of action of biotin-dependent enzymes in general, came from the studies of Northrop[606] and Northrop and Wood.[607] They carried out a detailed kinetic study on the biotin-dependent methyl-malonyl-CoA transcarboxylase [EC 2.1.3.1] which catalyzes the reversible transfer of carboxyl groups between methyl-malonyl-CoA and pyruvate. Their data were interpreted to suggest that transcarboxylase catalyzed a reaction involving a modified Ping-Pong mechanism where the active center consisted of two separate subsites. Northrop and Wood then introduced the concept that the two separate subsites were linked together by the covalently bound biotin acting as a mobile carboxyl-group carrier, shifting the activated carboxyl group from one subsite to the other. The idea of a flexible, covalently bound residue providing a shuttle system had not previously been considered to apply to the biotin-containing group of enzymes. A similar function had been postulated for lipoic acid, the prosthetic group of dihydrolipoyl transacetylase [EC 2.3.1.12], to transfer acetyl groups in the pyruvate dehydrogenase complex. Pantetheine was assigned a similar function in the acyl-carrier protein complex of the fatty-acid-synthesizing complex in yeast. However, the difference here is that in the latter two examples the carrier system is acting between enzymes, whereas the biotin is postulated to be acting within a single active center.

Since the original proposal was put forward by Northrop and Wood,[607] direct proof of the existence of separate subsites has been obtained from another biotin-dependent enzyme, acetyl-CoA carboxylase [EC 6.4.1.2] isolated from *Escherichia coli* (see Chapter 3). As a consequence of an active site consisting of two spatially distinct subsites, the classical approaches for studying the kinetic mechanism of action of an enzyme, namely initial velocity, product inhibition, and isotope exchange studies, are fraught with difficulties in terms of interpretation of experimental data. Nevertheless, over the last two to three decades considerable progress has been made in the elucidation of the reaction mechanisms of biotin-dependent enzymes. Several review articles covering advances in the general area have been published by Knappe,[450] Moss and Lane,[576] and Wood and Barden,[983] while investigations specifically related to pyruvate carboxylase [EC 6.4.1.1] have been discussed by Scrutton and Young[768] and Utter et al.[896]

II. THE OVERALL REACTION

In addition to the difficulties involved in elucidating the reaction mechanism for biotin-dependent enzymes, pyruvate carboxylase has its own special problems.

1. The enzyme catalyzes a reaction involving three substrates and three products, i.e.,

$$\text{Pyruvate} + HCO_3^- + \text{ATP} \rightleftharpoons \text{Oxaloacetate} + \text{ADP} + P_i \qquad (1)$$

2. In most cases, particularly with the enzyme isolated from vertebrate sources, it has two activators — acetyl-CoA and Mg^{2+}.
3. When either pyruvate or acetyl-CoA are the variable ligands in initial velocity studies, the enzyme exhibits non-Michaelis-Menten kinetic behavior.
4. The enzyme contains covalently bound biotin which acts as a mobile carboxyl group carrier. In most cases there is a stoichiometric ratio of approximately 4 mol of biotin per mole of protein in tetrameric forms of the enzyme[545,751,759,769] and 2 mol of biotin per mole of protein for the dimeric form of the enzyme isolated from *Pseudomonas citronellolis*.[856]

5. Except for the enzyme isolated from *Ps. citronellolis,* all pyruvate carboxylases examined so far contain a tightly bound transition metal[545,763,769,937] which was believed to be implicated in the catalytic process,[567,568,753,754] although subsequent findings[676,757] have cast considerable doubt on the original interpretations.

The overall reaction can be partitioned into two discrete steps or partial reactions:

$$\text{ENZ-biotin} + \text{MgATP}^{2-} + \text{HCO}_3^- \rightleftharpoons \text{ENZ-Carboxybiotin CO}_2^- + \text{MgADP} + \text{P}_i \quad (2)$$

$$\text{ENZ-Carboxybiotin CO}_2^- + \text{Pyruvate} \rightleftharpoons \text{ENZ-Biotin} + \text{Oxaloacetate} \quad (3)$$

The initial step involves the formation of a carboxybiotinyl complex, with the concurrent hydrolysis of MgATP to form MgADP and orthophosphate. When the enzyme is incubated with ATP, Mg^{2+}, $H^{14}CO_3^-$, and acetyl-CoA (in the case of acetyl-CoA-activated pyruvate carboxylases) an ENZ-[^{14}C]carboxybiotin intermediate can be isolated by Sephadex gel filtration.[752] In the second step, transfer of the carboxyl group to pyruvate occurs. It is assumed that activation of bicarbonate in the first partial reaction occurs at the first subsite while the transfer reaction is assumed to occur at the second subsite.

All species of the enzyme examined so far have exhibited isotopic exchange reactions between ATP and $^{32}P_i$, ATP, and [^{14}C]ADP, oxaloacetate and [^{14}C]pyruvate, and oxaloacetate and $H^{14}CO_3^-$.[48,157,544,752] The isotopic exchange between ATP and $^{32}P_i$ requires the presence of ADP, HCO_3^-, Mg^{2+}, and acetyl-CoA, but neither of the two components of the second partial reaction, pyruvate and oxaloacetate. Similarly, the exchange reaction between oxaloacetate and [^{14}C]pyruvate proceeds in the absence of any of the components of the first partial reaction. The occurrence and experimental conditions for these two exchange reactions is taken as further confirmation that the overall reaction proceeds via the two partial reactions as indicated above and that the ATP exchange reactions occur at one subsite while the pyruvate-oxaloacetate reaction occurs at the other subsite. On the basis of the isotopic exchange studies, the minimum mechanism represented by Equations 2 and 3, in the terminology of Cleland,[188] is described as a nonclassical Ping-Pong Bi-Bi Uni-Uni mechanism.

A. Steady-State Studies

A number of steady-state kinetic investigations involving initial velocity and product inhibition studies have been performed, beginning with Feir and Suzuki,[269] who studied the enzyme isolated from *Aspergillus niger.* Following this report, similar investigations were reported using the enzyme from rat,[546] chicken,[73] sheep,[28] and pig.[938] On the basis of the data obtained, it is reasonable to conclude that the enzyme isolated from these different sources catalyzes the reaction by the same basic mechanism. However, there is not general agreement on what that mechanism is. The earlier reports opted for a Ping-Pong or nonsequential pathway, even though the results indicated some inconsistencies with that interpretation. Later, a case was put forward for a sequential mechanism, but again that view did not explain all of the available data. Therefore, it is useful to trace back over the events leading up to the present time and to consider the cases for and against the two models.

1. The Nonclassical Ping-Pong Reaction Pathway

This mechanism was based mainly on the evidence of the isotopic exchange reactions at equilibrium and the isolation of the relatively stable ENZ-carboxybiotin complex[752] as mentioned above. However, the first report of an attempt to elucidate the kinetic mechanism of this multireactant enzyme using initial velocity and product inhibition studies was made by Feir and Suzuki.[269] These authors chose the enzyme isolated from

A. niger for their study because it did not have the property of being activated by acetyl-CoA. The results they obtained were interpreted as indicating that the active center consisted of three separate subsites. The model assumed that each subsite was specific for its particular substrate and could be occupied by the substrate independent of the addition of the other substrates to their respective subsites. The conclusions made from this pioneer study were later to be criticized, particularly by Warren and Tipton,[938] but they did create interest in the reaction mechanism, and reports of several other studies were to follow shortly after.

The first of these came from McClure et al.[546] using pyruvate carboxylase isolated from rat liver. They interpreted their initial velocity and product inhibition data as indicating a two-subsite Ping-Pong mechanism similar to that proposed by Northrop for transcarboxylase. However, for such a mechanism, double reciprocal velocity plots where either ATP or HCO_3^- is varied at different fixed levels of pyruvate should yield a family of parallel lines. In fact, this pattern was only obtained at low levels of pyruvate, and as the level of pyruvate was increased the lines intersected at the left of the ordinate. These authors rationalized this result by suggesting that the binding of pyruvate influenced the affinity of the enzyme for bicarbonate, but they did not explain how this occurred at high levels of pyruvate and not at low levels. Furthermore, this introduced a variation in the Northrop mechanism in that there must be interdependence between the two subsites at the active center. This was borne out by the fact that 2-oxobutyrate, which acted as a poor substrate and therefore must have bound at the second subsite, influenced the rate of exchange between ATP and ADP.[544]

Subsequently, the initial velocity and product inhibition study reported by Barden et al.[73] using chicken liver pyruvate carboxylase, tended to support a nonclassical Ping-Pong mechanism. In their investigation, these authors attempted to avoid the complication of the atypical kinetic behavior of the enzyme that is associated with pyruvate as mentioned above. However, in doing so, they unknowingly introduced another factor which could only lead to a misinterpretation of their data. This factor concerns the cause of the atypical kinetic response to various concentrations of pyruvate and will be discussed in a later section. By restricting their study to a low level and narrow concentration range of pyruvate, Barden et al.[73] were able to observe families of parallel lines in double reciprocal plots with either ATP or HCO_3^- as the variable and pyruvate as the fixed variable substrate. Nevertheless, the conclusion drawn from these studies was that the first partial reaction (Equation 2) involving MgATP, MgADP, HCO_3^-, and orthophosphate proceeded by a rapid equilibrium random Bi Bi mechanism and involved the formation of two abortive complexes, — P_i-ENZ-biotin-HCO_3^- and MgADP-ENZ-biotin-HCO_3^-. The carboxybiotin formed during this partial reaction then transferred its carboxy group to pyruvate to form oxaloacetate in a Uni-Uni reaction.

As pointed out by Barden et al.[73] in discussing their data, the classical concept of the active site of an enzyme catalyzing a reaction with a nonsequential mechanism is one which assumed that the active site functions as a single catalytic unit. For pyruvate carboxylase, this is shown diagramatically in Figure 1A. A nonclassical Ping-Pong mechanism assumes that the active site consists of two separate and functionally independent catalytic subsites linked together by a mobile carrier. A schematic representation of an active site of this type for pyruvate carboxylase is shown in Figure 1B. Since the nonclassical concept had been incorporated into the reaction mechanism of the biotin-dependent enzyme, methyl-malonyl-CoA transcarboxylase,[606] Barden et al.[73] extended their steady-state studies to determine whether or not pyruvate carboxylase exhibited a similar kinetic reaction mechanism. They used as a basis for their investigation the unique product inhibition patterns which are characteristic for each of the possible mechanisms (Figure 2) and compared these with their own experimental data

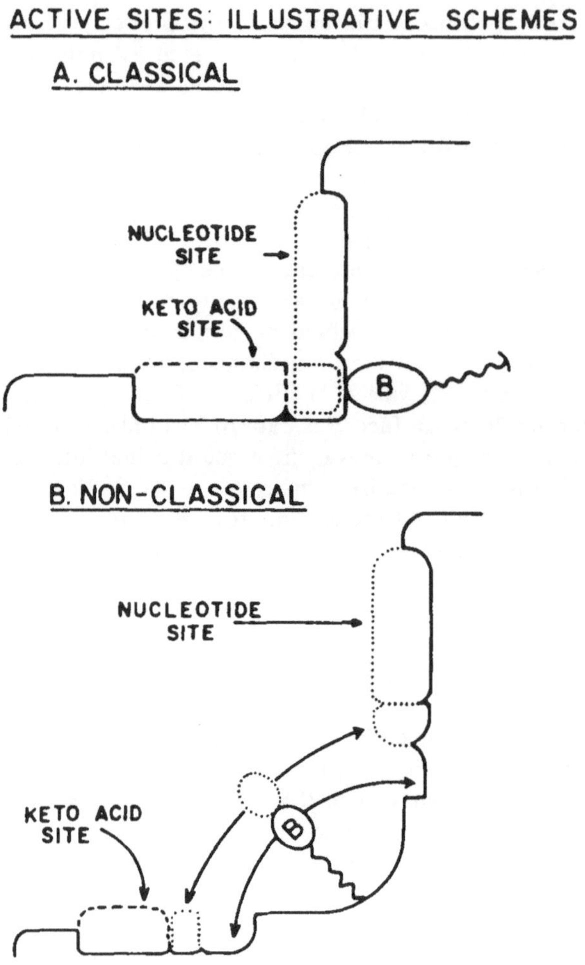

FIGURE 1. (A) a schematic illustration of an active site which functions as a single catalytic site. It is not assumed that the α-keto acids and the nucleotides are bound by an identical set of amino acids, only that there is an "overlap" between the α-keto acid and nucleotide sites sufficient to prevent the binding of α-keto acids to the enzyme-nucleotide complex, and vice versa; (B) a schematic illustration of an active site which is composed of separate catalytic sites for the reactants of each partial reaction. The two catalytic sites are assumed to be physically distinct to the extent that there is no "overlap," i.e., occupancy of one site does not prevent occupancy of the other. Yet these sites must be sufficiently close to each other so that biotin can function as carboxyl carrier between them. (From Cleland, W. W., *The Enzymes*, Vol. II, 3rd ed., Boyer, P. D., Ed., Academic Press, New York, 1970. With permission.)

(Table 1). Clearly, their data agreed very well with a product inhibition pattern for a nonclassical Ping-Pong mechanism involving the presence of abortive complexes.

2. The Sequential Reaction Pathway

In this reaction pathway, pyruvate, under steady-state conditions, must bind to the enzyme before ADP and P_i are released from the first subsite, implying an interdependence between the two subsites. The first suggestion of a sequential mechanism was made by Warren and Tipton[938] using pyruvate carboxylase isolated from pig liver. The results obtained by Warren and Tipton were not all that different from those reported

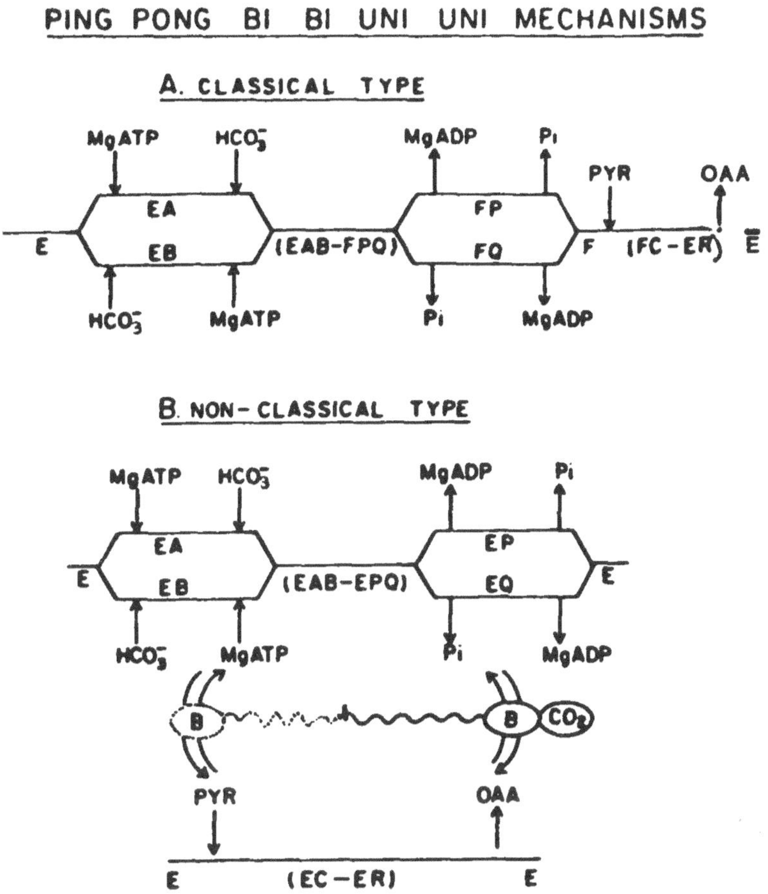

FIGURE 2. (A) A Ping-Pong Bi-Bi Uni-Uni mechanism based on the assumption that the active site functions as a single catalytic site. In this mechanism $MgATP^{2-}$, HCO_3^-, and oxaloacetate combine at a common site on E, which in this instance would be E-biotin, and phosphate, $MgADP^-$, and pyruvate combine at the same site but with a different form of the enzyme, F, which in this instance would be E-biotin-CO_2; (B) a Ping-Pong Bi-Bi Uni-Uni mechanism based on the assumption that the active site is composed of a separate catalytic site for each partial reaction. $MgATP^{2-}$, HCO_3^-, phosphate and $MgADP^-$, the reactions of the Bi-Bi partial reaction, combine at a common catalytic site of the enzyme. Pyruvate and oxaloacetate also share a common site, but this site is separate and distinct from that used by the reactants of the Bi-Bi partial reaction. Binding of reactants in this mechanism is assumed to be independent of whether E-biotin is carboxylated or not. Transfer of CO_2 between the two catalytic sites is presumably a function of the biotinyl residue which is essential for catalysis of the overall reaction. A rapid equilibrium Bi-Bi sequential mechanism is assumed for the Bi-Bi partial reaction in both A and B. (From Barden, R. E., Fung, C-H., Utter, M. F., and Scrutton, M. C., *J. Biol. Chem.*, 247, 1323, 1972. With permission.)

by the previous groups: however, these authors did point out some of the inconsistencies of interpretation made in the earlier publications. First of all, initial velocity studies of a three-substrate enzyme catalyzing a double-displacement mechanism must yield four out of a possible six primary plots comprising families of parallel lines. However, this was clearly not the case with pyruvate carboxylase. For the rat liver enzyme,[546] parallel lines were observed only at low levels of pyruvate while at high levels, all primary plots intersected to the left of the ordinate. For the yeast enzyme,[269] only two of the primary plots had families of parallel lines while for chicken enzyme, four sets

Table 1
COMPARISON OF PRODUCT INHIBITION PATTERNS PREDICTED BY FOUR DIFFERENT PING-PONG BI-BI UNI-UNI MECHANISMS TO EXPERIMENTALLY OBSERVED PATTERNS

Varied substrate	Product inhibitor	Classical ordered[a]	Classical rapid equilibrium	Nonclassical rapid equilibrium	Nonclassical with abortive complexes[b]	Observed
$MgATP^{2-}$	$MgADP^-$	UC	UC[c]	C	C	C
	Phosphate	NC	UC	C	C	C
	Oxaloacetate	C	C	NC	NC	NC
HCO_3^-	$MgADP^-$	UC	UC	C	NC	NC
	Phosphate	NC	UC	C	NC	NC
	Oxaloacetate					
Pyruvate	$MgADP^-$	C	C	UC	UC	UC
	Phosphate	UC	C	UC	UC	UC
	Oxaloacetate	NC	NC	C	C	C
$MgADP^-$	$MgATP^{2-}$	UC	UC	C	C	C
	HCO_3^-	NC	UC	C	NC	NC
	Pyruvate	C	C	NC	NC	
Phosphate	$MgATP^{2-}$	UC	UC	C	C	C
	HCO_3^-	NC	UC	C	NC	NC
	Pyruvate	NC	C	NC	NC	
Oxaloacetate	$MgATP^{2-}$	C	C	UC	UC	UC
	HCO_3^-	UC	C	UC	UC	NC
	Pyruvate	NC	NC	C	C	C

[a] Ordered Bi-Bi sequence assumed as $MgATP^{2-}$, HCO_3^-, P_i, and $MgADP^-$.
[b] The abortive complexes which form at the site of the Bi-Bi partial reaction are E-HCO_3^--P_i and E-HCO_3^--$MgADP^-$.
[d] NC designates noncompetitive; UC, uncompetitive,; C, competitive.

From Barden, R. E., Fung, C-H., Utter, M. F., and Scrutton, M. C., *J. Biol. Chem.*, 247, 1323, 1972. With permission.)

of parallel lines were obtained.[73] However, in this later study the concentration of pyruvate was restricted to less than 3 m*M* and, as pointed out elsewhere in this chapter, the reaction catalyzed by the enzyme isolated from vertebrate sources is not stoichiometric at low levels of pyruvate.

Second, the reaction sequence proposed for the rat and chicken liver enzymes based on the nonclassical Ping-Pong mechanism proposed for transcarboxylase[607] allows the formation of ternary complexes without affecting the parallel-line plots which characterize such a mechanism. Warren and Tipton[938] went on to point out that to achieve this situation, the binding sites for the two partial reaction components must bind their substrates at equilibrium and must be totally independent of each other. Clearly, this is not the case for the rat liver enzyme, where pyruvate is believed to affect the binding of HCO_3^- and the presence of oxaloacetate stimulates the rate of the isotopic exchange between ATP and ADP.[546]

The reaction pathway for the carboxylation of pyruvate, as proposed by Warren and Tipton[938] on the basis of their results obtained using pig liver pyruvate carboxylase in the presence of saturating concentrations of K^+ and acetyl-CoA, is shown in Scheme 1. In this scheme, free Mg^{2+} binds to the enzyme and remains bound during all further catalytic cycles. MgATP binds next, followed by HCO_3^-. (Warren and Tipton criticized the assumption of Barden et al.[73] that MgATP and HCO_3^- bind randomly to the enzyme. The basis of the Barden et al.[73] interpretation was that in the primary plots

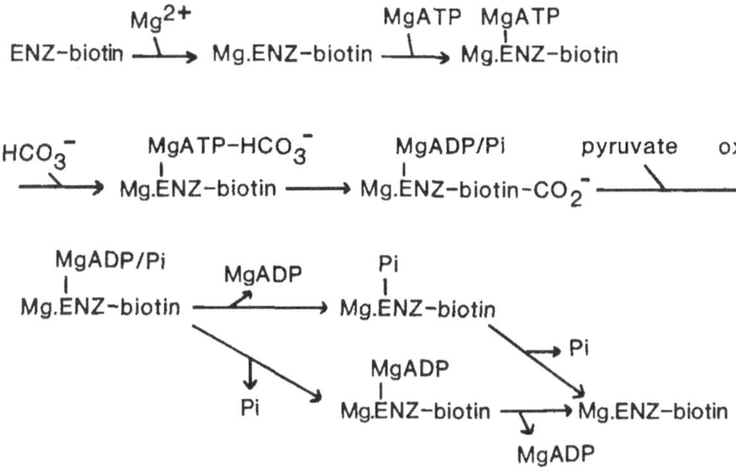

SCHEME 1

involving both ATP and HCO_3^-, intersecting lines were obtained. However, no evidence was presented to exclude an ordered sequence of binding.) Pyruvate binds and oxaloacetate is released before the random release of P_i and MgADP.

Support for a sequential mechanism came from Easterbrook-Smith et al.[253] who used two independent experimental approaches not previously applied to this enzyme. First, to avoid the variable stoichiometry associated with oxaloacetate formation when pyruvate is the variable substrate[251] (see Section II.D), the reaction rate was measured by following the rate of release of P_i. Initial velocities, when plotted as a function of varying concentrations of either MgATP or HCO_3^-, at fixed concentrations of pyruvate gave, in reciprocal form, a family of straight intersecting lines. As mentioned above, in previous publications where the rate of oxaloacetate formation was used as a measure of reaction velocity, and pyruvate was the variable substrate, families of straight, parallel lines were obtained only at low concentrations of pyruvate; at high concentrations the lines converged at a point to the left of the ordinate.

The second approach used by Easterbrook-Smith et al.[253] was to determine the reaction velocity using various concentrations of MgATP and using either pyruvate, 3-fluoropyruvate, or 2-oxybutyrate as alternative carboxy-group acceptor substrates. Under these experimental conditions, it was possible for these authors to predict that if the proposed Ping-Pong reaction sequence was correct, then plots of 1/v vs. 1/MgATP at fixed concentrations of either pyruvate or the alternative substrates should have identical slopes. Conversely, for a sequential reaction pathway, different slopes would be observed. When these experiments were carried out,[253] the slopes of the double reciprocal plots were significantly different.

Clearly, the results of both approaches supported the case for a sequential reaction pathway.

B. Equilibrium Studies

The conventional tools for studying enzyme-catalyzed reaction mechanisms have been initial velocity and product inhibition studies carried out under steady-state conditions. However, although these procedures describe the sequence of substrate binding and product release, they give very little information about the nature of the interactions which give rise to product formation. Therefore, other techniques, particularly isotopic exchange reactions performed under equilibrium conditions, are required to support a mechanism based solely on steady-state kinetic data and to provide evidence of intermediates and possible transition states.

This has been the approach used in studying the mechanism of pyruvate carboxylase. However, reconciling the data obtained from both steady-state and equilibrium conditions has not always been straightforward, particularly with regard to the first partial reaction.

1. Mechanism of the First Partial Reaction

In the first partial reaction (Equation 2) the hydrolysis of MgATP to MgADP and P_i is coupled to the fixation of CO_2 to form an ENZ-carboxybiotin complex. Although a number of possible mechanisms for coupling ATP hydrolysis to CO_2 fixation have been considered by various investigators, as indicated when discussing the overall reaction sequence, there is still a lack of general agreement on the sequence of events leading to the formation of the ENZ-carboxybiotin intermediate. One of the reasons for this uncertainty stems from the fact that most attention has been directed toward the process by which the biotin moiety is carboxylated. In its most stable configuration (Structure A), the ureido

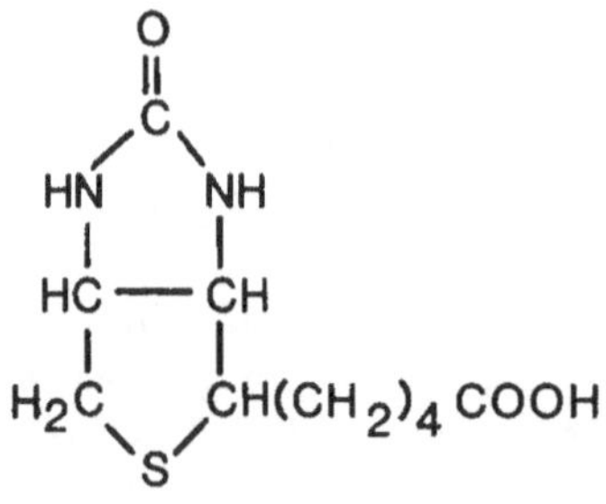

(A)

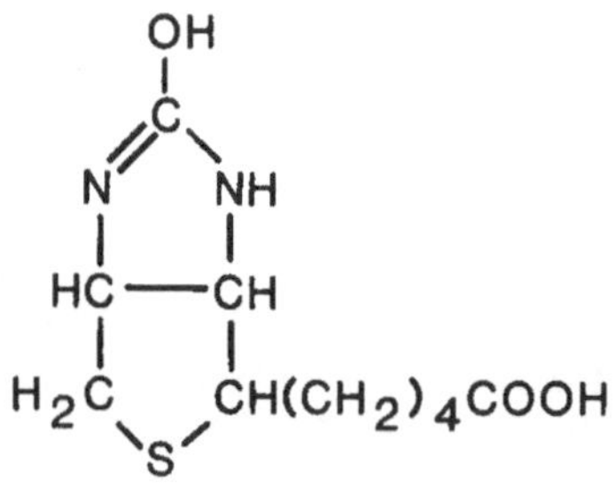

(B)

N-atom of biotin is an extremely poor nucleophile, but conversion of the keto form of the ureido group to its tautomeric form (Structure B) increases the nucleophilicity of the N-atom 10^{10}-fold.[390] Therefore, a great deal of debate has centered on (1) the site of carboxylation of biotin, and (2) the possible ways of increasing the nucleophilicity of the ureido group. However, what is also involved here is to clearly define the role of ATP in the reaction mechanism. In most other enzymic reactions involving ATP utilization, some compound is either phosphorylated, adenylated, or adenosylated, but in the case of the biotin-dependent enzymes, until recently, very little evidence for such compounds or intermediates has been found.

The mechanisms considered most likely to be involved in the formation of ENZ-carboxybiotin associated with ATP hydrolysis are (1) a concerted reaction which proceeds without the formation of a kinetically significant or detectable intermediate, (2) activation of the enzyme (or prosthetic group) by ATP to form an activated P_i-ENZ-biotin (or ADP-ENZ-biotin) complex which can then react with HCO_3^-, and (3) activation of HCO_3^- by formation of a phosphorylated intermediate. Each of the mechanisms and the evidence available to support each one will be considered.

a. The Concerted Mechanism

The first report suggesting that the first partial reaction proceeded via a concerted mechanism came from Kaziro et al.,[436] who were investigating the carboxylation of propionyl-CoA by propionyl-CoA-carboxylase. In their experiments using ^{18}O-labeled bicarbonate, they found that the equivalent of one ^{18}O atom was incorporated into orthophosphate for every two ^{18}O atoms incorporated into the carboxyl group of methyl-malonyl-CoA. This finding showed two points about the mechanism. First of all, it showed that bicarbonate rather than free CO_2 was the reactant — if CO_2 were the reactant only two atoms of ^{18}O would be expected to be utilized. Secondly, it implied the direct participation of HCO_3^- in the phosphoryl bond cleavage during the activation process. Consequently, a concerted reaction between the enzyme-bound biotin, HCO_3^-, and the γ-phosphoryl group of ATP was proposed.

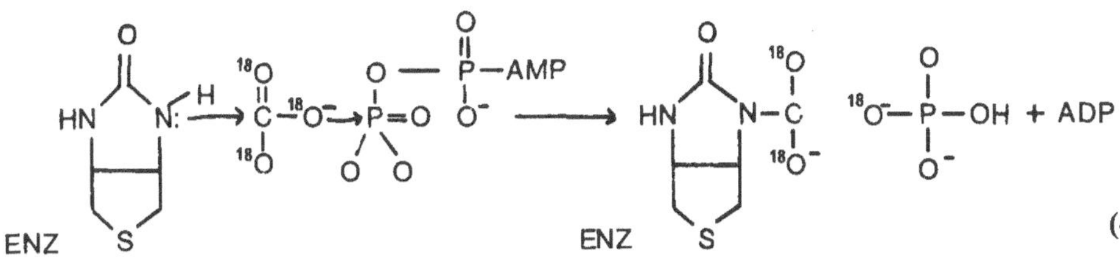

(4)

ENZ-biotin + HCO_3^- + ATP ⟶ ENZ-biotin-CO_2 + Pi + ADP

The possibility that pyruvate carboxylase might carboxylate biotin using the same mechanism received some support from the studies of McClure et al.[544] with the enzyme isolated from rat liver. These authors used identical assay conditions to study both the ATP:ADP and the ATP:P_i exchange reactions and showed that the rates of both reactions were the same, indicative of a concerted mechanism.

However, a concerted mechanism requires that both ADP and P_i are released after the formation of the ENZ-carboxybiotin complex and therefore, the exchange reaction between HCO_3^- and oxaloacetate should exhibit an absolute requirement for ATP. Scrutton and Utter[760] demonstrated that although both ADP and P_i were required for this exchange reaction to occur, it proceeded in the absence of ATP — a result which is inconsistent with a concerted mechanism. The most that can be said at this point is that there is direct participation of HCO_3^- in the hydrolysis of ATP and that the two exchange reactions involving ATP, i.e., ATP:ADP and ATP:P_i, must share a common pathway and some of the same intermediates.

b. A Mechanism Involving Activation of the Enzyme

In attempting to elucidate the reaction pathway, Scrutton and Utter[760] considered the possible existence of an activated P_i-ENZ-biotin complex. On the basis of the ATP:ADP exchange reaction, which required the presence of Mg^{2+} and the ATP:P_i exchange reaction which required Mg^{2+}, ADP, and HCO_3^-, they considered the possibility that the reaction might proceed as follows:

$$\text{ENZ-Biotin} + \text{MgATP} \rightleftharpoons P_i\text{-ENZ-Biotin} + \text{MgADP} \quad (5)$$

$$P_i\text{-ENZ-Biotin} + HCO_3^- \rightleftharpoons \text{ENZ-Carboxybiotin} + P_i \quad (6)$$

$$\text{ENZ-Carboxybiotin} + \text{Pyruvate} \rightleftharpoons \text{ENZ-Biotin} + \text{Oxaloacetate} \quad (7)$$

However, these authors discarded this sequence on the following evidence: (1) the exchange reaction between HCO_3^- and oxaloacetate required the presence of both ADP and P_i.[760] According to the above reaction pathway, the presence of ADP should not be required. They concluded that the ATP:ADP exchange was an abortive reaction and not on the main pathway and (2) they were unable to isolate the predicted P_i-ENZ-biotin complex.[760]

In retrospect, dismissing a mechanism involving activation of the enzyme using the above criteria was hasty. The attempt to isolate a phosphoryl-enzyme intermediate involved precipitation of the enzyme with trichloroacetic acid, thus assuming that the complex was acid stable. Furthermore, since the ATP:ADP exchange reaction only required the addition of Mg^{2+}, it was assumed that ADP must be released before HCO_3^- bound. However, the level of endogenous HCO_3^- in the reaction solution is relatively high (ca., 2.0 m*M* at pH 8.4) and it is conceivable the HCO_3^- could have been bound before ADP was released. If this were so, it would agree with the observation that ADP is essential for the exchange between HCO_3^- and oxaloacetate. This aspect will be considered in more detail in a later section.

c. A Mechanism Based on Substrate Activation

An alternate mechanism to "enzyme activation" is "substrate activation" where the enzyme reacts with an ATP-activated-bicarbonate species. This mechanism has received support from the work of Polakis et al.,[655,656] who showed that the biotin carboxylase component of acetyl-CoA carboxylase from *Escherichia coli* could catalyze the phosphoryl transfer from carbamoyl phosphate to ADP to form ATP:

$$\text{ADP} + NH_2\text{–}\overset{\overset{\displaystyle O}{\|}}{C}\text{–O–}PO_3^{2-} \rightarrow H_2NC\begin{matrix}\nearrow O \\ \searrow O^-\end{matrix} + \text{ATP} \qquad (8)$$

The reaction required the presence of divalent cation (Mg^{2+}) and in the presence of free *d*-biotin and ethanol, the rate of phosphoryl transfer was to 40% of the rate observed for the carboxylation of free *d*-biotin. These authors proposed that carbamoyl phosphate (Structure C) was an analog of carboxyphosphate (Structure D) and therefore this reaction was of interest, since it indicated that either a carbonyl- or a carboxyphosphate might be an intermediate in the ATP-dependent carboxylation of biotin.

$$H_2N\text{–}\overset{\overset{\displaystyle O}{\|}}{C}\text{–O–}PO_3^{2-} \qquad HO\text{–}\overset{\overset{\displaystyle O}{\|}}{C}\text{–O–}PO_3^{2-} \qquad HO\text{–}\overset{\overset{\displaystyle O}{\|}}{C}\text{–}CH_2\text{–}PO_3^{2-}$$

(C) (D) (E)

Although the presence of free *d*-biotin stimulated the rate of phosphoryl transfer, it did not appear to participate in the reaction. This was shown by the fact that chemical substitution of the 1-N position of biotin, i.e., at the site of enzymic carboxylation reduced, but did not abolish, the ability of this derivative to support phosphoryl transfer. Therefore, Polakis et al.[655,656] suggested that the presence of biotin may reorient the binding sites for ATP and HCO_3^- in such a way that the immediate transfer of the carboxy group to the adjacent biotin moiety would restrict the hydrolysis of the unstable carboxyphosphate. Furthermore, the premature splitting of ATP would be forestalled until the "CO_2" acceptor, i.e., the biotin prosthetic group, is in the carboxylation site.

Ashman and Keech[28] observed a similar conversion of carbamoyl-phosphate and ADP to ATP using pyruvate carboxylase isolated from a sheep kidney. Unlike acetyl-CoA carboxylase from *E. coli,* the subunits of pyruvate carboxylase from vertebrate sources are multifunctional polypeptides, and it was not possible to examine the effects of *d*-biotin on the rate of this reaction. However, Ashman and Keech[28] showed that phosphonoacetate (Structure E) was a potent inhibitor of pyruvate carboxylase and since the ionic groups of carboxyphosphate ($^-OOC\text{–}O\text{–}PO_3^{2-}$) and phosphonoacetate ($^-OOC\text{–}CH_2\text{–}PO_3^{2-}$) are identical, they considered that this provided additional evidence that carboxyphosphate might be a transitory intermediate.

In summary, by the late part of the 1970s, although there was general agreement that the mechanism of the first partial reaction for all biotin-dependent carboxylases was the same, there was not universal consensus on the sequence of events leading to the carboxylation of the covalently bound biotin. Evidence was available from various sources to support three possible mechanisms. However, as pointed out by Polakis et al.,[655] on the basis of isotopic exchange experiments and because convincing evidence to the contrary was lacking, the first partial reaction was generally visualized as a concerted process. Nevertheless, the fact that an exchange between HCO_3^- and oxaloacetate occurs in the absence of ATP[760] excludes this mechanism. Furthermore, as pointed out by Polakis et al.,[655] a step-wise chemical mechanism which involves carboxyphosphate as a tightly bound enzyme intermediate is also consistent with previous findings.

More recently, Phillips et al.,[651] studying the process in pyruvate carboxylase, proposed a step-wise sequence of events which satisfied all of the known information and proceeded as follows:

$$\text{ENZ-Biotin} + \text{ATP} \rightleftharpoons [\text{ATP-ENZ-Biotin}] \tag{9}$$

$$[\text{ATP-ENZ-Biotin}] + HCO_3^- \rightleftharpoons \left[\begin{matrix} \text{Pi} \\ \vdots \\ CO_2^- \end{matrix} \Big\rangle \text{ENZ-Biotin}\right] + \text{ADP} \tag{10}$$

$$\left[\begin{matrix} P_i \\ \vdots \\ CO_2^- \end{matrix} \Big\rangle \text{ENZ-Biotin}\right] \xrightleftharpoons{\text{Acetyl CoA}} \text{ENZ-Carboxybiotin} + P_i \tag{11}$$

Equation 9 to 11 give, as a sum,

$$\text{ENZ-Biotin} + \text{ATP} + HCO_3^- \rightleftharpoons \text{ENZ-Carboxybiotin} + \text{ADP} + P_i$$

In this mechanism, under nonsteady-state conditions, both ATP and HCO_3^- bind to the enzyme before ADP is released leaving a

$$\begin{matrix} P_i \\ \vdots \\ CO_2^- \end{matrix} \Big\rangle \text{ENZ-biotin}$$

form of the enzyme. As pointed out above, there is sufficient endogenous HCO_3^- in the reagents to permit an exchange reaction between ATP and ADP without further addition. Since it was not possible to completely remove HCO_3^- from all reagent solutions, Phillips et al.[651] were not able to demonstrate an absolute dependence for HCO_3^-. However, the velocity profile for this exchange reaction exhibits a substrate inhibition pattern when the HCO_3^- concentration is raised above an optimum level, suggesting that at high levels of HCO_3^- there is an increased level of

$$\begin{matrix} P_i \\ \vdots \\ CO_2^- \end{matrix} \Big\rangle \text{ENZ-biotin}$$

and a corresponding decrease in the amount of the other forms of the enzyme. This in turn would make it more difficult for ATP to participate in the exchange reaction, resulting in a lower rate of exchange. Similar comments apply to the ATP:P_i exchange with the additional complication that acetyl-CoA appears to be involved at this point. From the ^{18}O-labeling experiments of Kaziro et al.,[436] it appears that at some point during the reaction, nucleophilic attack by the bicarbonate ion occurs at the γ-phosphoryl phosphorous atom. Therefore, it is reasonable to conclude that HCO_3^- binds to the enzyme before hydrolysis of ATP occurs.

Several other features of this mechanism are worthy of comment. First of all, HCO_3^- is shown to bind to the enzyme and during hydrolysis of the ATP it is activated to a putative carboxyphosphate derivative (Equation 10). Then, under the influence of acetyl-CoA, the carboxyl group is shifted to the N′-atom of the ureido group of the biotin molecule. This two-step process was proposed by Phillips et al.,[651] who were able to isolate a

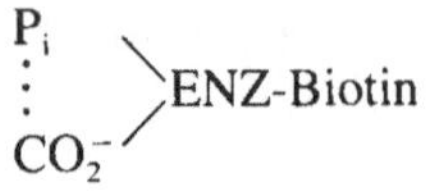

complex by gel filtration from a solution containing ENZ-biotin, Mg^{2+}, ATP, and HCO_3^-. However, they were unable to transfer the carboxyl group to pyruvate unless acetyl-CoA was present in the transferring medium. It appears that at least one of the roles of acetyl-CoA is to facilitate the transfer of the activated carboxyl group from the enzyme to the biotin to form N-carboxybiotin. It is not known whether the carboxyl and phosphoryl groups remain attached to each other in the absence of acetyl-CoA and it raises the question as to how the enzyme functions in the absence of acetyl-CoA. Preliminary evidence[651] indicates that K^+ functions in the same way as acetyl-CoA and appears to be involved in the transferral of the carboxyl group to biotin.

The proposed mechanism is also consistent with the evidence of Polakis et al.[655,656] and Ashman and Keech[28] discussed above, and agrees with the data of Scrutton and Utter,[760] who showed that the HCO_3^-:oxaloacetate exchange reaction required both ADP and P_i.

d. Participation of Biotin in the First Partial Reaction

The literature concerning the carboxylation of biotin in the first partial reaction has been reviewed extensively by Moss and Lane[576] and Wood and Barden[983] and therefore, only the salient features will be considered. From a mechanistic point of view, the two main points of interest are determining the site of carboxylation on the biotin (Structure A) and how the carboxylation is achieved. Therefore, the discussion will focus on the three functional groups in the molecule, i.e., the carboxyl group of the valeric acid sidechain, the thioether, and the ureido group. From the available evidence, biotin is covalently attached to each of its apoenzymes through its carboxyl group by an amide link to the ε-amino group of a specific lysyl residue in the polypeptide chain. The first piece of evidence along these lines came from the detection of ε-*N*-biotinyl-L-lysine (biocytin) in a cell-free extract.[986] Later, Kosow and Lane[466] isolated ^{14}C-biocytin from a pronase digest of rat liver propionyl-CoA carboxylase and since then the evidence has accumulated which suggests that the biotin in all biotin-dependent enzymes is attached in the same way. Thus, it appears that the carboxyl group has no other function and the enzymic process is more concerned with the other two functional groups.

During the first partial reaction, carboxybiotin is formed, presumably through a the same way. Thus, it appears that the carboxyl group has no other function and the enzymic process is more concerned with the other two functional groups.

During the first partial reaction, carboxybiotin is formed, presesumably through a nucleophilic attack of the ureido ring on an electrophilic center of a carboxy donor which may be either bicarbonate or an activated carboxyl group. This first report on the site of the enzymically catalyzed carboxylation of biotin came from Lynen et al.[516] using a β-methyl crotonyl-CoA carboxylase. They were able to show that this enzyme was able to carboxylate free biotin in the absence of its natural substrate. The relatively unstable carboxybiotin was stabilized by converting it to its dimethyl ester with diazomethane and then shown to be identical with the dimethyl ester of *N*-carboxybiotin formed by chemical synthesis[452,516] (Structure F). Later, it was conclusively demonstrated by X-ray crystallographic analysis[116] that the methoxy carbonyl group of the methylated carboxylation product was located on the 1′-*N*-position. Subsequently, it was confirmed that the site of carboxylation of the covalently bound biotin in enzymes such as β-methyl crotonyl-CoA carboxylase,[451,453] propionyl-CoA carboxylase,[485] transcarboxylase,[984] and acetyl-CoA carboxylase[608,609] was also in the 1′-*N*-position. It is assumed that the same situation applies for pyruvate carboxylase.

However, some controversy has arisen concerning the assignment of the 1′-*N* carboxyl as the reactive group. From their model studies, Bruice and Hegarty[133] indicated that the carbonyl oxygen was the most nucleophilic site in the biotin ring system. They suggested that carboxylation occurred at the ureido-*O* and that during the methylation process used to stabilize the carboxybiotin, *O*-carboxylated biotin underwent a rearrangement that transferred the carboxyl group to the ureido-*N*. The debate was extended by Guchhait et al.,[347] who showed quite conclusively that the 1′-*N*-form of carboxybiotin is the reactive species in catalysis. Using authentic 1′-*N*-carboxybiotin, ADP and P_i, and the biotin carboxylase component of acetyl-CoA carboxylase, they found that ATP was synthesized. Furthermore, using 1′-*N*-carboxybiotinol with the carboxytransferase component, they demonstrated efficient carboxylation of acetyl-CoA to form malonyl-CoA.

A proposal based on model studies, but more in keeping with the experimental evidence obtained from enzymic catalysis, was put forward by Kluger and Adawadkar.[448,449] On the basis of the hydrolysis of a synthetic cyclic phosphourea derivative, they suggested that the urea moiety of biotin is nucleophilic toward phosphate derivatives. Therefore, with ATP, HCO_3^-, and biotin bound in close proximity at the first subsite, they proposed that the formation of *N*-carboxybiotin could occur via a mechanism involving *O*-phosphobiotin as an intermediate (Structure G). A similar role for *O*-phosphobiotin in the carboxylation of biotin was

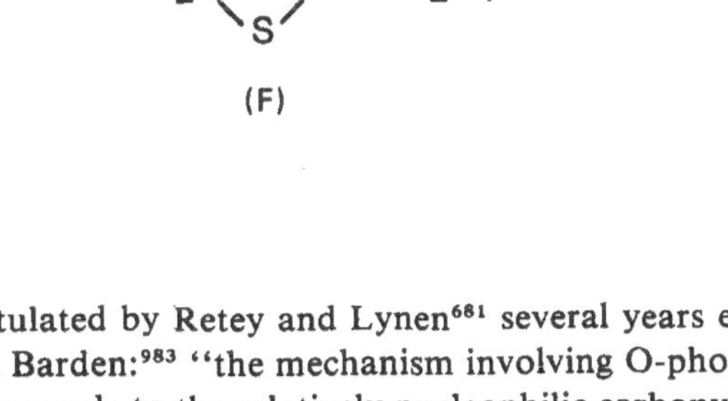

(F)

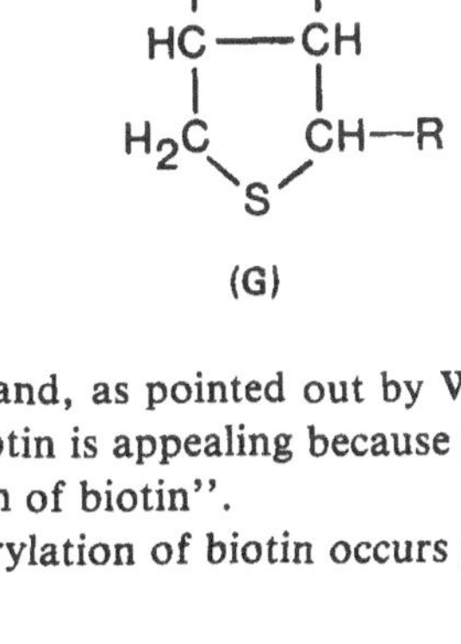

(G)

postulated by Retey and Lynen[681] several years earlier, and, as pointed out by Wood and Barden:[983] "the mechanism involving O-phosphobiotin is appealing because it assigns a role to the relatively nucleophilic carbonyl oxygen of biotin".

A drawback with this mechanism is that the phosphorylation of biotin occurs prior to the reaction with HCO_3^-, which is therefore not involved in the hydrolysis of ATP. Although phosphorylation of the biotin at the ureido-O would increase the nucleophilicity of the ureido-N, this same objective might be achieved by the interaction of protein residues in the first subsite with biotin to induce tautomerization; such a sytem, which is an extension of one proposed by Wood and Barden,[983] is

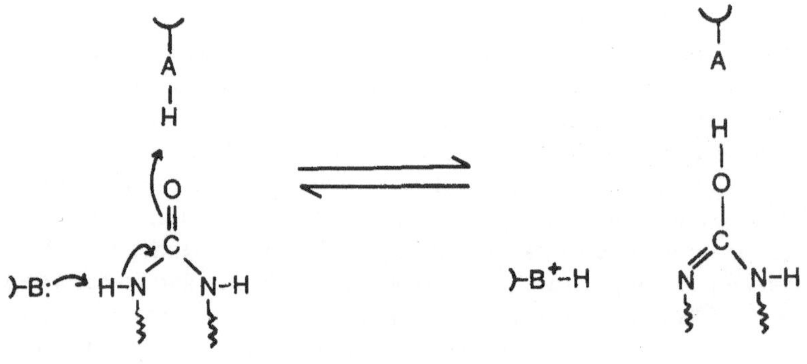

(H)

At the present time there does not appear to be any direct involvement of the sulfur atom contained in the tetrahydrothiophene ring in the carboxylation reaction. The importance of the thioether is beyond question since dethiobiotin and biotin sulfoxides are inactive,[516] and on the basis of the data obtained by Sigel et al.,[789] it is reasonable to conclude that a possible function of the sulfur atom is to complex with metal ions.

C. Factors that Influence the Translocation of the N-Carboxybiotin Moiety Between the Two Subsites

Once the carboxybiotin complex has been formed in the first subsite, it has to move to the second subsite to complete the reaction. The first piece of information available concerning the translocation process came from a study by Easterbrook-Smith et al.[251] These authors showed that pyruvate, the substrate of the second partial reaction (Equation 3), acted as a signal to summon the activated carboxybiotin moiety to the second subsite in readiness for the carboxylation of pyruvate. In a subsequent study specifically focusing on determining the factors that influence the location and movement of the carboxybiotin, Goodall et al.[323] showed that the presence of Mg^{2+} favored the holding of the complex in the first subsite. The half-life of the isolated ENZ-carboxybiotin complex increased from 50 to 700 min in the presence of 10 mM $MgCl_2$. From this observation one would predict that high concentrations of Mg^{2+} would inhibit both the overall reaction and the pyruvate:oxaloacetate exchange reaction, and there are indications in the literature to suggest that this prediction is correct.[439,752]

If the scheme of Goodall et al.[323] is correct, the implication is that Mg^{2+} does not remain bound at the first subsite throughout the catalytic cycle as proposed by Warren and Tipton.[938] The scheme shows that Mg^{2+} is released before carboxybiotin moves to the second subsite and must therefore rebind subsequent to the carboxylation of pyruvate.

Goodall et al.[323] confirmed the previous findings of Easterbrook-Smith et al.[251] and showed that pyruvate overcame the Mg^{2+} effect and induced the complex to move into the second subsite. During their investigation, Goodall et al.[323] used a number of compounds possessing structural features resembling pyruvate to determine what features of the pyruvate molecule affect the release of carboxybiotin from the first subsite and the translocation to the second subsite. The outcome of their investigation was that pyruvate and fluoropyruvate caused the release of the carboxybiotin at a rate too fast to measure in the absence of rapid reaction techniques. The addition of a methyl group at the C-3 position (as in 2-oxobutyrate) drastically decreased the efficiency of the process, but nevertheless, the binding of 2-oxo-butyrate did release the carboxybiotin from the first subsite. Compounds such as propionate, lactate, and acetaldehyde that lack either the carboxy or oxo group of pyruvate are neither substrates nor inhibitors and are incapable of releasing the carboxybiotin. Thus, both the carboxy and oxo group appear to be essential for both binding and initiating the transfer process.

D. Transfer of the Carboxyl Group from Carboxybiotin to Pyruvate

In discussing the sequence of events that occur at the second subsite, it is necessary to consider both the evidence relating to the mechanistic aspects and the nonclassical kinetic behavior associated with the second partial reaction. As mentioned previously, a consistent feature of the data obtained in most initial velocity studies has been that when pyruvate is the varied substrate, double reciprocal plots are concave from below.[163,545,773,860] For a long while, very few attempts were made to understand the cause or significance of this phenomenon and in some initial velocity studies, experimental conditions were chosen to avoid nonlinear double reciprocal plots.[73,938] As a result, some misinterpretation of kinetic data has been made.

The original postulates advanced to explain the so-called "substrate activation" were not adequately supported by experimental data. Taylor et al.[860] suggested negative cooperativity[464] with respect to pyruvate binding as a possible interpretation for the atypical kinetic behavior, but the demonstration by Nakashima et al.[592] that the active protomeric form of the rat liver enzyme exhibited nonlinear double reciprocal plots, negated any explanation based on subunit interactions. The suggestion that the nonclassical kinetic profile was due to a pyruvate-induced interaction between subsites[546] overlooked the fact that biphasic pyruvate saturation curves were observed for the exchange reaction between pyruvate and oxaloacetate,[752] where apparently only one subsite is involved.

Subsequently, Easterbrook-Smith et al.[251] examined the atypical kinetic phenomenon and found that the stoichiometry of the sheep liver pyruvate carboxylase reaction was not constant under all experimental conditions. Previously, Utter and Keech[899] had demonstrated a one-to-one relationship for all substrates and products, but this exercise was performed using "saturating conditions" for all substrates. Thereafter, it was tacitly assumed that this relationship held under all experimental conditions, but from the result obtained by Easterbrook-Smith et al.,[251] this assumption was not justified. It was pointed out by these authors that in all studies reported in the literature on pyruvate carboxylase, enzymic activity had been assayed by measuring the rate of production of oxaloacetate, either by spectrophotometric or isotopic means. Therefore, in their study, they measured the reaction velocity in terms of both oxaloacetate produced and orthophosphate liberated. In double reciprocal form, the plot obtained using oxaloacetate production as a measure of reaction velocity was nonlinear with downward curvature, as had been reported previously.[163,545,773,860] However, the plot obtained using the rate of orthophosphate release was a straight line (Figure 3). This illustration clearly indicated that only at saturating levels of pyruvate was the ratio of the amount of oxaloacetate produced and the amount of orthophosphate liberated, one-to-one, while at low concentrations of pyruvate this stoichiometric relationship no longer held; at low concentrations of pyruvate, significantly more orthophosphate was liberated than oxaloacetate formed.

The postulate advanced by Easterbrook-Smith et al.[251] to account for this variable stoichoimetry, was that part of the carboxybiotin formed at the first sub-site underwent a hydrolytic breakdown in the second subsite (Schemes 2 and 3). The proposed mechanism suggested that the binding of pyruvate induced the carboxybiotin to move into the second subsite, and the efficiency of the transfer of the carboxyl group to the acceptor depended on whether or not the pyruvate binding site was occupied. If, as

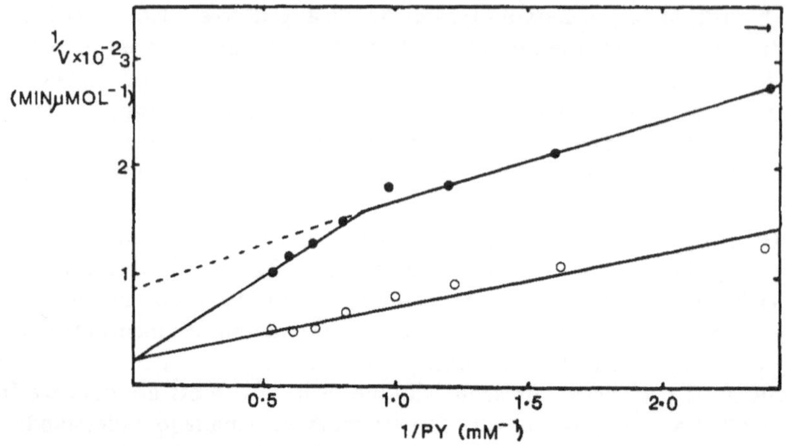

FIGURE 3. Reciprocal plots of the velocity of the pyruvate carboxylase reaction plotted as a function of pyruvate concentration; the reaction velocity was measured either as orthophosphate released (○—○) or as oxaloacetate produced (●—●). (From Easterbrook-Smith, S. B., Campbell, A. J., Keech, D. B., and Wallace, J. C., *Biochem. J.*, 179, 497, 1979. With permission.)

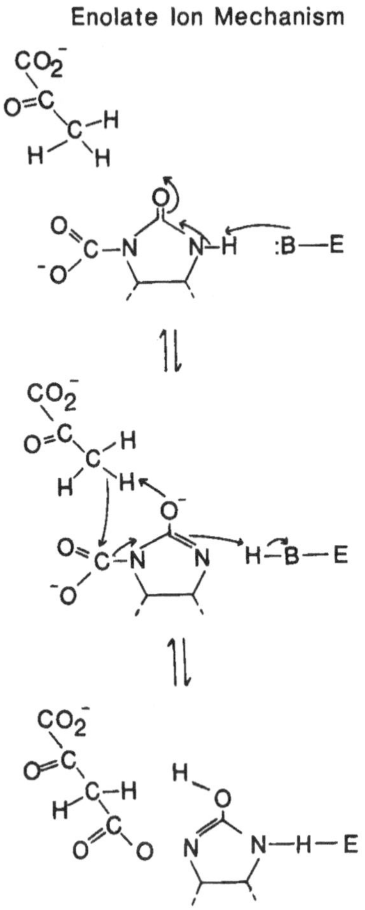

SCHEME 2

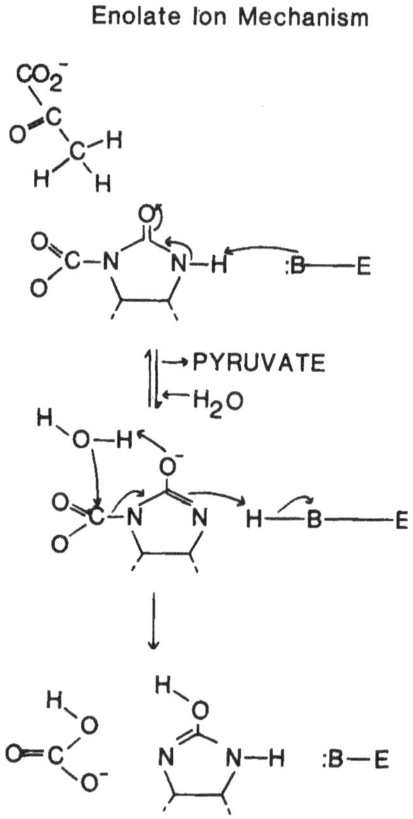

SCHEME 3

seems likely at high concentrations of pyruvate, the pyruvate binding site will be occupied most of the time, and when the carboxybiotin arrives, the efficiency of transfer will be high, i.e., the carboxyl group will be transferred to pyruvate and oxaloacetate will be formed. However, if the pyruvate dissociates and the binding site is not occupied when the carboxybiotin arrives — a situation which is quite feasible at nonsaturating levels of pyruvate — then, if the microenvironment is unfavorable for carboxybiotin, the carboxyl group will be transferred to water to form HCO_3^-.

This explanation for the variable stoichiometry is supported by several lines of evidence. For example, during their nuclear magnetic resonance (NMR) studies, Mildvan and Scrutton [567] showed that, in the case of the chicken liver pyruvate carboxylase, pyruvate bound to the enzyme in a rapid equilibrium manner; they showed that pyruvate moved into and out of the second subsite at rates which were two orders of magnitude faster than the rate of the overall reaction. Therefore, if the binding of pyruvate is the signal for the carboxybiotin to move into the second subsite, there is no guarantee that it will still be there when the carboxybiotin arrives. Furthermore, when studying the rate of oxaloacetate decarboxylation, i.e., the reversal of Equation 3, Easterbrook-Smith et al.[251] found that in the presence of nonsaturating levels of pyruvate there was a very significant rate of decarboxylation. If, however, pyruvate was removed from the system using lactate dehydrogenase and NADH, the rate of decarboxylation was negligible, again suggesting that pyruvate was a necessary catalyst in labilizing the carboxybiotin. Finally, the above explanation is also consistent with the conclusion of Cheung and Walsh,[182] who studied the kinetic isotope effect of tritiated pyruvate on both the catalytic step involving the carboxylation of pyruvate and the overall reaction. As a result of their studies, one of the conclusions these authors made was that "after the pyruvate molecule binds to the enzyme, it is only about 50% committed to catalysis, i.e., it will dissociate without reacting one out of two times it binds."

Whether or not any physiological significance can be attached to such a "slipping clutch" mechanism involving an excessive use of ATP is not known at present. Although energetically inefficient at low concentrations of pyruvate, it does permit a single protein to serve a multipurpose function. Easterbrook-Smith et al.[251] point out that even at the lowest concentration of pyruvate normally found in the cell, sufficient oxaloacetate can be synthesized to maintain the essential anaplerotic and gluconeogenic roles. However, in response to elevated levels of pyruvate, e.g., following muscular activity, there is a compensatory increase in enzymic efficiency to remove the accumulated substrate. Thus, the energy expenditure of the substrate-induced variation in stoichoimetry of the pyruvate carboxylase reaction is somewhat analogous to the futile cycles[292,598] for which a regulatory function has been proposed.[186,599,762]

In discussing the events occurring during the second partial reaction it is tacitly assumed that the transfer of the carboxyl group from carboxybiotin to acceptor molecules is the same for all biotin-dependent enzymes. Therefore, it is necessary to consider any evidence relating to the mechanism obtained from all of these enzymes. However, any mechanism that is proposed for pyruvate carboxylase must take into account the "hydrolytic leak" (described above) that occurs at low levels of pyruvate and the fact that the abortive breakdown of carboxybiotin does not appear to occur with other biotin-dependent enzymes.

At the molecular level, two mechanisms, namely a concerted mechanism and a mechanism involving the formation of a carbanion, have been proposed.

1. The Concerted Mechanism

In 1965, Retey and Lynen[681] proposed that the transfer of the carboxyl group from carboxybiotin to propionyl-CoA, as catalyzed by propionyl-CoA carboxylase, occurred by a concerted reaction involving a six-membered ring complex (Structure I). This proposal was based on their finding that carboxylation of propionyl-CoA occurs with retention of configuration, i.e., a proton is removed from the substrate and the carboxyl group is added to the same face. Later, Prescott and Rabinowitz[658] obtained data which were consistent with the proposed mechanism. They found that the rate of release of ^3H from [2-^3H] propionyl-CoA was the same as the rate of the overall reaction. Furthermore, ^3H from ^3H_2O was not incorporated into propionyl-CoA in the absence of the reverse reaction.

Additional support for a concerted mechanism came from studies using pyruvate carboxylase[568] and transcarboxylase.[702] Both of these enzymes catalyze carboxylation reactions with retention of configuration, and loss of a proton from the substrate does not occur in the absence of other components of the overall reaction. In addition, Rose et al.[702] found that with transcarboxylase there was a small but significant amount of transfer of hydrogen between the substrates of the two partial reactions, i.e., ^3H from [3-^3H] pyruvate to propionyl-CoA and ^3H from [2-^3H] propionyl-CoA to pyruvate. These authors suggested that a proton carrier functions between the two subsites of

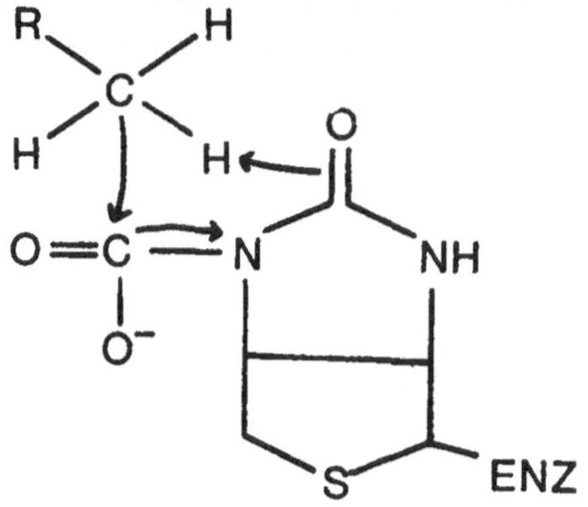

(I)

transcarboxylase and that the transfer is most likely to be a base on the biotin carboxyl carrier protein. Despite its rapid equilibrium with the medium, they suggest this could be the 2′0 of the ureido group of biotin.

In proposing a concerted mechanism for the carboxylation of pyruvate by pyruvate carboxylase, Mildvan et al.[568] suggested that a transannular interaction between the carbonyl group and the sulfur atom of the thiophane ring could contribute to the polarization of the carbonyl group of biotin. This in turn would assist the removal of a proton from the substrate molecule. However, Bowen et al.[119] have since shown that such an interaction would have a negligibly small effect.

Mildvan et al.[568] also proposed that the tightly bound Mn^{2+} in the enzyme coordinated with the carbonyl oxygen of pyruvate and the carboxyl group of carboxybiotin, thereby facilitating the departure of a proton from the methyl group of pyruvate and assisting the formation of a carbonium ion from carboxybiotin. However, Fung et al.[303] have since shown that the pyruvate binding site is too far away from the Mn^{2+} to permit direct interaction; Fung and collaborators[302] suggested that the bound metal ions in pyruvate carboxylase and transcarboxylase promote the acidity of an inner-sphere water ligand which forms a hydrogen bond with the carbonyl oxygen of pyruvate, thereby assisting proton abstraction from the methyl group. Subsequently, these authors[301] proposed two alternative concerted mechanisms which incorporate this suggestion.

A deficiency of a simple concerted mechanism, such as that illustrated in Structure I, is that model studies[151,152,457,817,914] have, in general, indicated that carboxybiotin is a relatively stable intermediate. For example, Caplow[152] found that *N*-carboxyimidazolidone, a model for carboxybiotin, lost its carboxy group by an intramolecular cyclic decarboxylation process and had no tendency to transfer it to a nucleophile. Furthermore, the crystal structure analyses of biotin[229] and 1′*N*-methoxy-carbonyl methyl ester[817] show that the carboxylation of biotin at the 1′*N*-position restores the double-bond character found in biotin to the new bond between 1′*N* and the carboxyl C. Stallings et al.[817] suggest that this effect may contribute to the stability of carboxybiotin; the stability of carboxybiotin is essential for it to survive the translocation between the two subsites. As a result of these studies, the idea has evolved that the enzymatic process involves more than just aligning the carboxybiotin to the acceptor molecule as required in a concerted mechanism.

2. A Mechanism Involving Carbanion Formation

Recently, as a result of the work of Stubbe and Abeles[830] and Stubbe et al.,[831] a stepwise process involving the formation of a carbanion intermediate has been proposed.[831] The evidence advanced to support this mechanism came from investigations using β-fluoropropionyl-CoA as a substrate in both the propionyl-CoA carboxylase and transcarboxylase catalyzed reactions. In the case of propionyl-CoA carboxylase, they found that with β-fluoropropionyl-CoA as "substrate", HF and acrylyl-CoA were formed (Equation 12). For this reaction to occur, ATP was required

$$CH_2FCH_2COSCoA + ATP \xrightarrow{HCO_3^-} CH_2 = CHCOSCoA + ADP + P_i + HF \qquad (12)$$

(HCO_3^- was also present) and was hydrolyzed to ADP and P_i, although no carboxylation products of β-fluoroporpionyl-CoA were observed. The rate of ATP hydrolysis was the same as the rate of acrylyl-CoA formation.

Using transcarboxylase, Abeles and collaborators[831] showed that oxaloacetate was required for the formation of acrylyl-CoA and like propionyl-CoA carboxylase, the rate of pyruvate formation was equal to the rate of acrylyl-CoA formation. Furthermore, the formation of acrylyl-CoA with either enzyme only occurred after carboxybiotin had been formed.

These authors considered several possible mechanisms, but concluded that the most likely sequence of events was the abstraction of the α-hydrogen of β-fluoropropionyl-CoA to form a carbanion which then eliminated F^- and formed acrylyl-CoA (Equation 13):

$$\begin{array}{ccccc}
\begin{array}{c} \mathrm{H\ \ H\ \ O} \\ |\ \ \ |\ \ \ \| \\ \mathrm{F{-}C{-}C{-}C{-}SCoA} \\ |\ \ \ | \\ \mathrm{H\ \ H} \end{array} & \rightarrow &
\begin{array}{c} \mathrm{H\ \ H\ \ O} \\ |\ \ \ |\ \ \ \| \\ \mathrm{F{-}C{-}C{-}C{-}SCoA} \\ |\ \ \ \ominus \\ \mathrm{H} \end{array} & \rightarrow &
\begin{array}{c} \mathrm{H\ \ H\ \ O} \\ |\ \ \ |\ \ \ \| \\ \mathrm{C{=}C{-}C{-}SCoA} \\ | \\ \mathrm{H} \end{array} \\
\beta\text{-Fluoropropionyl CoA} & & \text{Carbanion} & & \text{Acrylyl CoA}
\end{array} \qquad (13)$$

They pointed out that if the carbanion mechanism is accepted, then it is most likely that the normal carboxylation catalyzed by these two enzymes proceeds via the same pathway and a concerted mechanism is not involved.

Perhaps the most significant feature of this work was that it demonstrated that proton abstraction could occur without concomitant carboxyl group transfer from carboxybiotin to the acceptor substrate. Also, the proposed mechanism postulated the presence of a base in the catalytic subsite of the enzyme. However, in terms of a general mechanism for all biotin-dependent carboxylases (in particular, pyruvate carboxylase), it does not take into account the fact that carboxybiotin (of pyruvate carboxylase) is unstable in the second subsite[251] (i.e., carboxybiotin is capable of reacting with water in the second subsite). Nor does it account for the instability of carboxybiotin in propionyl-CoA carboxylase in the presence of β-fluoropropionyl-CoA.

3. An Enolate Ion Relay Mechanism

Additional information relevant to the sequence of events occurring at the second subsite was obtained by Goodall et al.[323] during their investigation of the factors that influence the translocation of carboxybiotin between the two subsites of pyruvate carboxylase. They showed that the presence of compounds such as hydroxypyruvic acid (Structure J), oxamic acid (Structure K), and glyoxylic acid (Structure L) readily caused the decarboxylation of carboxybiotin. This was interpreted to indicate that these compounds were effective signals for attracting the carboxybiotin complex out of the first subsite and into the second subsite, and since these compounds were not carboxylated, the carboxybiotin spontaneously decarboxylated. In the case of oxamate and glyoxylate, carbanion formation could not have occurred.

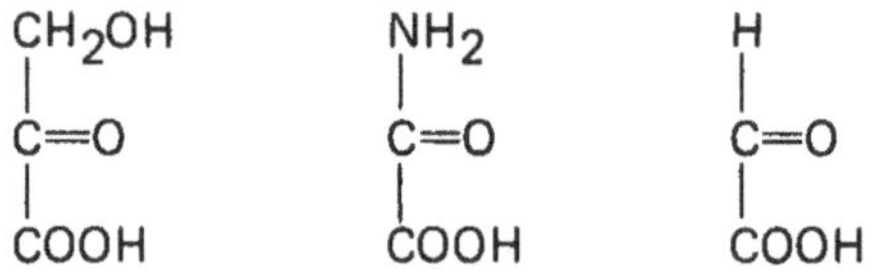

(J) (K) (L)

Therefore, to account for all of the available evidence on the transfer of the carboxyl group from carboxybiotin to acceptor substrate molecules, Goodall[322] proposed the mechanism illustrated below.

In this mechanism, the binding of the acceptor substrate at the second subsite induces a conformation change in the enzyme with the result that the carboxybiotin translocates from the first subsite and binds at the second subsite. In the second subsite a base is positioned adjacent to the 3′-N atom of the carboxybiotin and abstracts a proton producing an enolate anion of carboxybiotin. A cyclic concerted reaction similar to the one originally proposed by Retey and Lynen[681] then occurs, except that it is the enolate anion of carboxybiotin that reacts rather than the ureido form. Once the carboxylated product is formed, it dissociates from the enzyme and the enol-biotin returns to the first subsite, in most cases tautomerizing to the ureido form on the way. As the step in which the product is formed is concerted, the proposed mechanism is consistent with the observation of retention of configuration, lack of exchange between substrate and water protons when the enzyme is not carboxylated, and transfer of tritium between the substrates of transcarboxylase.

The abortive hydrolysis of carboxybiotin at low concentrations of pyruvate could occur if the pyruvate molecule, having induced the carboxybiotin to bind at the second subsite, dissociates from the enzyme. This would allow a water molecule access to the enolate anion of carboxybiotin and a cyclic reaction would occur with water as the substrate as shown below. Pyruvate analogs such as hydropyruvate, oxamate, and glyoxylate would cause decarboxylation in the same way.

The mechanism proposed by Goodall[322] also accounts for the observations made by Stubbe et al.[830,832] that propionyl-CoA carboxylase and transcarboxylase catalyze the elimination of F^- from β-fluoropropionyl-CoA with concomitant decarboxylation of carboxybiotin. In this case, the binding of the acceptor molecule, β-fluoropropionyl-

CoA, induces carboxybiotin to move into the second subsite where the enolate anion of carboxybiotin is formed. The anion abstracts a proton from the C-2 atom of the substrate analog (as it would in the normal reaction), but instead of the C-2 atom attacking the carboxyl group of carboxybiotin, fluoride is eliminated from fluoropropionyl-CoA. (Stubbe et al.[830,831] suggest several reasons why fluoride elimination occurs in preference to carboxylation.) The net result is formation of acrylyl-CoA and enol carboxybiotin; when the acrylyl-CoA dissociates, the carboxybiotin reverts to its ureido form and water now has access to the carboxybiotin which hydrolyzes by the mechanism already described. It should be remembered that fluoride elimination only occurs when the biotin is in the carboxybiotin form.

Final proof of the mechanism proposed by Goodall[322] would depend on the demonstration of the existence of a base in the second subsite.

III. THE EFFECT OF ACETYL-CoA

A. Variation in the Dependence of Enzymic Activity on the Presence of Acetyl-CoA

Although the basic reaction mechanisms of pyruvate carboxylases are very similar or even identical, the degree of dependence among the species on the presence of acyl-CoA compounds varies considerably. For example, among the prokaryotes, the activity of the enzyme isolated from *Pseudomonas citronellolis*,[772,856] *Aspergillus niger*,[111,269] and *Azotobacter vinelandii*[758] has no dependence on the presence of acetyl-CoA under any conditions, while the enzyme isolated from *Arthrobacter globiformis*[128] and *Bacillus (stearothermophilis)*[163] appears to exhibit almost complete dependence. On the other hand, the yeast enzyme is active in the absence of acetyl-CoA but is stimulated two- to three-fold by its presence.[158,203] In fact, this enzyme is activated by CoASH, palmitoyl-CoA, and acetoacetyl-CoA, as well as acetyl-CoA. Within the group of enzymes isolated from vertebrate sources, there appears to be a variation in the allosteric activator dependence also, although it is not quite as marked. The rat[545,764] and sheep[29] enzymes have an appreciable activity under appropriate assay conditions minus the activator, but again show a several-fold stimulation of activity on its addition. The chicken enzyme, on the other hand, has an almost absolute requirement for the presence of acetyl-CoA.[898]

B. Desensitization of the Enzyme Against Acetyl-CoA Activation

The fact that pyruvate carboxylases catalyze an acetyl-CoA-independent synthesis of oxaloacetate and yet exhibit a significant stimulation of enzymic activity in the presence of acetyl-CoA has been of considerable interest. The ability to separate the two activities has made it possible to probe the relationship between the catalytic and allosteric effector sites of the enzyme by chemical modification techniques. The first experiments along these lines were carried out by Keech and Farrant,[438] who showed that the enzyme from sheep possessed an essential lysyl residue. They showed that chemical modification of the enzyme using amino-group specific reagents such as 1-fluoro-2,4-dinitrobenzene, 2,4,6-trinitrobenzene sulfonic acid, and potassium cyanate resulted in loss of catalytic activity. Since the presence of acetyl-CoA protected the catalytic activity against the inhibitory action of these amino-group reagents, Keech and Farrant[438] suggested that the ε-amino group of the lysine side chain might be involved in the enzyme-acetyl-CoA interaction. Subsequently, Ashman et al[30] and Scrutton and White[765] extended these observations.

Ashman et al.,[30] using the enzyme isolated from sheep, presented evidence showing that the chemical modification of the lysyl group by 2,4,6-trinitrobenzene sulfonic acid actually desensitized the enzyme against acetyl-CoA stimulation. First of all, they confirmed the earlier work of Keech and Farrant,[438] and in addition, they showed that the acetyl-CoA stimulation of the exchange reactions between ATP and P_i and pyru-

Table 2
SPECIFICITY OF ACTIVATION OF PYRUVATE CARBOXYLASE AT pH 7.8

Activator	Activator constant K_a[a] (*M*)	Relative V_{max}
Formyl-CoA	1.9×10^{-4}	11
Acetyl-CoA	3.3×10^{-5}	100
Propionyl-CoA	1.1×10^{-4}	94
Crotonyl-CoA	1.35×10^{-4}	82
Isobutyryl-CoA	4.1×10^{-4}	59
n-Butyryl-CoA	2.8×10^{-4}	4
n-Valeryl-CoA	2.7×10^{-4}	1.3
CoA	10^{-2}	0.5

[a] K_a is the concentration of activator required to produce 50% maximal activation.

From Scrutton, M. C. and Utter, M. F., *J. Biol. Chem.*, 242, 1723, 1967. With permisssion..

vate and oxaloacetate was abolished by trinitrophenylation of the enzyme. The interesting aspect of their findings was that the chemical modification of the enzyme caused a significant increase in the rate of the acetyl-CoA-independent activity. So far, no adequate explanation for this observation has been advanced.

Scrutton and White[765] used the enzyme isolated from rat liver and yeast. The results they obtained confirmed that incubation of these enzymes with 2,4,6-trinitrobenzene sulfonic acid caused the loss of the acetyl-CoA-stimulated activity, although they did not report a concomitant increase in the acetyl-CoA-indendent activity. Interestingly, the pyruvate carboxylase purified from *Azotobacter vinelandii* is not activated by acetyl-CoA and was not effected by incubating with this amino-group reagent.

Although it had been known for sometime that acetyl-CoA was an allosteric activator of pyruvate carboxylase, these two reports were the first to provide direct evidence for the separation of the catalytic and modifier sites on the enzyme.

C. Specificity of the Activation Process

The first report on the ability of various acetyl-CoA derivatives to replace acetyl-CoA as activators of pyruvate carboxylase was made by Keech and Utter.[439] They showed that propionyl-CoA, crotonyl-CoA, and *n*-butyryl-CoA could replace acetyl-CoA. Later, Utter and Scrutton[761,904] extended this list of compounds to include formyl-CoA, isobutyryl-CoA, and *n*-valeryl-CoA, and at the same time they were able to assess the relative effectiveness of each compound in terms of the V and K_a values. These data are presented in Table 2. As pointed out by Utter and Scrutton,[904] the structural requirements for optimal activation appear to be relatively stringent and only a small group of compounds ranging from acetyl-CoA to isobutyryl-CoA are effective activators. These authors reported that carboxylated analogs of the acyl-CoA activators such as malonyl-CoA and methylmalonyl-CoA are relatively weak competitive inhibitors, while modification of the CoA-SH moiety in any way (e.g., removal ot the 3′-phosphate from the ribose molecule or replacement of the $-NH_2$ group by an $-OH$ group at the 6-position of the adenine ring) resulted in the conversion of these compounds to inhibitors of the enzyme. In summary then, an effective activator must possess both an intact CoA moiety and an acyl substituent that contains no more than four carbon atoms and of the correct steric configuration. Even now, the significance of these data is not understood.

D. The Deacylation Reaction

The variation in the degree of dependence on acetyl-CoA exhibited by pyruvate carboxylases isolated from different sources is only one of a number of interesting features concerning the association between the enzyme and its allosteric activator. For example, preparations of the enzyme of high specific activity, isolated from either sheep or chicken liver mitochondria, catalyze the slow deacylation of acetyl-CoA to CoASH and acetate.[29,761] The deacylase activity copurifies with pyruvate carboxylase during both ion-exchange chromatography and gel filtration, and is stimulated by the presence of substrates of the carboxylation reaction.[761] These data imply that both the deacylase and carboxylase activities are associated with pyruvate carboxylase and that the deacylase activity is not due to a separate contaminating enzyme. The end result is that the enzyme is capable of destroying its own allosteric activator and although the significance of this function is not known, it does have some bearing on experimental design and interpretation as will be seen later on.

E. Effect of Acetyl-CoA on Pyruvate Carboxylase During Cold Stress and on Dilution

In addition to its more obvious effects on the catalytic activity of pyruvate carboxylase, acetyl-CoA also exerts an effect on some of the more unusual properties of the enzyme. For example, the enzyme from chicken liver is cold labile below 15°C,[426,759] while the enzyme from all vertebrate sources tested so far is inactivated when diluted to less than 3 to 4 enzyme units per milliliter.[29] The presence of saturating levels of acetyl-CoA protects the enzyme against both phenomena. Strangely, although saturating levels of acetyl-CoA protect the enzyme against the adverse effects of both processes, it is not able to reactivate the enzyme. In the case of cold inactivation, warming at 23°C is sufficient to restore activity while recent studies[1003] have shown that Mg^{2+} can reactivate dilution-inactivated enzyme. It has also been observed that nonionic substances such as sucrose and polyethylene glycol reduce the rate of both dilution and cold inactivation while polar solutes such as KCl increase the rate.[190]

In terms of the effect of dilution on the kinetic properties of the enzyme, from unpublished observations the following points may be made. In addition to the loss of overall enzymic activity, both ATP:P_i and HCO_3^-:oxaloacetate exchange reaction activities are also lost, while the rates of both the ATP:ADP and the pyruvate:oxaloacetate exchange reactions appear to be largely unaffected. The interpretation placed on these observations is that when the enzyme is diluted below a certain critical level there is a change in the normal microenvironment surrounding the first subsite. This change could be a slight shift in its normal locus or a change in its configuration as a result of decreased protein-protein interaction. No matter how the change occurs, the indications are that biotin can no longer react at the first subsite. This may be caused by an inability of acetyl-CoA to bind at the first subsite after the configuration change, which would mean that the carboxylation of biotin could not occur. From the mechanisms proposed by Phillips et al.[651] (see Equations 9, 10, and 11), both the ATP:P_i and HCO_3^-:oxaloacetate exchange reactions involve the participation of biotin at the first subsite. On the other hand, the ATP:ADP exchange reaction proceeds without the involvement of the biotin moiety.

Since the pyruvate:oxaloacetate exchange reaction is unaffected by dilution, it is assumed that the second subsite remains unchanged and the biotin has ready access to the pyruvate binding site.

This explanation could indicate that biotin is not symmetrically located between the two subsites but is placed nearer the second subsite. However, it is consistent with the observation covering many enzyme preparations of different specific activities that the ratio of the rate of the pyruvate:oxaloacetate exchange and the rate of the overall reaction is extremely variable. The exchange reaction continues to occur long after the overall activity has decreased or even been lost.

F. The Sigmoidal Velocity Response Curve

Another interesting facet of the interaction between pyruvate carboxylase and acetyl-CoA has been the observed sigmoidal velocity response to increasing concentrations of acetyl-CoA. This phenomenon was first reported by Barritt et al.[77] using sheep kidney pyruvate carboxylase and later shown to be a property of other pyruvate carboxylases.[203,761,771] With Hill coefficients (nH) approaching 3.0 for the chicken enzyme and 2.0 for the rat and sheep enzymes, it was concluded that acetyl-CoA bound to these enzymes in a highly cooperative manner.[896] Despite a vast amount of research, this claim remained unsubstantiated and for a long time there were many more questions than answers. What was the mechanism of action of acetyl-CoA in stimulating the activity of pyruvate carboxylase? Was the "apparent" cooperativity a part of the mechanism? Why did the enzyme isolated from some sources require acetyl-CoA while others did not? What was the significance of the deacylation reaction?

The view that acetyl-CoA bound to the enzyme in a cooperative manner[896] was based entirely on kinetic data. No direct evidence was available until Frey and Utter,[287] using a modified version of the rapid flow dialysis method of Colowick and Womack,[201,444] presented evidence that the binding curve for acetyl-CoA was also sigmoidal, although markedly less sigmoidal than the activity profile. Thus, the case for cooperative binding appeared to be complete.

However, unequivocal interpretation of the binding data was difficult. As pointed out above, purified preparations of pyruvate carboxylase deacylate acetyl-CoA. When investigating the kinetic properties of the enzyme, the deacylase activity is of little consequence since low concentrations of enzyme and short incubation times are used. Binding studies, on the other hand, required high enzyme concentrations and longer time intervals. When discussing the data of Frey and Utter[287] in a subsequent publication, Easterbrook-Smith et al.[250] calculated that, under the conditions used for the binding studies and from the V and known K_m values, 35% of the [1-^{14}C]acetyl-CoA was hydrolyzed to [1-^{14}C]acetate by the time the unlabeled acetyl-CoA was added. Since the rate of diffusion of a molecule in solution is inversely proportional to the average radius, the [1-^{14}C]acetate would be expected to diffuse across the cell membrane many times faster than [1-^{14}C]acetyl-CoA. As a result, this would lead to an underestimate of the binding of acetyl-CoA by pyruvate carboxylase, particularly at low concentrations of the ligand. This, in turn, would give rise to a sigmoidal-shaped binding curve.

In order to resolve this dilemma, Easterbrook-Smith et al.[250] investigated the kinetic response of the deacylation reaction to increasing concentrations of acetyl-CoA and showed that it exhibited classical Michaelis-Menten behavior. This observation indicated very strongly that acetyl-CoA did not bind to the enzyme in a cooperative manner. It also suggested that the sigmoidal velocity curve obtained with the carboxylation reaction must be due to some special features of the assay system and a number of possible factors were considered. The phenomenon of irreversible inactivation of the enzyme on dilution below approximately 4 enzyme units per milliliter (discussed above) was considered by Easterbrook-Smith et al.[250] to be a significant contributing factor to the apparent atypical kinetic response. Ashman et al.[29] showed that although acetyl-CoA protected the enzyme against dilution inactivation, below 40 μM acetyl-CoA a rapid loss of enzymic activity occurred and was not recovered by the addition of more acetyl-CoA. As a result, the effect of dilution inactivation on the velocity curve is that at low concentrations of acetyl-CoA a falsely reduced rate of enzymic activity is observed.

A second factor which these investigators[250] considered may influence the shape of the velocity curve concerned the effect of acetyl-CoA on the degree of saturation of the other reaction components. Utter et al.[896] reported that variations in the concentration of acetyl-CoA had very little effect on the apparent K_m values of the other substrates. However, no mention was made of any precautions being taken to prevent dilution inactivation from occurring and only a narrow range of acetyl-CoA concentrations was examined (2 to 20 μM). In another study, Ashman et al.,[29] using sheep pyruvate carboxylase, varied the concentration of acetyl-CoA from 0 to 260 μM and reported an eightfold decrease in the K_m value for pyruvate in the presence of acetyl-CoA and an even greater decrease in the K_m value for HCO_3^-. Scrutton and White[764] reported similar changes for HCO_3^- using the enzyme isolated from rat liver. When considering these observations, Easterbrook-Smith et al.[250] reasoned that when acetyl-CoA was the variable ligand, the fixed concentration of pyruvate (and HCO_3^-), although saturating at high concentrations of acetyl-CoA, would become nonsaturating as the concentration of acetyl-CoA decreased. Hence, they predicted that the degree of sigmoidicity, as reflected by the Hill coefficient, would depend, at least to some extent, on the pyruvate concentration used in the assay solution. The data presented in Figure 4 and Table 3 show that this reasoning was correct.

These factors prompted Easterbrook-Smith et al.[250] to suggest that the observed reaction velocity was in fact, the sum of two velocities, i.e., the acetyl-CoA-dependent activity catalyzed by those enzyme molecules which are saturated with acetyl-CoA and an acetyl-CoA-independent activity catalyzed by those molecules devoid of acetyl-CoA. Under normal acetyl-CoA-dependent assay conditions, the enzyme (0.05 enzyme units per milliliter) is fully saturated with acetyl-CoA and all molecules are fully active. As the concentration of acetyl-CoA is decreased to the point where some enzyme molecules do not have acetyl-CoA bound, they rapidly undergo dilution inactivation. In addition, the acetyl-CoA-deficient molecules which have not undergone irreversible inactivation cannot catalyze the reaction at a maximum rate because they are no longer fully saturated with pyruvate and HCO_3^-. As a result, the observed reaction velocity at low concentrations of acetyl-CoA is no longer first order with respect to acetyl-CoA concentration and a sigmoidal velocity response curve is observed. However, if the reaction is carried out under acetyl-CoA-independent assay conditions, i.e., at high enzyme concentrations (5 enzyme units per milliliter) to avoid dilution inactivation and high pyruvate and HCO_3^- concentrations so that the enzyme is saturated with these substrates irrespective of the acetyl-CoA concentration, then the only effect of varying the acetyl-CoA concentration would be to vary the proportion of enzyme molecules catalyzing the acetyl-CoA-dependent and -independent activities, i.e.,

$$v = \frac{V^D}{1 + \frac{Ka}{A}} + \frac{V^I}{1 + \frac{A}{Ka}} \quad (14)$$

where V^D and V^I are the maximum velocities for the dependent and independent activities, A is the acetyl-CoA concentration, and K_a is the dissociation constant between the enzyme and activator.

In order to test this proposal, Easterbook-Smith et al.[250] carried out such an experiment and the sigmoidal velocity curve was completely eliminated and in double reciprocal form, the velocity plot was concave downward (Figure 5) — the opposite of the parabola expected for positive cooperative binding. Furthermore, theoretical curves generated from Equation 14 by calculating the dependence of reaction velocity on acetyl-CoA concentration (Figure 6) were similar to the plots of the experimental data, suggesting that the above proposal is a plausible explanation for the nonclassical velocity profile obtained with this ligand.

G. Orientation of the Acetyl-CoA Molecule at the Active Site

Unlike most other allosteric effectors, the binding of acetyl to pyruvate carboxylase

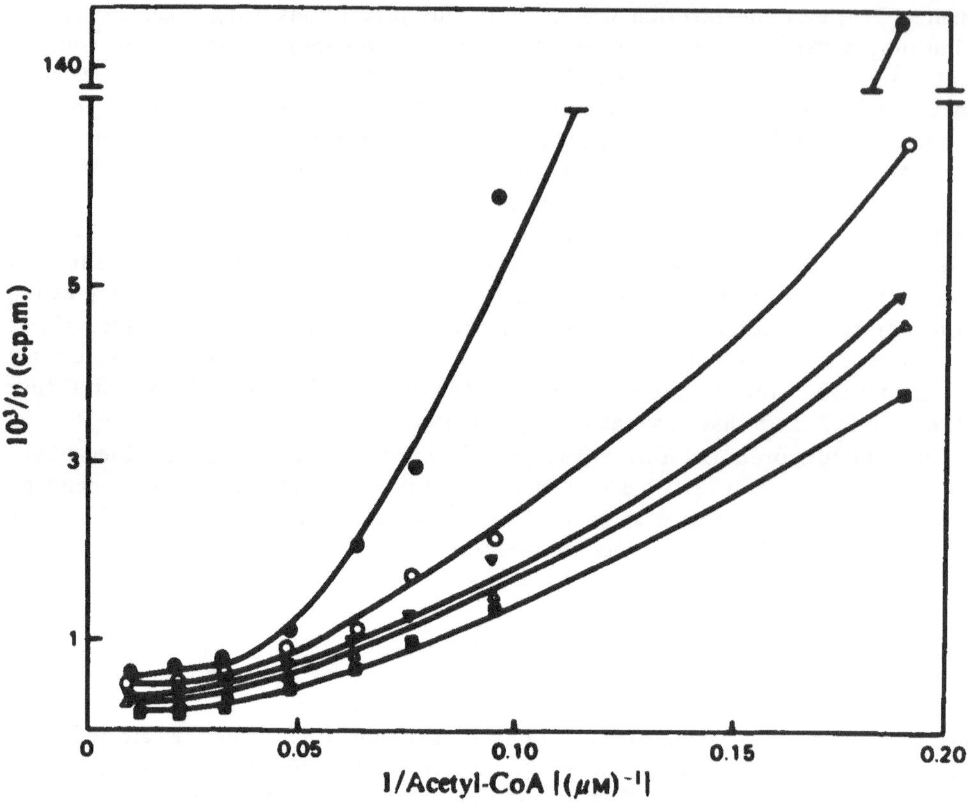

FIGURE 4. Effect on sheep pyruvate carboxylase of various acetyl-CoA concentrations at different fixed concentrations of pyruvate. The concentrations of pyruvate were (m*M*): 1.0 (●); 3.0 (○); 5.0 (▼); 10 (△); 20 (■). (From Easterbrook-Smith, S. B., Campbell, A. J., Keech, D. B., and Wallace, J. C., *Biochem J.*, 179, 497, 1979. With permission.)

Table 3
EFFECT OF VARIOUS PYRUVATE CONCENTRATIONS ON THE HILL COEFFICIENT (h) FOR ACETYL-CoA

Conc. of pyruvate (m*M*)	h	
	Sheep enzyyme	Chicken enzyme
1.00	3.40 ± 0.05	4.20 ± 0.19
3.00	2.77 ± 0.23	3.48 ± 0.17
5.00	2.15 ± 0.15	—
10.00	1.82 ± 0.04	2.24 ± 0.21
20.00	1.95 ± 0.15	2.22 ± 0.14
30.00	—	1.84 ± 0.20

From Easterbrook-Smith, S. B., Campbell, A. J., Keech, D. B., and Wallace, J. C., *Biochem. J.*, 179, 497, 1979. With permission.)

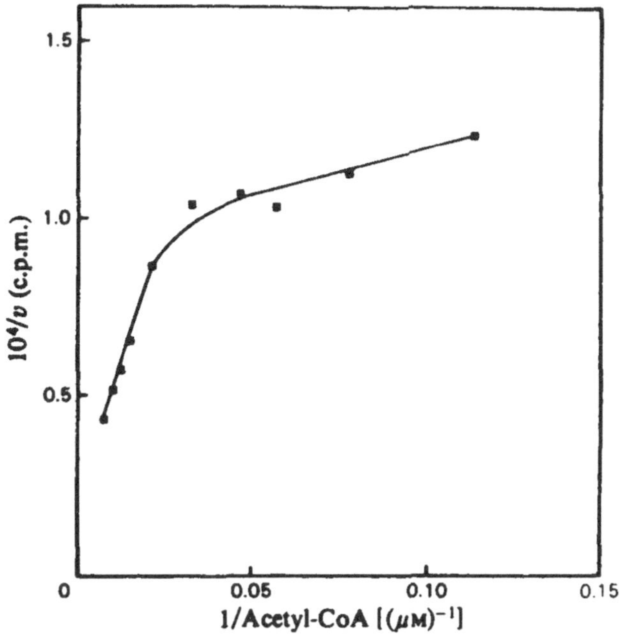

FIGURE 5. Reciprocal of reaction velocity vs. reciprocal acetyl-CoA concentration at saturating concentrations of all other reaction components with sheep pyruvate carboxylase. The assay contained, in a final volume of 0.5 m*l* (μmol): Tris chloride (pH 8.4), 40; ATP, 4; $MgCl_2$, 17; pyruvate, 36; NH_4Cl, 83; $NaH^{14}CO_3$ (9.86×10^5 c.p.m./μmol), 38; 2.6 units of enzyme. The reaction time was 30 sec. (from Easterbrook-Smith, S. B., Campbell, A. J., Keech, D. B., and Wallace, J. C., *Biochem J.*, 179, 497, 1979. With permission.)

does more than just convert the enzyme from a weakly active to a most-active form. The known effects are: (1) it increases the overall reaction rate, (2) it decreases the apparent K_m value for both HCO_3^- and pyruvate,[29] (3) it is required for the ATP:P_i exchange reaction, (4) it stimulates the rate of pyruvate:oxaloacetate exchange reaction, and (5) it protects the enzyme against dilution inactivation[29] and cold inactivation.[426,759] The evidence available to support the suggestion that these effects result from subtle conformational change within the enzyme is of two types. First of all, it was reported by Scrutton and Utter[761] that the presence of acetyl-CoA increased the rate of enzyme inactivation by avidin. However, the effect of acetyl-CoA on the rate of inactivation was of a biphasic form; concentrations of acetyl-CoA up to 200 μ*M* caused an increase in rate while higher concentrations actually reduced the rate of inactivation until, at 3 m*M* acetyl-CoA, the rate was the same as for the free enzyme. Therefore, any interpretation of an acetyl-CoA-induced conformational changes based on this data would have to be treated with caution, particularly since there was no change in the inactivation rate in the range of the K_a value (14 μ*M*[87]) where the enzyme is operating at half-maximum activity. Second, McGurk and Spivey[554] studied the occurrence of ligand-induced conformational changes using the fluorescent probe 8-anilinonaphthalene sulfonic acid and showed that the binding of acetyl-CoA quenched the fluorescence of the probe by 10%. Again, it is difficult to associate this change with an effect on enzymic activity since the concentration of acetyl-CoA used was 70-fold greater than the dissociation constant.

On the other hand, other attempts to demonstrate an acetyl-CoA-induced conformational change have not been successful. These include the use of electron micros-

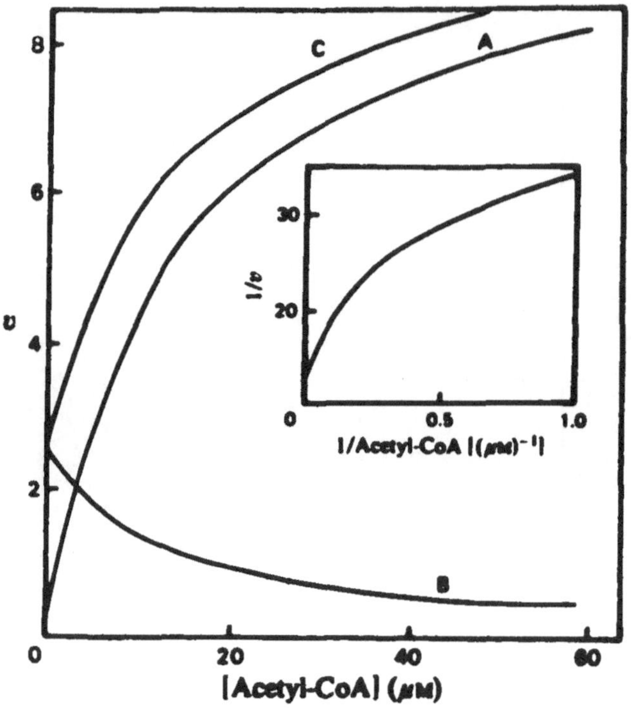

FIGURE 6. Theoretical curves generated from Equation 14 in the text by calculating the dependence of reaction velocity on acetyl-CoA concentration. (Curve A) the simulation of the acetyl-CoA-dependent term alone; (Curve B) the simulation of the acetyl-CoA-independent term alone; (Curve C) the simulation of the sum of both terms. The values of K_a, V^D (i.e., the maximum velocity of the acetyl-CoA-dependent reaction, and V^I (the maximum velocity of the acetyl-CoA-independent reaction) were 12 μM, 10 and 2.5 μM, respectively. Inset: the double reciprocal plot of Curve C. (From Easterbrook-Smith, S. B., Campbell, A. J., Keech, D. B., and Wallace, J. C., *Biochem. J.*, 179, 497, 1979. With permission.)

copy, optical rotary dispersion, and protein fluorescence. Small UV spectral changes were observed by Frey and Utter.[287] Therefore, in view of the lack of convincing evidence to support the concept, other models to explain the multiple effects of acetyl-CoA need to be considered.

In an earlier section concerning the mechanism of the first partial reaction, the reaction sequence put forward by Phillips et al.[651] (Equations 9—11) indicated that the role of acetyl-CoA in the reaction mechanism was to facilitate the shifting of the activated carboxyl group from the enzyme carboxyphosphate complex to the N′-atom of the ureido group of biotin. Although the precise mechanism by which this process is achieved is not known, this role for acetyl-CoA could account for the increase in the rate of the overall reaction and its requirement in the ATP:P_i exchange reaction.

Previous studies on the role and locus of action of acetyl-CoA[752,901] showed that although the exchange reaction between pyruvate and oxaloacetate proceeded in its absence, the exchange reaction between ATP and P_i exhibited an absolute requirement for the activator. This led to the conclusion that the locus of action of acetyl-CoA was at the first partial reaction subsite. Later, acetyl-CoA was shown to influence events occurring at the second subsite. Ashman et al.[29] demonstrated that the activator decreased the K_m value for pyruvate while Seubert et al.,[773] McClure et al.,[544] and Ashman et al.[30] all showed that the exchange reaction between pyruvate and oxaloacetate was stimulated two- to three-fold by acetyl-CoA. These observations provided compelling evidence that the influence of acetyl-CoA was not restricted to the first partial reaction but that both subsites were affected. (N.B.: It has been reported that with chicken liver pyruvate carboxylase, acetyl-CoA has no affect on the pyruvate:oxaloacetate exchange rate. However, in this laboratory it has been observed that the effect of acetyl-CoA is related to the concentration of enzyme used in the assay. At high enzyme concentrations, the effect of acetyl-CoA on the exchange rate is barely detectable while at low concentrations of enzyme, a threefold stimulation of the exchange rate is observed.)

The binding studies of Frey and Utter[287] showed that there were four binding sites for acetyl-CoA on the tetrameric form of the enzyme and so it can be assumed that only 1 mol of acetyl-CoA is bound at each active center. Therefore, to account for the fact that the effector molecule influences events occurring at both subsites, it is not unreasonable to propose that acetyl-CoA must span across the active site, forming a bridge between the two subsites. Using this concept of acetyl-CoA binding, it is possible to propose a mechanism to account for the protection of the enzymic activity on dilution. It was proposed in the discussion on dilution inactivation that diluting the enzyme in the absence of acetyl-CoA permitted a change in either the microenvironment or location of the first subsite, resulting in a loss of enzymic activity. However, if acetyl-CoA binds at both subsites, it could hold the active site in a more rigid conformation, preventing a conformational change from occurring on dilution. This idea is contrary to the current view of the role of an allosteric effector, which is assumed to bind and induce a conformational change. However, the above proposal does account for the known effects of the activator on the enzyme and provides an explanation for the inability to demonstrate an acetyl-CoA-induced conformational change in the concentration range known to affect enzymic activity.

IV. THE ROLE OF INORGANIC CATIONS

A. The Role of the Freely Dissociable Divalent Ions

Utter and Keech[898] first demonstrated a requirement for Mg^{2+} in the pyruvate carboxylase assay solution and at that time it was assumed that the Mg^{2+} complexed with the ATP^{4-} to form the true substrate, $MgATP^{2-}$. If this was the sole function of the added Mg^{2+}, then Mg^{2+} and ATP^{4-} would be required in approximately stoichiometric amounts with respect to each other. However, it was pointed out by Keech and Barritt[437] that for optimal assay conditions of biotin-dependent carboxylating enzymes, the total divalent metal ion concentration must be in excess of the total ATP^{4-} concentration.[340,435,507] This observation suggested that besides being necessary for the formation of the substrate, i.e., $MgATP^{2-}$, free Mg^{2+} may also be required to activate the $MgATP^{2-}$-dependent enzymes. For sheep kidney pyruvate carboxylase, Keech and Barritt[437] and Blair,[107] who reexamined their results, both concluded that Mg^{2+} and $MgATP^{2-}$ were required for activity. Later, similar conclusions were reached for pyruvate carboxylases isolated from yeast,[159] chicken liver,[74] and *Aspergillus niger*.[269] However, in the case of the rat liver enzyme, McClure et al.[545] concluded that because at infinite $MgATP^{2-}$ concentration the Michaelis-Menten constant was zero, there was no absolute requirement for free Mg^{2+}.

Although the requirement for free-divalent cations appears to be characteristic of all pyruvate carboxylases, different specificities are observed. Mg^{2+}, Mn^{2+}, and Co^{2+} activate the enzyme isolated from chicken liver[439] and rat liver,[545] but only Mn^{2+} is effective as an activator of the enzyme from *Bacillus stearothermophilus*[838] and only Mn^{2+} and Mg^{2+} for the yeast enzyme.[514] Other divalent metal ions, e.g., Ca^{2+}, Zn^{2+}, Cu^{2+} and Ni^{2+} act as competitive inhibitors with respect to free Mg^{2+}.[545,755]

The initial work of Keech and Barritt[437] was later extended by Bais and Keech.[48] They confirmed that the enzyme has an absolute requirement for Mg^{2+} and that Mg^{2+} and $MgATP^{2-}$ combine with the enzyme in a random manner to form an activated complex consisting of Mg^{2+}-ENZ-$MgATP^{2-}$. Furthermore, from two separate lines of evidence, they obtained data which clearly indicated that the binding of Mg^{2+} induced a significant conformational change in the enzyme. However, it was left to Goodall et al.[323] to assign a role for Mg^{2+} in the reaction mechanism. Earlier, Easterbrook-Smith et al.[251] had provided evidence that the binding of pyruvate apparently acts as a signal to induce carboxybiotin to move into the second subsite, but in this location the carboxybiotin is unstable and if the carboxyl group is not transferred immediately to pyruvate, it reacts with water to form HCO_3^-. Goodall et al.[323] were able to show that at least one of the roles of Mg^{2+} is to hold the carboxybiotin in the first subsite where it increases the half-life of the complex at least tenfold. The mechanism by which this is achieved is not yet understood.

B. The Effect of Monovalent Metal Ions

The first demonstration of an effect by univalent metal ions on the activity of pyruvate carboxylase was made by Bloom and Johnson,[111] who reported a stimulation of catalytic activity of the enzyme isolated from *Aspergillus*. Following on from this, Ruiz-Amil et al.[710] observed a concentration-dependent stimulation of activity with respect to increasing the concentrations of K^+. In addition, they showed that Na^+ inhibited enzymic activity of yeast pyruvate carboxylase. However, the first extensive investigation carried out on the effect of the monovalent inorganic cations on the vertebrate enzyme was performed by McClure et al.[545] These authors showed that rat liver pyruvate carboxylase required the presence of a monovalent cation for activity and showed a hyperbolic response to increasing concentrations of KCl. Unlike the effect observed by Ruiz-Amil et al.[710] for the enzyme from yeast, the rat liver enzyme was activated to varying extents by all of the alkali metal cations and NH_4^+.

McClure et al.[545] concluded that their results provided strong evidence that the univalent cations specificity of the rat liver enzyme appeared to be influenced by the nonhydrated volume (or size) of the cation and by the anionic field strength of the combining site on the enzyme-substrate complex. (It seems strange for these authors to use as a criterion, the nonhydrated form of the cations. Under biological conditions or with the experimental conditions used, the nonhydrated ionic radii would not apply. On the other hand, the hydrated ionic radii decrease in magnitude in going from Li^+ to Cs^+, i.e., the opposite order of the nonhydrated form). By examining these two factors and by comparing the order of decreasing cationic selectivity displayed by rat liver pyruvate carboxylase, McClure et al.[545] deduced that either a phosphate or a carboxyl group was the anionic site responsible for complexing the cation to the enzyme-substrate complex. This particular aspect has never been extended beyond this point.

Included in the investigation carried out by McClure et al.[545] was an examination of the relationship of K^+ activation to the saturation of the other substrates. The most striking feature they observed was the large enhancing effect HCO_3^- and K^+ have on the binding of one another to their respective steady-state forms of the enzyme. A similar effect was observed between $MgATP^{2-}$ and K^+, although the magnitude of the effect was considerably less, while there appeared to be very little interaction between pyruvate and K^+. Barden and Scrutton[74] obtained similar results with the enzyme isolated from chicken liver, except that when K^+ was the variable reactant, the double reciprocal plots obtained at a series of changing fixed concentrations of HCO_3^- intersected on the ordinate. These authors pointed out that such a pattern is characteristic of an equilibrium-ordered mechanism of addition with HCO_3^- adding first. They also pointed out that "the binding of the negatively charged substrate to the enzyme before the positively charged activator is exactly opposite to that expected for the formation of an enzyme-metal-substrate bridge complex, which has been involved to explain the activation by both univalent and divalent cations in several other systems."

These findings prompted Barden and Scrutton[74] to suggest that a possible role for K^+ in stimulating the pyruvate carboxylase reaction may be attributable at least in part to a restoration of charge balance in the catalytic site. At the same time, they were careful to point out that this postulate did not necessarily imply direct interaction between substrate and activator on the enzyme surface. However, this suggestion could only be a partial explanation since McClure et al.[545] had previously shown that a K^+-induced stimulation of isotopic exchange reactions occurred at both subsites. Therefore, for the Barden and Scrutton suggestion to be correct, K^+ has to restore a charge balance with respect to HCO_3^- at the first subsite and pyruvate at the second subsite. Alternatively, it may be that K^+ maintains a charge balance with the carboxybiotin moiety as it moves from the first subsite to the second subsite. In other words, K^+ could facilitate the movement of carboxybiotin through a hydrophobic region in the active site. In this way, a charge balance would be maintained at both subsites of the active center.

An interesting aspect of both of these studies was the fact that in double reciprocal plots of initial velocity data, where K^+ was the fixed variable reactant, families of intersecting lines were obtained when any of the three substrates (i.e., $MgATP^{2-}$, HCO_3^-, or pyruvate) was the variable ligand. It is difficult to explain these results in terms of a nonclassical Ping-Pong mechanism. However, if K^+ binds to the enzyme after HCO_3^- binds (as clearly demonstrated by Barden and Scrutton[74]) followed by the binding of pyruvate before MgADP and P_i are released, then this evidence supports an overall sequential mechanism. Furthermore, it places the binding of K^+ in the reaction mechanism in precisely the same position as Phillips et al.[651] have indicated that acetyl-CoA participates and exerts its effect. This in turn tempts one to suggest that the role of K^+ is closely related to the role of acetyl-CoA in converting the putative carboxyphosphate derivative to carboxybiotin in preparation for its carboxylation of pyruvate.

V. CONCLUSION

The present evidence points towards the steady-state mechanism of the overall reaction as being sequential. This means that there is interaction between the subsites of the active site in that pyruvate binds at the second subsite before the products of the first partial reaction, ADP and P_i are released from the first subsite. A recent report by Beaty and Lane[89] suggests that such a mechanism also applies to acetyl-CoA carboxylase.

In the first partial reaction, HCO_3^- is directly involved in the hydrolysis of ATP and it is possible that a carboxyphosphate-type intermediate is formed. The major role of acetyl-CoA is to mediate the transfer of the carboxyl group from such an intermediate onto the ureido-N of biotin to form carboxybiotin. The function of Mg^{2+} in the overall reaction is to hold the carboxybiotin in the first subsite, where it is stable.

The binding of pyruvate acts as a signal to call the carboxybiotin into the second subsite, thus disturbing the equilibrium between bound and free carboxybiotin of the first subsite. During the subsequent reaction, pyruvate is deprotonated, possibly by the enolate anion of biotin. This only occurs if the biotin is carboxylated and thus a cyclic concerted carboxylation of pyruvate is likely. As pyruvate binds in a rapid equilibrium manner, it may dissociate before carboxylation can occur; in this case, the carboxybiotin can react with water to form HCO_3^- and biotin.

Chapter 5

REGULATION OF ENZYMATIC ACTIVITY

G. J. Barritt

TABLE OF CONTENTS

I. GLUCONEOGENIC AND LIPOGENIC TISSUES OF VERTEBRATES

A. Evidence that Pyruvate Carboxylase Catalyzes a Rate-Limiting Step in the Gluconeogenic and Lipogenic Pathways

Pyruvate carboxylase catalyzes the first step in the pathway for the formation of glucose from pyruvate and other three-carbon precursors in the liver and kidney cortex[507,899,904] (Figure 1), and an early step in the conversion of pyruvate to fatty acids in the liver, adipose tissue and mammary gland[51,341,459,532,696,904,940] (Figure 2). Evidence that this enzyme is subject to metabolic regulation includes observations of the properties of the enzyme and the reaction which it catalyzes,[79,599,759,762,816,901] assessment of the tissue concentration of metabolites of the gluconeogenic pathway under different physiological conditions,[265,277,550,770,880,961,963,964,970,1000] and the observation that the activity of the enzyme is altered by changes in diet and certain hormones (Sections I.B and I.C.6 below).

The arguments which provide indirect evidence that pyruvate carboxylase catalyzes a rate-limiting step in the pathways of gluconeogenesis and lipogenesis have recently been summarized.[79] More direct evidence which indicates that the enzyme catalyzes a rate-limiting step in the gluconeogenic pathway has been provided by comparison of the concentrations of tissue metabolites measured in the liver under basal conditions with those present when the rate of gluconeogenesis is increased or decreased, and application of the "cross-over" theorem.[265,277,550,770,880,961,963,964,970,1000] Evidence for a cross-over between pyruvate and phosphoenolpyruvate (PEP) (Figure 3) has been obtained in experiments conducted using livers in vivo, perfused livers and isolated liver cells in which the rate of the overall gluconeogenic pathway was altered by treatment with glucagon,[265,961] long-chain fatty acids,[963,964,1000] 4-pentenoic acid,[227,880,970] or starvation.[277,770] These results are consistent with the conclusion that the reaction catalyzed by pyruvate carboxylase is a rate-limiting step in the gluconeogenic pathway.[964] However, interpretation of data from cross-over plots of this type is complicated.[698] The possibilities that the cross-over may be due to a change in the rate of the reaction catalyzed by PEP-carboxykinase[264] or to the removal or formation of a metabolite of the gluconeogenic pathway catalyzed by an enzyme in another pathway cannot be completely eliminated, although they are probably unlikely.

B. Regulation of the Total Amount of Enzyme Protein

1. Methods Employed in Estimation of the Total Amount of Pyruvate Carboxylase Present in Tissue Extracts

Changes in the total amount of pyruvate carboxylase protein, through alterations in the rate of synthesis, constitute an important mechanism for the long-term regulation

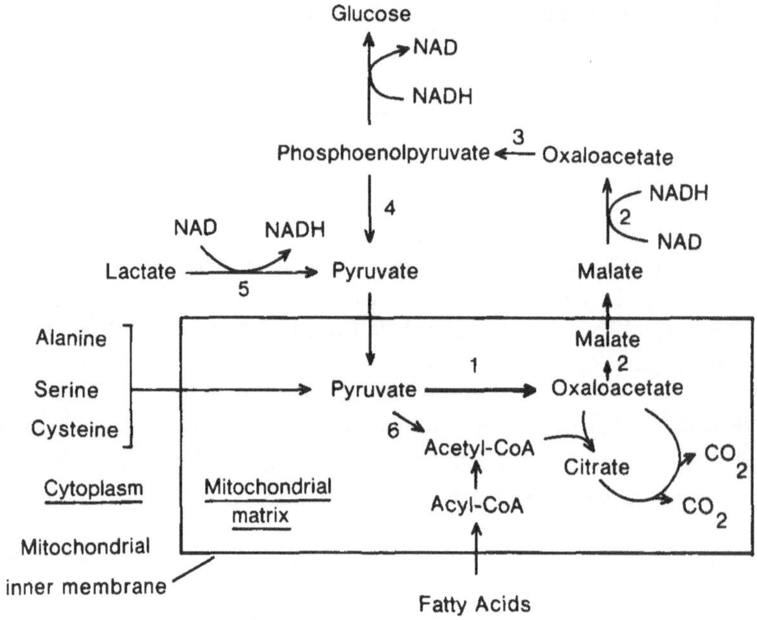

FIGURE 1. The role of pyruvate carboxylase in the pathway of gluconeogenesis from three-carbon precursors in mammalian liver and kidney, and its relationship to pyruvate and fatty acid oxidation. The enzymes specifically indicated are (1) pyruvate carboxylase, (2) malate dehydrogenase, (3) PEP-carboxykinase, (4) pyruvate kinase, (5) lactate dehydrogenase, and (6) pyruvate dehydrogenase. The scheme has been adapted from Denton and Halestrap.[223] In avian and guinea pig livers, PEP-carboxykinase is located exclusively or partly, respectively, in the mitochondrial matrix leading to the formation of all or some, respectively, of PEP in this compartment (reviewed by Utter[893]).

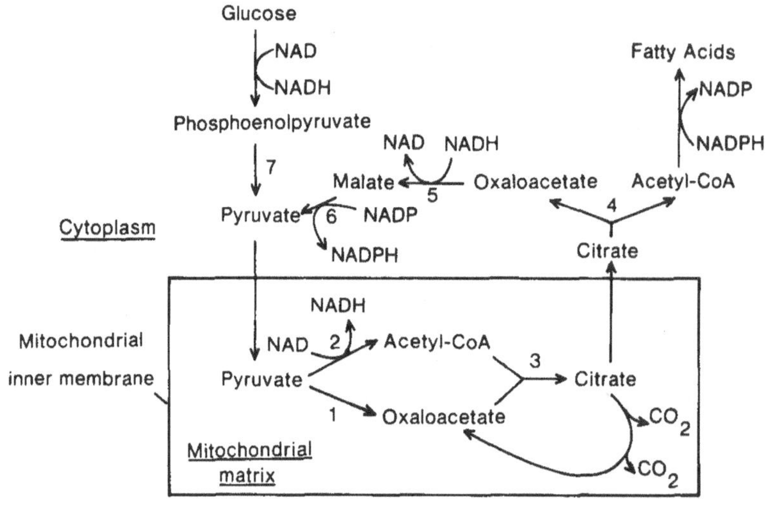

FIGURE 2. The role of pyruvate carboxylase in the pathway of lipogenesis (fatty acid synthesis) from glucose in mammalian liver, adipose tissue, and mammary gland. The enzymes specifically indicated are (1) pyruvate carboxylase, (2) pyruvate dehydrogenase, (3) citrate synthetase, (4) citrate lyase, (5) malate dehydrogenase, (6) malic enxyme, and (7) pyruvate kinase. (From Denton, R. M. and Halestrap, A. P., *Essays Biochem.*, 15, 37, 1979. With permission.)

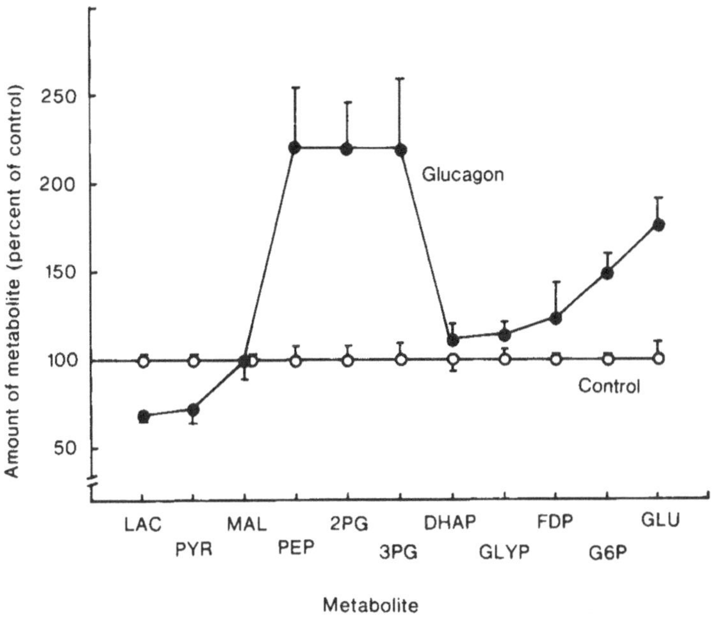

FIGURE 3. A "cross-over" plot for the effect of glucagon on the concentrations of intermediates of the gluconeogenic pathway in perfused livers from starved rats. The amount of each intermediate present in control livers (O) and livers perfused with glucagon (●) is expressed as a percentage of the control value. The abbreviations are : LAC, lactate; PYR, pyruvate; MAL, malate; PEP, phosphoenolpyruvate; 2PG, 2-phosphoglycerate; 3PG, 3-phosphoglycerate; DHAP, dihydroxyacetone phosphate; GLYP, glyceraldehyde-3-phosphate; FDP, fructose-1,6-biphosphate; G6P, glucose-6-phosphate; GLU, glucose. (From Exton, J. H. and Park, C. R., *J. Biol. Chem.* 244, 1424, 1969. With permission.)

of metabolic flux from pyruvate to oxaloacetate in the liver, kidney, and adipose tissue.[79] These changes are associated with the increases in the rate of gluconeogenesis during starvation and diabetes (reviewed by Ruderman et al.[709] and Owen et al.[621]), periods of enhanced cellular metabolism induced by thyroid hormones[285,560,587] and neonatal development,[58,649,936,992] and increases in the rate of lipogenesis in developing adipocytes.[334,520] Changes in the total amount of pyruvate carboxylase activity are also associated with alterations in the concentration of other key gluconeogenic[770] and lipogenic[520] enzymes.

In the majority of studies reported, increases or decreases in the total amount of pyruvate carboxylase in a given tissue have been described by estimating the total amount of enzyme activity present in tissue homogenates, or in extracts of isolated mitochondria. This should, in principle, provide a reliable estimate of the intracellular concentration of the enzyme. However, a number of factors may influence the amount of enzyme activity observed in these extracts or the difference in activities between two physiological conditions. These factors include the degree of extraction of the enzyme from the mitochondrial matrix, increased fragility of mitochondria in tissues from starved or diabetic animals[305,862] variations in the concentrations of substrates and modifiers present in crude extracts, and inadequate quantitative estimation of all secondary products of oxaloacetate metabolism. In some recent studies, uncertainties in estimation of the total amount of enzyme activity have been overcome by measurement of the total amount of enzyme protein by immunoprecipitation[172,289,944,945] and by direct measurement of the rates of synthesis and degradation of the enzyme.[944,945]

Some variation in the relative amounts of pyruvate carboxylase activity observed

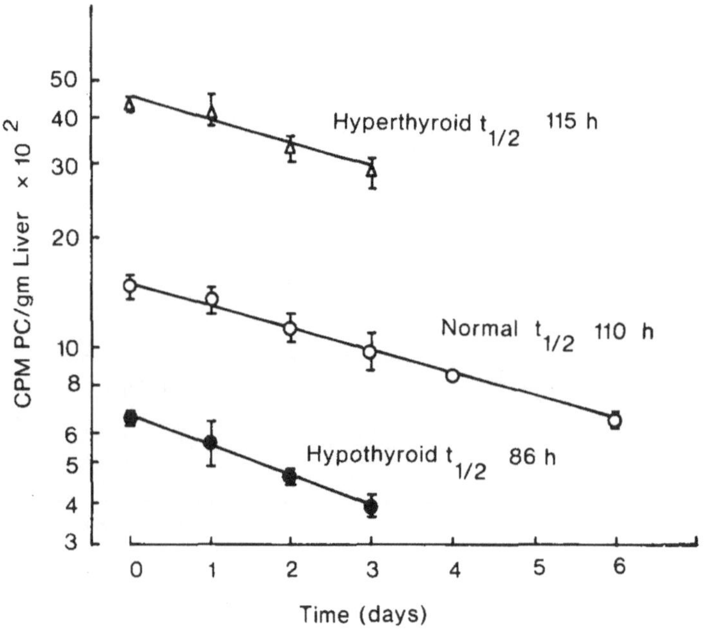

FIGURE 4. Rates of degradation of pyruvate carboxylase in livers from normal, hyperthyroid, and hypothyroid rats. Rats were injected with L[4,5-^{3}H] leucine and the amount of radioactivity present in immunoprecipitable pyruvate carboxylase measured at the times indicated. (From Weinberg, M. B. and Utter, M. F., *J. Biol. Chem.* 254, 9492, 1979. With permission.)

under different physiological conditions which have been reported by different laboratories may also be due to the way in which the activity of pyruvate carboxylase has been related to the gluconeogenic capacity of the tissue or animal. Most results have been expressed in micromoles oxaloacetate formed per minute (units) per gram wet weight of tissue. In some earlier studies, expression of enzyme activity in units per milligram protein may have been misleading because the amount of protein extracted may vary between different physiological states. In order to relate the amount of pyruvate carboxylase to the total capacity of a given tissue to synthesize glucose or lipids, the activity of the enzyme has been related to total body weight[945] or the DNA content[945,976] of the tissue.

2. *Adult Liver and Kidney*

a. *Turnover Time of Pyruvate Carboxylase*

In the liver of the adult rat, the half-time for the turnover of pyruvate carboxylase has been estimated to be 4.6 days (110 hr)[944,945] (Figure 4). This is slightly longer than the value of 3.8 days for the average turnover time of mitochondrial protein, but significantly shorter than that of 8.1 days for pyruvate dehydrogenase,[944] the concentration of which does not appear to change under different physiological conditions.[944,945] The turnover time of apopyruvate carboxylase is similar to that for the holoenzyme.[288] The rate constants for the synthesis and degradation of rat liver holopyruvate carboxylase have been estimated to be 44μg of protein per milligram of DNA per day and 0.15 days^{-1}, respectively.[944]

b. *Starvation and Refeeding*

In most species investigated, the total activity of hepatic pyruvate carboxylase (generally expressed as units per gram wet weight of liver or per milligram of extracted

Table 1
EFFECTS OF STARVATION AND DIABETES ON THE TOTAL AMOUNT OF PYRUVATE CARBOXYLASE ACTIVITY IN THE LIVERS OF RATS

	Pyruvate carboxylase activity (μmol/min at 30°C)	
Condition of rat	Per gram of liver	Per 100 μg DNA
Normal fed	3.45 ± 0.15	1.85 ± 0.10
Normal starved	6.95 ± 0.42[a]	2.60 ± 0.17[b]
Alloxan-diabetic fed	13.05 ± 0.31[a]	5.14 ± 0.15[a]
Streptozotocin-diabetic fed	10.25 ± 0.66[a]	5.49 ± 0.48[a]

[a] $p < 0.001$.
[b] $p < 0.01$.

From Wimhurst, J. M. and Manchester, K. L., *Biochem. J.*, 120, 95, 1970. With permission.

protein) increases two- to three-fold after a period of starvation (Table 1). These species include the rat[286,397,770,805,808,976] (but see Krebs[472] and Filsell et al.[278]), sheep,[278,862] cow[56] guinea pig,[805,808] and pigeon.[805] The activity of hepatic (and renal) pyruvate carboxylase has also been reported to increase in dogs fed a carbohydrate-free diet.[91] In the rabbit[305,673] and chicken,[776] no change in the total amount of hepatic pyruvate carboxylase is observed on starvation. Little information is available on the effects of refeeding starved animals on hepatic pyruvate carboxylase activity. The refeeding of starved chickens results in a decrease in the total amount of enzyme activity present.[776]

In the kidney of the rat,[278] sheep,[278] and chicken,[776] a small increase in total pyruvate carboxylase activity is observed on starvation, although some authors did not detect a change in activity in the starved state.[3,397] After a period of refeeding, pyruvate carboxylase activity in the kidney of the chicken was found to return to control levels[776] and that in rats refed in a high-protein diet to values below control levels.[3]

c. Diabetes

Experimental diabetes is associated with an increase of two- to fourfold (Table 1) in the total amount of pyrvuate carboxylase in rat liver[171,286,397,661,829,918,945,947,976] and kidney,[829] and in sheep liver.[278,862] Increased amounts of the enzyme in the livers of genetically obese rodents, compared to livers of lean controls, have also been reported.[197,401] No change in total pyruvate carboxylase activity was observed in the kidneys of diabetic sheep.[278] As in the case of starvation, no change in pyruvate carboxylase activity was observed in the livers of diabetic rabbits.[673] It has been suggested that the absence of a change in hepatic pyruvate carboxylase in the diabetic rat reported by some authors[126,263,472,991] may have been due, at least in part, to the method employed for the assay of pyruvate carboxylase activity in liver homogenates.[945]

In a reinvestigation of the question of whether diabetes alters the intracellular concentration of hepatic pyruvate carboxylase, Weinberg and Utter[945] have shown that in the diabetic rat, a two-fold increase in the total activity of hepatic pyruvate carboxylase in extracts of mitochondria is accompanied by a similar increase in the total amount of immunoprecipitable enzyme protein (Table 2). This was shown to be due to a 2- to 2.5-fold increase in the rate of synthesis of the enzyme with little change in the rate of degradation (Table 2). In comparison, no change in the rate of synthesis of pyruvate dehydrogenase was observed in the diabetic liver (Table 2).

In contrast to the increase in the total amount of pyruvate carboxylase activity, the rate of pyruvate carboxylation catalyzed by intact mitochondria isolated from the livers of diabetic rats has been reported to be lower than that in control mitochondria.[556] This may be due to other factors (e.g., the intramitochondrial concentration of acetyl-CoA) which may limit the rate of pyruvate carboxylation when this is assayed in isolated mitochondria.

d. Hormonal Mechanisms in Starvation and Diabetes

In contrast to knowledge of the hormonal regulation of the intracellular concentration of PEP-carboxykinase (reviewed by Tilghman et al.[874]), the hormonal mechanisms which regulate the total amount of pyruvate carboxylase activity in liver and kidney during starvation, refeeding, or in diabetes have not been clearly defined. It is likely that changes in the plasma concentrations of glucagon, insulin, and the glucocorticoids play an important part, but there is little evidence to support this. Increases in the activity of hepatic pyruvate carboxylase during the starvation of rats have been correlated with increases in the plasma concentration of glucagon and glucocorticoids.[770] Although the administration of glucocorticoids to fed sheep,[278] rabbits,[673] and rats[643,829] has been reported to yield no change or a decrease in pyruvate carboxylase in the liver and/or kidney, these hormones have been shown to induce an increase in pyruvate carboxylase in adrenalectomized rats.[397,834] There is one report[395] that the administration of cortisol to fed rats increased the concentration of pyruvate carboxylase in rat liver.

The increase in pyruvate carboxylase activity in experimental diabetes is likely to be due to an increase in the ratio of the plasma glucagon/insulin concentrations. However, attempts to demonstrate this have proven inconclusive. In one test of this proposal, the treatment of diabetic rats with insulin has been reported to lead to a decrease in hepatic pyruvate carboxylase activity to values present in control livers[661] or to a much smaller decrease,[171] or no change[945] in enzyme activity. Weinberg and Utter[945] have suggested that the failure of treatment of diabetic rats with insulin for a relatively short period of time (5 days) to completely reverse the effects of diabetes may be due to the slow turnover rate of pyruvate carboxylase.

e. Thyroid Hormones

Pronounced changes in the amount of hepatic pyruvate carboxylase are associated with alterations in the plasma concentrations of the thyroid hormones. The administration of thyroxin to normal rats (hyperthyroid state) leads to a two-fold increase in the activity of pyruvate carboxylase in extracts of the liver while removal of the thyroid gland (hypothyroid state) decreases enzyme activity to about half that present in control livers.[117,560,579,944] Administration of thyroid hormone to the hypothyroid rat restores the amount of pyruvate carboxylase activity to a value equal to, or slightly greater than control values.[117,944]

These alterations in enzyme activity are associated with parallel changes in the total amount of pyruvate carboxylase protein.[944] Weinberg and Utter[944] have shown that increases in pyruvate carboxylase activity which follow the administration of L-thyroxine to hypothyroid or normal rats are due to an increase in the rate of synthesis of pyruvate carboxylase. Little change in the rate of degradation of the enzyme was observed (Figure 4). The observation that the increase in pyruvate carboxylase activity, which follows the administration of thyroid hormone to thyroidectomized rats, is inhibited by actinomycin C[117] is consistent with the conclusion that the increase in pyruvate carboxylase activity is the result of *de novo* synthesis of the enzyme. Thus, hepatic

Table 2
EFFECT OF STREPTOZOTOCIN-INDUCED DIABETES ON THE TURNOVER OF PYRUVATE CARBOXYLASE AND PYRUVATE DEHYDROGENASE IN RAT LIVER

Enzyme	Rats	Enzyme protein (μg/mg of liver DNA)	Enzyme activity (units/ mg of liver DNA)	Apparent rate constant of synthesis (μg of protein/mg DNA per day)	Apparent rate constant of degradation ($days^{-1}$)	Half-life (days)
Pyruvate carboxylase	Normal	225 ± 23	5.5 ± 0.7	34.1	0.15	4.6
	Diabetic	417 ± 25	10.2 ± 1.4	64.1	0.15	4.5
	Insulin-treated diabetic	327 ± 21	7.5 ± 0.7			
Pyruvate dehydrogenase	Normal	3.3 ± 0.2	2.1 ± 0.2	3.6	0.085	8.1
	Diabetic	4.1 ± 0.5	2.4 ± 0.1	4.8	0.093	7.5

From Weinberg, M. B. and Utter, M. F., *Biochem. J.*, 188, 601, 1980. With permission.

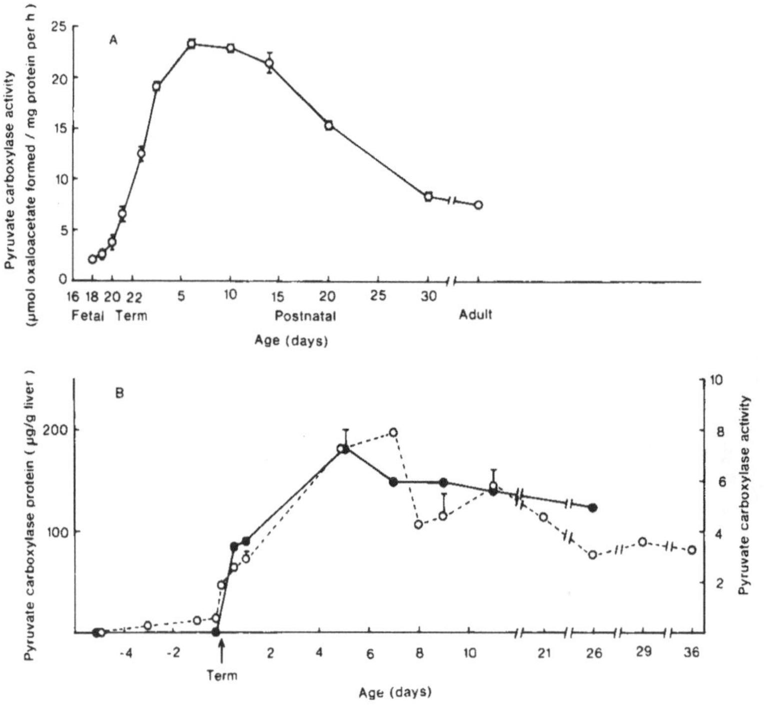

FIGURE 5. Changes in pyruvate carboxylase activity and the amount of enzyme protein in rat liver during neonatal development. (A) The increase and decline in the total amount of pyruvate carboxylase activity measured in homogenates of rat liver during neonatal development. (From Yeung, D., Stanley, R. S., and Oliver, I. T., *Biochem. J.*, 105, 1219, 1967. With permission.) (B) Comparison of the total amount of immunoprecipitable pyruvate carboxylase protein (●) with the total amount of enzyme activity (○) during neonatal development of the rat (From Chang, L. O., *Pediatr. Res.*, 11, 6, 1977. With permission.)

pyruvate carboxylase appears to be one of a number of specific enzymes, the synthesis of which is induced by thyroid hormones (reviewed by Sterling[822]).

f. Correlation with Rates of Gluconeogenesis

Changes in the total amount of hepatic pyruvate carboxylase activity are consistent with observed alterations in the rate of gluconeogenesis from three- or four-carbon precursors in starvation,[709] experimental diabetes,[621] and the hyperthyroid state[285,560,587] (but see Bartels and Sestoft[80]). Therefore, it may be concluded that changes in the amount of enzyme protein, which appear to be principally due to alterations in the rate of synthesis with little change in the rate of degradation[944,945] constitute a long-term mechanism for the regulation of gluconeogenesis under these physiological conditions.

3. Developing Liver and Kidney

Rates of gluconeogenesis in a variety of mammalian species[58,649,936,992] have been shown to be low in the fetal liver but to increase markedly within a few hours after birth. Early in the gestation period the total amount of pyruvate carboxylase detected in mitochondrial extracts from fetal livers is very low but rises slowly as gestation progresses (Figure 5). A rapid increase in activity occurs just before birth (Figure 5). During the early neonatal period the amount of enzyme activity continues to increase to a maximum at about 5 days after birth. At this time, the activity is about two- to three-fold greater than that present in adult liver (Figure 5).

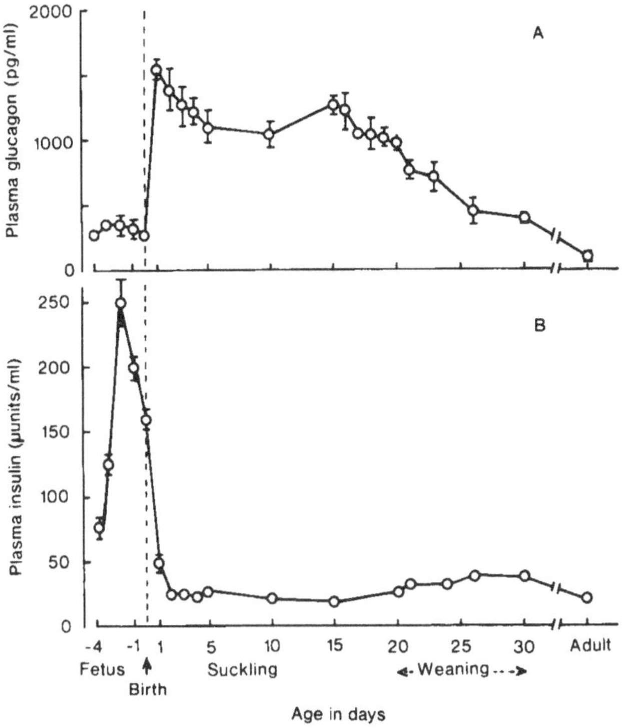

FIGURE 6. The concentration of glucagon (A) and insulin (B) in the plasma of the rat during development. (From Girard, J. R., Ferre, P., Kervran, A., Pegorier, J. P., and Assan, R., in *Glucagon: Its Role in Physiology and Clinical Medicine,* Foa, P. P., Bajaj, J. S., and Foa, N. L., Eds., Springer-Verlag, New York, 1977, 563. With permission.)

This general pattern (Figure 5) has been observed in livers of the developing rat,[52,172,993] guinea pig,[433,685] rabbit,[148] pig,[391,566] dog,[184] cow,[662,663] sheep,[255] and man.[531] In the guinea pig, the concentration of pyruvate carboxylase reached in the period before birth is higher than it is in other species at this stage of development.[433] Chang[172] has shown that changes in total pyruvate carboxylase activity in neonatal rat liver are associated with similar changes in the quantities of immunoprecipitable pyruvate carboxylase (Figure 5B), confirming that the increase and decline in enzyme activity does indeed reflect changes in the total amount of enzyme protein.

The increase in the total amount of hepatic pyruvate carboxylase during the neonatal period correlates with an increase in the rate of [^{14}C]HCO_3^- fixation by intact mitochondria.[184,799] Such a correlation would be expected if the increase in the amount of enzyme is part of the mechanism by which gluconeogenesis increases during the neonatal period.

In the livers of rats and guinea pigs born prematurely, the marked increase in pyruvate carboxylase activity normally observed at birth (Figure 5) does not occur, and the total amount of pyruvate carboxylase activity remains significantly lower than normal for up to 72 hr after birth.[172,692,699] It has been proposed that this may be primarily responsible for hypoglycemia in the neonatal guinea pig.[699]

The mechanism which mediates the increase in hepatic pyruvate carboxylase activity at birth has not been clearly defined, but is likely to involve the marked changes in the plasma concentrations of glucagon and insulin which occur during this period (Figure 6).[316] There is some evidence that glucocorticoids are not responsible for the rapid

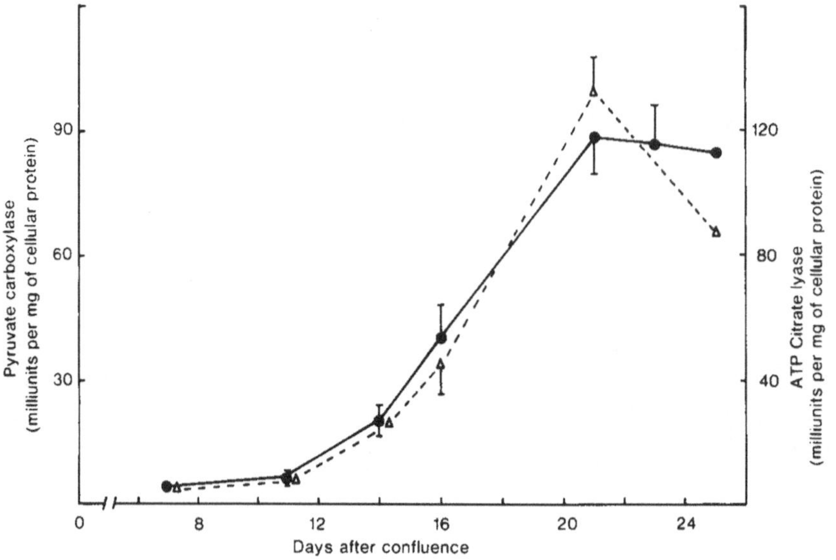

FIGURE 7. Increases in the total amount of pyruvate carboxylase (●) and ATP-citrate lyase (Δ) activities in 3T3-L1 cells during differentiation. (From Mackall, J. C. and Lane, M. D., *Biochem. Biophys. Res. Commun.*, 79, 720, 1977. With permission.)

increase in pyruvate carboxylase activity at birth,[993] although a role for these hormones has not been eliminated.

In contrast to the liver, total pyruvate carboxylase activity in the kidneys of sheep, guinea pigs, and chickens increases slowly during gestation and the early neonatal period. During the neonatal period, the amount of pyruvate carboxylase present in the kidney is much lower than that in kidneys of the adult. Only a small increase is observed at birth,[256,433,685,686,775] in comparison with the sharp rise in enzyme activity observed in the liver.

In addition to changes in the activity of pyruvate carboxylase in the livers of animals during the neonatal period, increases in total hepatic pyruvate carboxylase activity have also been reported in cows and rats during pregnancy[278] and lactation,[56,797] although Smith[797] has reported a decrease in hepatic pyruvate carboxylase activity in the pregnant rat. These changes are consistent with evidence which indicates that the rate of hepatic gluconeogenesis increases during pregnancy.[402] Furthermore, the administration of estradiol to rats has been shown to increase the activity of hepatic pyruvate carboxylase.[213]

4. *Adipose Tissue*

One of the best experimental systems in which changes in the total amount of pyruvate carboxylase activity have been studied is the 3T3-L1 preadipocyte cell line. In the presence of insulin, glucocorticoids, and other factors, 3T3-L1 cells differentiate to form cells with many of the morphological and biological characteristics of mammalian adipocytes.[334,481] Differentiation is associated with the accumulation of triglyceride and increases in the amount of several enzymes of the lipogenic pathway.[334,520] The total amount of pyruvate carboxylase activity (expressed as units per milligram cellular protein) increases by about 20-fold (Figure 7).[198,289,290,518] Precipitation of pyruvate carboxylase with antibodies and quantitative polyacrylamide gel electrophoresis (PAGE) have been used to show that this increase is due to an increase in the amount of pyruvate carboxylase protein.[289] Angus et al.[16] have presented evidence to indicate that these changes are due to an increase in the concentration of translatable pyruvate

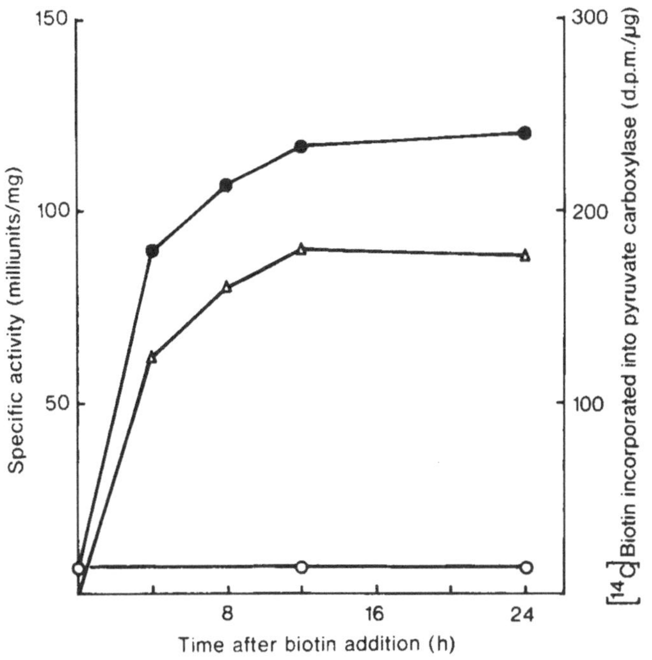

FIGURE 8. Effect of biotin on the recovery of pyruvate carboxylase activity in biotin-deficient 3T3-L1 cells. Cells treated with avidin were incubated for a period of 16 days after confluency. At this time, [^{14}C]-biotin was added to the culture medium of the cells under test. At the times indicated, the cells were harvested and the specific activity of pyruvate carboxylase (●,○), and the amount of [^{14}C]-biotin (Δ) associated with immunopreciptiable pyruvate carboxylase protein were determined; (●), the specific activity of pyruvate carboxylase in cells incubated in the presence of biotin; (○), the specific activity of the enzyme in cells incubated in the absence of added biotin. (From Freytag, S. O. and Utter, M. F., *Proc. Natl. Acad. Sci. U.S.A.*, 77, 1321, 1980. With permission.)

carboxylase messenger RNA, and hence an increase in the rate of synthesis of the enzyme.

When 3T3-L1 preadipocytes were permitted to differentiate in the presence of avidin, an increase in the amount of pyruvate carboxylase protein, but no change in enzyme activity was observed.[289] Subsequent addition of excess biotin led to an increase in the specific activity of the enzyme (Figure 8), indicating that in the absence of biotin the apoenzyme of pyruvate carboxylase accumulates in the cell.[289]

Relatively little is known about possible changes in the amount of pyruvate carboxylase activity present in adipose tissue in vivo. Under conditions in which lipogenesis from glucose and acetate in adipose tissue of sheep was increased by elevating the proportion of solid carbohydrate in the diet, no increase in adipose tissue pyruvate carboxylase was observed, in contrast to large increases in ATP-citrate lyase and NADP-malate dehydrogenase.[50] On the other hand, total pyruvate carboxylase activity was found to be increased in adipose tissue from limit-fed genetically obese Zucker rats, although the amount of enzyme present in adipose tissue from obese rats fed a normal diet did not differ from that present in lean litter mates.[812]

C. Control of Enzyme Activity

1. Intramitochondrial Concentration of Pyruvate

In the previous section alterations in the total amount of enzyme activity arising from a change in the rate of synthesis, or possibly degradation, of pyruvate carboxylase were discussed. Mechanisms which modify the activity of the enzyme without changing the total amount of enzyme protein will now be examined. These processes are rapid in onset and chiefly involve changes in the concentration of substrates, products, or metabolites which are modifiers of enzyme activity. Changes in the concentration of the substrate, pyruvate, will be considered first.

The concentration of pyruvate would influence the activity of pyruvate carboxylase if the steady-state intramitochondrial concentration of this substrate fluctuated around a value equal to the apparent K_m of pyruvate carboxylase for pyruvate. This possibility has been tested experimentally by (1) comparing the estimated intramitochondrial concentration of pyruvate with the apparent K_m of the isolated enzyme for this substrate and (2) measuring the rates of gluconeogenesis or pyruvate carboxylation under conditions in which the rate of pyruvate transport across the mitochondrial inner membrane is altered.

Interpretation of the apparent K_m values of purified pyruvate carboxylase enzymes for pyruvate in physiological terms is complicated by the observation, discussed in Chapter 4, that double reciprocal plots of the rate of oxaloacetate formation as a function of pyruvate concentration for pyruvate carboxylase from sheep kidney,[251,860] rat liver,[545,773] and chicken liver[897] are nonlinear and best described by two lines with different slopes (Figure 3). From the linear plot obtained at low pyruvate concentrations, the apparent K_m value for pyruvate is estimated to be in the range of 0.05 to 0.3 m*M*. On the other hand, using the complete curves of velocity as a function of pyruvate different slopes (Chapter 4, Figure 3). From the linear plot obtained at low pyruvate concentration, the apparent K_m value for pyruvate is estimated to be in the range of 0.05 to 0.3 m*M*. On the other hand, using the complete curves of velocity as a function of pyruvate concentration, the concentration of pyruvate which gives half-maximal stimulation of the rate of oxaloacetate formation is much higher, of the order 1 to 10 m*M*. Thus, in physiological terms, the activity of pyruvate carboxylase might be expected to respond to changes in pyruvate concentration over a wide range (0.05 to 10 m*M*).

The intramitochondrial concentration of pyruvate can be estimated to be in the range of 0.05 to 1 m*M* from data for the amounts of pyruvate present in freeze-clamped livers,[45,54,339,897] assuming that pyruvate is evenly distributed within the cell. Recently, Siess et al.[785] have made a more precise measurement of the mitochondrial pyruvate concentration by measuring the amount of pyruvate present in the mitochondrial fraction obtained after the disruption of hepatocytes with digitonin. For cells prepared from the livers of starved rats and incubated in the presence of lactate, the intramitochondrial concentration of pyruvate was found to be 1 m*M*. Thus, the estimated values for the intramitochondrial concentration of pyruvate are within the range which is expected to alter the activity of pyruvate carboxylase.

In the second approach to the question of whether the intramitochondrial concentration of pyruvate contributes to regulation of the activity of pyruvate carboxylase in vivo, the rate of delivery of pyruvate to the mitochondrial matrix has been altered by the action of hormones,[4,386] or by the use of inhibitors of mitochondrial pyruvate transport.[363,873]

Adam and Haynes[4] showed that the administration of glucagon to rats leads to stimulation of both pyruvate carboxylation and decarboxylation in intact mitochondria subsequently isolated from the liver. It was also shown that the apparent K_m for extramitochondrial pyruvate of pyruvate carboxylation in intact mitochondria (0.04 to 0.08 m*M*) is lower than that (0.22 m*M*) observed when the rate of the reaction is measured in disrupted mitochondria. These results suggested that the delivery of pyruvate to the mitochondrial matrix may be a rate-limiting step in pyruvate carboxylation.[4,386] The observation that α-cyano-4-hydroxycinnamate, an inhibitor of mitochondrial pyruvate transport, inhibits pyruvate carboxylation in intact liver mitochondria, but has little effect on the activity of isolated pyruvate carboxylase[363] is consistent with this conclusion.

In a more rigorous investigation, Thomas and Halestrap[873] have used α-cyano-4-hydroxycinnamate to examine the role of mitochondrial pyruvate transport in the regulation of gluconeogenesis in the liver. Measurement of the rate of gluconeogenesis from lactate in isolated liver cells at different concentrations of α-cyano-4-hydroxycinnamate led to the conclusion that the transport of pyruvate from the cytosol to the mitochondrial matrix is a rate-limiting step in the conversion of three-carbon precursors to glucose. This is consistent with the argument developed above that in the mitochondria the reaction catalyzed by pyruvate carboxylase is limited by a steady-state intramitochondrial concentration of pyruvate which is low relative to the apparent K_m of the enzyme for pyruvate. However, Thomas and Halestrap[873] have pointed out that a situation in which gluconeogenesis is limited by the activities of pyruvate kinase and/or PEP carboxykinase is also compatible with their data.

2. Adenine Nucleotides

In addition to the requirement of pyruvate carboxylase for $MgATP^{2-}$ as a substrate, steady-state kinetic studies have shown that the enzyme is inhibited by the reaction product, $MgADP^-$ (Chapter 4, Table 1). When assayed under more physiological conditions (37°C, pH 7.3, and the presence of 85 m*M* KCl), the degree of inhibition of isolated pyruvate carboxylase by $MgADP^-$ was found to depend on the concentration of pyruvate.[543] Significant inhibition was observed at 5 m*M* pyruvate, but $MgADP^-$ had a much smaller effect at 0.1 m*M* pyruvate.[543]

Walter and co-workers have conducted a detailed study of the effects of changes in the concentration of ATP and ADP on the rate of pyruvate carboxylation in isolated rat liver mitochondria.[123,833,917,931] These workers have shown that under a variety of experimental conditions designed to alter the intramitochondrial concentration of ATP and ADP, changes in the rate of pyruvate carboxylation are positively correlated with changes in the ratio of intramitochondrial ATP:ADP (Figure 9). On the basis of these observations, Walter and colleagues have proposed that the ratio of intramitochondrial ATP:ADP is an important parameter which regulates the activity of pyruvate carboxylase under physiological conditions.[123,833,917,931]

In similar experiments, Davis-van Thienen and Davis[215] employed a range of steady-state pyruvate concentrations and showed that the effects of changes in the ratio of ATP:ADP (altered by varying the concentration of an exogenous purified mitochondrial F_1 ATPase) on flux through pyruvate carboxylase in isolated rat liver mitochondria are most pronounced at low extramitochondrial pyruvate conentrations and in the absence of an exogenous source of acetyl-CoA. Using a slightly different approach, it has been shown that the rate of pyruvate carboxylation in isolated rat liver mitochondria is decreased under conditions of increased demand for intramitochondrial ATP imposed either by the addition of hexokinase[669] or by promotion of the activity of carbamoyl phosphate synthetase,[138] which also utilizes ATP as a substrate.

Two other groups of experiments have provided further evidence which indicates that in liver mitochondria pyruvate carboxylase activity is regulated in part by the steady-state concentration of ATP. In the first, the stimulation of pyruvate carboxylation by glucagon has been shown to involve an increase in the concentration of intramitochondrial ATP.[78,137,166,664,879] As discussed below, it has been proposed that this leads directly to an activation of pyruvate carboxylation.

The second group of experiments concerns changes in liver mitochondria of the newborn. During the first few hours after birth, the total adenine nucleotide content and the concentration of ATP in the matrix of intact mitochondria isolated from the livers of the rat and rabbit increases markedly,[18,19,21,127,593,711,844] and is associated with an

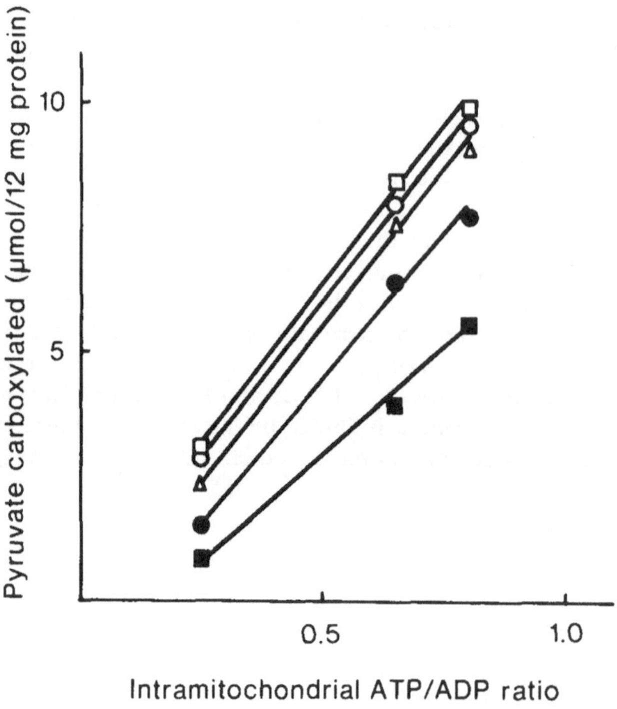

FIGURE 9. Effect of increases in the ratio of intramitochondrial ATP:ADP concentrations on the rate of pyruvate carboxylation in isolated rat liver mitochondria at different fixed concentrations of acetyl-CoA. The concentrations of acetyl-CoA were (□) 0.5, (○) 0.42, (△) 0.33, (●) 0.25, and (■) 0.17 nmol/mg protein. The concentrations of adenine nucleotides and acetyl-CoA were varied by the addition of creatine kinase plus creatine, acetyl carnitine, or arsenite. (From Von Glutz, G. and Walter, P., *FEBS Lett.*, 72, 299, 1976. With permission.)

increase in the rate of pyruvate carboxylation (Figure 10)[21,127,711] and gluconeogenesis.[127]

This has been attributed to an increase in the intramitochondrial concentration of ATP.[21,127,711] Consistent with this conclusion is the observation that an increase in the rate of pyruvate carboxylation was also observed when the total adenine nucleotide content of mitochondria isolated from rat liver at birth was increased by incubation of mitochondria with ATP in vitro.[21]

From the results of the different types of experiments described it may be concluded that changes in the concentrations of ATP and ADP in liver mitochondria contribute to regulation of the activity of pyruvate carboxylase under some physiological conditions. These include the stimulation of gluconeogenesis by hormones (associated with increased turnover of ATP and ADP) and marked changes in the intramitochondrial concentration of ATP such as those which occur immediately following birth. Under other physiological conditions in which the steady-state concentrations of the nucleotides do not change markedly, the roles of ATP and ADP in the regulation of pyruvate carboxylation appear to be less clear.

e. Acetyl-CoA

a. Association of Elevated Cellular Concentrations of Acetyl-CoA with Enhanced Gluconeogenesis and Pyruvate Carboxylase Activity

On the basis of the discovery that acetyl-CoA stimulates the activity of isolated

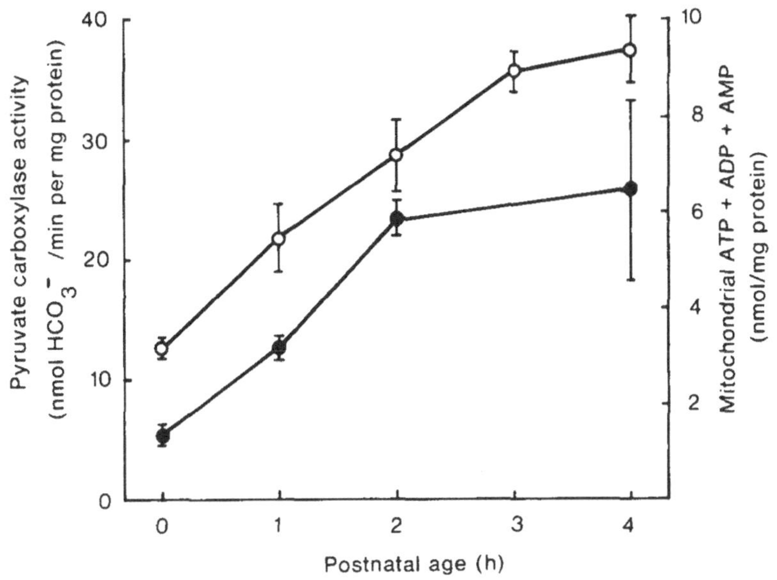

FIGURE 10. Changes in pyruvate carboxylase activity (●) and total adenine nucleotide content (O) in mitochondria isolated from rats of different postnatal ages. Rat pups were delivered by Caesarean section on day 21 of gestation and either killed immediately (0 hr) or placed in a humidified incubator at 32°C and killed at the times indicated. (From Aprille, J. R., Yaswen, P., and Rulfs, J., *Biochem. Biophys. Acta*, 675, 143, 1981. With permission.)

chicken liver pyruvate carboxylase, Keech and Utter[439] proposed that this acyl-CoA derivative is a major physiological regulator of the activity of pyruvate carboxylase. It was further pointed out that acetyl-CoA would provide a link between enhanced fatty acid oxidation and increased of gluconeogenesis (Figure 1).[439,901] Soon after the initial discovery of the activation of pyruvate carboxylase by acetyl-CoA, Kornberg[460] proposed that this provides a mechanism by which the supply of oxaloacetate to the citric acid oxidation and increased rates of gluconeogenesis (Figure 1).[439,901] Soon after the initial discovery of the activation of pyruvate carboxylase by acetyl-CoA, Kornberg[460] proposed that this provides a mechanism by which the supply of oxaloacetate to the citric acid cycle is coordinated with the availability of acetyl-CoA. This is presumably true for pyruvate carboxylase in both the gluconeogenic and lipogenic tissues, as well as in other vertebrate tissues and in invertebrates where the enzyme has a more general anaplerotic role (Sections II and III).

Since the proposal that acetyl-CoA acts as a physiological regulator of pyruvate carboxylase was first made,[439] it has taken some time to test whether the activity of the enzyme is regulated by acetyl-CoA *in situ*. Although some details are still not fully explained, there is now a large body of evidence consistent with the conclusion that changes in acetyl-CoA concentration do play a major role in the regulation of pyruvate carboxylase activity in cells of gluconeogenic tissues. In contrast, relatively little is known of the role of acetyl-CoA in the regulation of pyruvate carboxylase activity in adipocytes where the enzyme participates principally in lipogenesis.[51,459,532,696]

In searching for a physiological connection between the observation that isolated pyruvate carboxylase is activated by acetyl-CoA and the mechanisms which regulate the rate of gluconeogenesis, Krebs et al.,[477] using slices of kidney cortex, were one of the first groups to show that metabolites which are precursors of acetyl-CoA, including acetate, acetoacetate, and short-chain fatty acids, stimulate the rate of gluconeogenesis while increasing the tissue concentration of acetyl-CoA. Subsequently, a large number of studies have shown that the perfusion of livers,[559,630,807,828,865,960,963-966,968,1000] incubation of isolated hepatocytes[129,432,475,788] or kidney cortex slices[703] in the presence of fatty acids or their carnitine derivatives, or the administration of triacylglycerol to newborn rats[277] stimulates the synthesis of glucose from lactate or pyruvate and raises the total tissue concentration of acetyl-CoA.[964,965] In other experiments, 4-pentenoic acid[277,560,880,965,970] or branched-chain α-keto acids,[925] which inhibit the formation of acetyl-CoA from fatty acids or pyruvate, respectively, were found to decrease the rate of gluconeogenesis from pyruvate and lactate in perfused livers, newborn cats, or isolated hepatocytes. This change was associated with a decrease in the tissue concentration of acetyl-CoA.[880,925,970] In contrast to these, the stimulation by glucagon of the gluconeogenic pathway is associated with little or no change in the concentration of acetyl-CoA[4,129,785,964] (Section I.C.6). These results indicate that there is a correlation between the rate of gluconeogenesis and the total tissue concentration of acetyl-CoA.

b. Discrepancy Between Measured Concentrations of Intramitochondrial Acetyl-CoA and Apparent K_a for Pyruvate Carboxylase

The intramitochondrial concentration of acetyl-CoA in liver cells has been calculated to be in range of 0.05 to 1 m*M* using data obtained for intact liver tissue or isolated mitochondria.[79,917,925,931,963] In a more direct estimation of the concentration of mitochondrial acetyl-CoA *in situ*, Siess et al.[785] have reported that the quantity of mitochondrial acetyl-CoA present in isolated hepatocytes, treated with digitonin in order to distinguish between cytosolic and mitochondrial metabolites, is about 0.6 m*M*.

Comparison of the estimated intramitochondrial concentrations of acetyl-CoA with the values of about 2, 20, and 15 μM obtained for the apparent K_a for acetyl-CoA of pyruvate carboxylase purified from chicken liver,[897] rat liver,[87,545] and sheep, liver,[250] respectively, showed that the estimated total concentration of acetyl-CoA in the mitochondrial matrix appears to be about 5 to 10 times higher than the apparent K_a of the purified enzymes for acetyl-CoA.[79,543,917,931,963] This suggested that in these tissues, pyruvate carboxylase would always be saturated with this activator and would be unlikely to respond to changes in the concentration of acetyl-CoA.[79,809,894,897,904,963]

A resolution to this problem was sought by measuring the activity of pyruvate carboxylase and the concentration of acetyl-CoA in intact isolated mitochondria under conditions designed to yield different concentrations of intramitochondrial acetyl-CoA. In the presence of arsenite, which inhibits the formation of acetyl-CoA from pyruvate, and precursors of acetyl-CoA (acetyl-carnitine and fatty acids), the rate of pyruvate carboxylation was shown to increase with increasing concentrations of acetyl-CoA in isolated chicken[79] and rat[917] liver mitochondria (Figure 11). The apparent K_a of this reaction for acetyl-CoA was estimated to be 320 and 210 to 290 μM for mitochondria from chicken[79] and rat[917] liver, respectively. This is about 25- and 10-fold higher, respectively, than the apparent K_a value of the isolated enzymes for this activator.

A double reciprocal plot of the data obtained for the activation of pyruvate carboxylation by acetyl-CoA in intact chicken liver mitochondria was linear (Figure 11B). When the data were expressed in the form of a Hill plot, the slope was found to be 1.5,[79] compared to a value of about 3 for the Hill coefficient for the activation of purified pyruvate carboxylase from chicken liver by acetyl-CoA.[761] It was suggested that this discrepancy may be due to the limited range of acetyl-CoA concentrations employed in studies with intact mitochondria, and the fact that the total concentration of acetyl-CoA rather than the concentration of free acetyl-CoA was measured.[79] However, another explanation may be that the enzyme is not inactivated by dilution[250] in intact mitochondria.

The reasons for the 10- to 20-fold difference between the apparent K_a values for the activation of purified pyruvate carboxylases by acetyl-CoA and concentration of in-

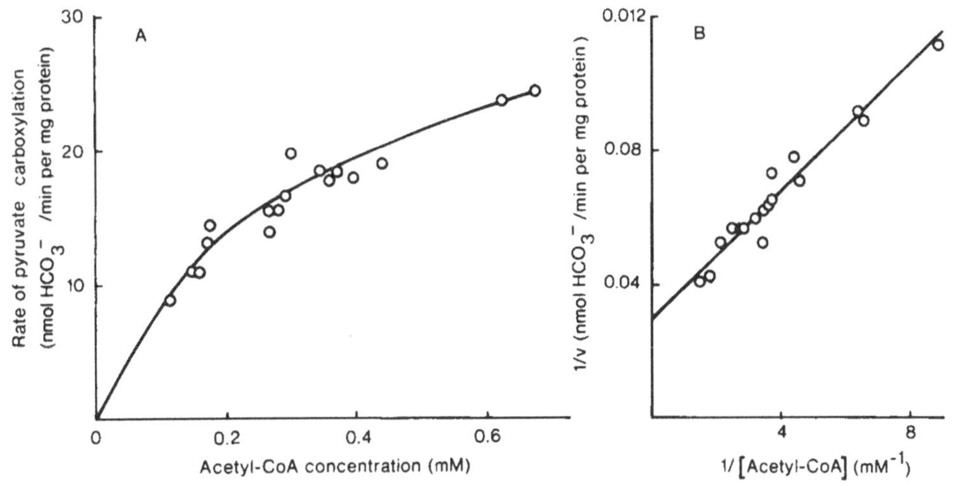

FIGURE 11. (A) The rate of pyruvate carboxylation in isolated chicken liver mitochondria plotted as a function of the measured concentration of intramitochondrial acetyl-CoA. The concentration of acetyl-CoA was varied by the addition of different concentrations of acetyl carnitine to mitochondria incubated in the presence of arsenite. (B) A double reciprocal plot of the data in A. (From Barritt, G. J., Zander, G. L., and Utter, M. F., *Glyconeogenesis*, Hanson, R. W. and Mehlman, M. A., Eds., John Wiley & Sons, New York, 1976, 3. With permission.)

tramitochondrial acetyl-CoA required to give half-maximal activity of pyruvate carboxylation in intact mitochondria have not been fully resolved. Three main explanations have been advanced. These are

1. The proposal that in the mitochondrial matrix the concentration of free acetyl-CoA is significantly less than that of total acetyl-CoA.[79,750,809,904]
2. Observations which suggest that under more physiological conditions of ionic composition and concentration, the apparent K_a of purified pyruvate carboxylase for acetyl-CoA is significantly greater than that for isolated pyruvate carboxylase assayed under standard in vitro assay conditions.[79,744,750,897]
3. The possibility that free acetyl-CoA in the mitochondrial matrix is distributed at different concentrations in more than one compartment.[79,750,917] Each of these explanations will now be briefly considered.

c. Association of Acetyl-CoA with Enzymes of the Mitochondrial Matrix

The concentration of pyruvate carboxylase in the matrix of chicken and rat liver mitochondria has been calculated to about 20 mg/m*l* from the measured amount of pyruvate carboxylase activity in these mitochondria, the specific activity of the purified enzyme, and the estimated volume of the matrix space.[79] Similar calculations have shown that the apparent concentrations of a number of other enzymes of the mitochondrial matrix are also near this value.[79,809] Barritt et al.[79] calculated that significant quantities of acetyl-CoA are likely to be bound to pyruvate carboxylase, citrate synthetase, and hydroxymethylglutaryl-CoA synthetase, so that the estimated concentration of free acetyl-CoA is about ten times lower than that of total acetyl-CoA.[79,750] It was estimated that for chicken liver mitochondria the apparent K_a for acetyl-CoA is about 150 μM when expressed as total acetyl-CoA, compared with a value of 13 μM expressed as free acetyl-CoA (Figure 12).[79,750]

d. Kinetic Properties of Pyruvate Carboxylase Under More Physiological Conditions

The apparent K_a values of purified pyruvate carboxylase enzymes for acetyl-CoA

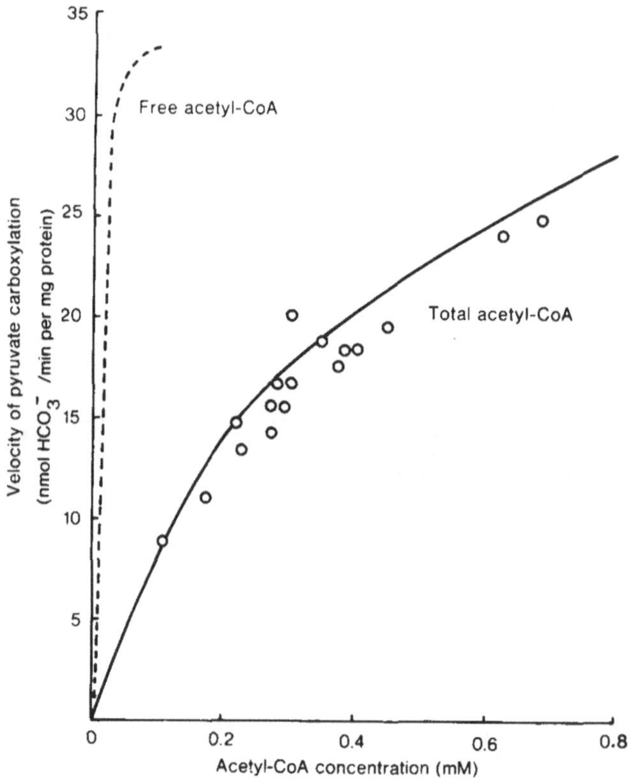

FIGURE 12. Effect of the combination of acetyl-CoA with enzymes of the mitochondrial matrix on the observed K_a value of pyruvate carboxylase for acetyl-CoA. The calculated curve of pyruvate carboxylase activity as a function of the concentration of total acetyl-CoA (—) is compared with data for the rate of pyruvate carboxylation in isolated chicken liver mitochondria plotted as a function of the measured concentration of intramitochondrial acetyl-CoA (O) and the response of purified pyruvate carboxylase to changes in the concentration of free acetyl-CoA (----). Calculations of pyruvate carboxylase activity as a function of total acetyl-CoA concentration were made using the estimated intramitochondrial concentrations of enzymes which bind acetyl-CoA (pyruvate carboxylase, citrate synthetase, acetyl-CoA-acyltransferase, hydroxymethylglutaryl-CoA synthetase, pyruvate dehydrogenase, carnitine palmitoyltransferase, α-ketoglutarate dehydrogenase, and succinyl-CoA synthase) and the apparent K_m values of these enzymes for acetyl-CoA, assuming that the Hill coefficient for the interaction of acetyl-CoA with pyruvate carboxylase is 1.0[750] (From Scrutton, M. C. and Griffiths, J. R., *Short-Term Regulation of Liver Metabolism,* Hue, L. and Van de Werve, G., Eds., Elsevier/North-Holland, New York, 1981, chap 11. With permission.)

have generally been determined at very low concentrations of the enzyme, a pH of about 8, temperatures between 20 and 30°C, low concentrations of monovalent ions, pyruvate concentrations of about 5 m*M*, and in the absence of modifiers of enzyme activity.[250,545,897] These conditions contrast with those likely to be present in the mitochondrial matrix, including high concentrations of pyruvate carboxylase,[79] a pH below 8,[409] a temperature of 37° C, about 100 m*M* K^+ (calculated from the data of Halestrap[360] assuming that the volume of the mitochondrial matrix is about 1 μl/ mg protein[376]), a pyruvate concentration of about 1.0 m*M*,[785] and the presence of other modifiers of enzyme activity. Although the concentration of pyruvate carboxylase in the mitochondrial matrix is much higher than that employed in kinetic studies,[79] no convincing evidence that this causes changes in the kinetic properties of the enzyme has been provided.

McClure and Lardy[543] have shown that at 5 m*M* pyruvate, the apparent K_a of purified rat liver pyruvate carboxylase for acetyl-CoA is 60 μM when measured in the presence of 85 m*M* KCl at pH 7.3 and 37°C compared with a value of 20 μM obtained at 20 m*M* KCl, pH 8.0, and 25°C. Furthermore, Scrutton[744] has shown that for rat liver pyruvate carboxylase, the apparent K_a for acetyl-CoA increases from 50 to 55 μM at 10 m*M* pyruvate to 120 μM at 0.1 m*M* pyruvate. The apparent K_a for acetyl-CoA of chicken liver pyruvate carboxylase has also been shown to be increased by succinyl-CoA and CoA-SH at concentrations likely to be present in the mitochondrial matrix in vivo.[79,894,897] These results indicate that a number of differences between the chemical environment of pyruvate carboxylase employed in studies with the isolated enzyme and that present in the mitochondrial matrix may contribute to the observation that the apparent K_a for acetyl-CoA in intact mitochondria is higher than that observed for purified pyruvate carboxylase.

e. Compartmentation of Acetyl-CoA in the Mitochondrial Matrix

The existence of more than one compartment of acetyl-CoA in the matrix of rat liver mitochondria was first proposed by Delisle and Fritz[220] on the basis of a number of considerations, including the observation that the addition of pyruvate to an incubation medium containing mitochondria caused a decrease in $^{14}CO_2$ formation, but an increase in [^{14}C]-acetoacetate formation from [^{14}C]-palmitate. The results of further experiments conducted by Mullhofer and Kuntzen[580] and Von Glutz and Walter[916] were also interpreted as evidence for the presence of two compartments of acetyl-CoA in rat liver mitochondria.

Other investigations, however, have led to the conclusion that acetyl-CoA is distributed homogeneously within the mitochondrial matrix.[296,511] Interpretation of data from the experiments described by Delisle and Fritz[220] and Von Glutz and Walter[916] may be complicated because factors other than possible compartmentation of acetyl-CoA may have contributed to the final specific activity of the radioactive label in a given metabolite.[511] Thus, it may be concluded on the basis of present knowledge that the possibility that compartmentation of acetyl-CoA in the mitochondrial matrix contributes to differences between the K_a of isolated pyruvate carboxylase for acetyl-CoA, and the much higher concentration of acetyl-CoA observed in the mitochondrial matrix appears to be unlikely.

f. Conclusion

While the question of the apparent discrepancy between the range of acetyl-CoA concentrations which activate isolated pyruvate carboxylase on the one hand and pyruvate carboxylation in intact mitochondria on the other has not been fully resolved, it seems most likely that the binding of acetyl-CoA to enzymes in the mitochondrial matrix and the presence of ions, metabolites, and inhibitors of the enzyme activity which are present in the mitochondrial matrix can account for most of the difference. It is of interest to note that activation of pyruvate carboxylase by acetyl-CoA represents one of the few examples of mammalian regulatory enzymes in which extensive experiments have provided reasonable evidence to indicate that the effects of the apparent "allosteric" modifier (acetyl-CoA) on the purified enzyme do indeed constitute an important physiological mechanism for the regulation of enzyme activity in vivo.

4. Other Metabolites Which may Affect Pyruvate Carboxylase Activity Under Physiological Conditions

A number of acyl derivatives of coenzyme A as well as CoA-SH itself have been

Table 3
OTHER EFFECTORS OF HEPATIC PYRUVATE CARBOXYLASE ACTIVITY: COMPARISON OF THE APPARENT K_a AND K_i VALUES FOR ACTIVATION OR INHIBITION OF ISOLATED PYRUVATE CARBOXYLASE ENZYMES WITH THE ESTIMATED INTRAMITOCHONDRIAL CONCENTRATION OF THE EFFECTOR

Activator	Apparent K_a value[a] (μM)	Estimated concentration in mitochondrial matrix (μM)
Propionyl-CoA	14 (chicken)[761]	5—60[b] (rat liver)[806]
	155 (rat)[744]	400—1200[d] (rat liver)[139,796]
n-Butyryl-CoA	280 (chicken)[761]	
Inhibitor	**Apparent K_i value[a] (μM)**	**Estimated concentration in mitochondrial matrix (μM)**
CoA-SH	500 (chicken)[79,299]	30—100[b] (rat liver, sheep liver, and kidney)[351,800]
		3,100[c] (rat hepatocytes)[785]
		100—2000[d] (rat and chicken liver)[79,139,601,796]
Succinyl-CoA	250 (chicken)[79,299]	5—10[b] (rat liver)[786]
	1580 (rat)[744]	600—1600[d] (rat and chicken liver)[79,139,796]
Malonyl-CoA	470 (chicken)[904]	1—15[b] (rat liver)[352,551-553]
Acetoacetyl-CoA	200 (chicken)[79,299]	50[d] (chicken liver)[79]
	1800 (rat)[748]	
L-Methylmalonyl-CoA	480 (chicken)[299,904]	240[f] (rat liver)[540]
	1770 (rat)[744]	
Hydroxymethyl-glutaryl-CoA	140 (chicken)[299,904]	12—17[g] (rat liver)[444,446]
α-Ketoglutarate	5000—10,000 (chicken)[766]	3200[c] (rat hepatocytes)[129,785]
		7—36[e] (rat liver)[632]
Glutamate	5000—20,000 (chicken and rat)[766]	8000—15,500 (rat hepatocytes)[766,785]
		900—6100[e] (rat liver)[632,962]
Citrate	1000 (chicken and rat)[766]	4800—6000[c] (rat hepatocytes)[766,785]
		1600—12,900[e] (rat liver)[632,962]

[a] The apparent K_a or K_i value is defined as the concentration of activator or inhibitor which gives half-maximal activation or inhibition, respectively, of pyruvate carboxylase activity under the conditions tested.

[b] Estimated from the amount of metabolite present in whole tissue or isolated cells (nmol/g wet wt) assuming that the metabolite is distributed evenly throughout the tissue (1 m*l* volume corresponds to 1 g wet wt).

[c] Measured in the mitochondrial fraction obtained after treatment of isolated hepatocytes with digitonin[785] or after subfractionation in a nonaqueous medium.[803]

[d] Calculated from the measured amount of intramitochondrial metabolite (nmol/mg mitochondrial protein) in an incubation of isolated mitochondria assuming that the volume of the matrix space is 1 μl/mg protein.[376]

[e] The distribution of metabolite between mitochondria and cytoplasm was calculated from the measured concentrations of this and other metabolites using mass action equilibria equations.

[f] The apparent K_m value of methylmalonyl-CoA mutase for methylmalonyl-CoA.

[g] The apparent K_m value of hydroxymethylglutaryl-CoA reductase for hydroxymethylglutaryl-CoA.

shown to alter the activity of purified pyruvate carboxylase. These are summarized in Table 3 of Chapter 4 and in References 79, 761, 894, and 904. Those compounds which may be physiologically important are listed in Table 3.

Concentrations of propionyl-CoA in the livers of nonruminants are comparable with the apparent K_a of pyruvate carboxylase for this activator (Table 3), suggesting that propionyl-CoA, which in nonruminants is derived from the oxidation of short-chain fatty acids, may play a role in the regulation of pyruvate carboxylase activity. Williamson et al.[964] have proposed that the ratio of propionyl-CoA to methylmalonyl-CoA, an intermediate formed in the conversion of propionyl-CoA to succinyl-CoA, may contribute to the regulation of pyruvate carboxylase under conditions where the concentration of acetyl-CoA remains constant. If the apparent K_m value of the enzyme methylmalonyl-CoA mutase for methylmalonyl-CoA reflects the mitochondrial concentration of this metabolite, it would appear that the concentration of methylmalonyl-CoA is comparable with the apparent K_i value for inhibition of pyruvate carboxylase (Table 3) and hence that methylmalonyl-CoA can act as an inhibitor of pyruvate carboxylase in vivo. The observations that propionate inhibits gluconeogenesis from pyruvate in the perfused rat liver[167] and induces hypoglycemia in patients with a congenital deficiency of methylmalonyl-CoA mutase provides further evidence which is consistent with this conclusion.

In the ruminant liver, propionate and butyrate, derived from fermentation reactions in the rumen, are readily converted to propionyl-CoA and butyryl-CoA, respectively.[97] Although little is known about the intramitochondrial concentration of these compounds in the ruminant liver, it has been proposed that they are physiological activators of pyruvate carboxylase in the livers of these animals.[97]

Avian liver pyruvate carboxylase appears to be considerably more sensitive to inhibition by acyl derivatives of coenzyme A than the enzyme from rat liver.[543,748] Comparison of the apparent K_i values of purified pyruvate carboxylase enzymes for CoA-SH, succinyl-CoA, malonyl-CoA, acetoacetyl-CoA, and hydroxymethylglutaryl-CoA with the estimated total intramitochondrial concentrations of these compounds (Table 3) indicates that CoA-SH and succinyl-CoA are the only metabolites in this group which are likely to be physiologically important inhibitors of pyruvate carboxylase in vivo. Barritt et al.[79] have drawn attention to some problems associated with the measurement of CoA-SH and succinyl-CoA in tissue or mitochondrial extracts, and have suggested that estimates of the intramitochondrial concentrations of these metabolites, particularly CoA-SH, should be interpreted with caution. These authors have also shown by calculation that significant amounts of CoA-SH and succinyl-CoA may be bound to enzymes in the mitochondrial matrix. However, the concentrations of free CoA-SH and succinyl-CoA, which have been estimated to be 50 and 90%, respectively, of the total concentrations,[79] are comparable with the apparent K_i values for inhibition of isolated pyruvate carboxylase.

It has been suggested that succinyl-CoA may play a role in controlling the rate of pyruvate carboxylation in relation to the requirement for the anaplerotic synthesis of four-carbon intermediates from three-carbon precursors.[79] More direct experimental support for the possible regulation of pyruvate carboxylase by succinyl-CoA has come from the work of Siess et al.,[786] who showed that glucagon causes a two-fold decrease in the content of succinyl-CoA in rat livers or isolated hepatocytes. It was proposed that this may contribute to the stimulation of pyruvate carboxylation by glucagon.

While the inhibition of pyruvate carboxylase by malonyl-CoA would theoretically provide a feed-back mechanism to control the rate of transfer of acetyl-CoA from mitochondria to the cytoplasm during lipogenesis,[79] the concentration of malonyl-CoA in the mitochondrial matrix appears to be too low to inhibit pyruvate carboxylase (Table 3). Moreover, no mechanism for the rapid transport of this acyl-CoA from the cytoplasm to the mitochondrial matrix has been described.

The measured concentration of mitochondrial acetoacetyl-CoA is much lower than its apparent K_i value (Table 3), and it has been estimated that the free concentration of this metabolite is very much lower than the total concentration.[79] Thus, it is unlikely that acetoacetyl-CoA, or the ratio of 3-hydroxybutyryl-CoA to acetoacetyl-CoA (as proposed by Utter and Fung,[897]) plays a physiological role in the regulation of pyruvate carboxylase activity.

Scrutton and colleagues have investigated the effects of di- and tricarboxylic acids on hepatic pyruvate carboxylase in order to test the proposal that one or more of these compounds regulates the activity of the enzyme in relation to its role in the provision of four-carbon precursors for anaplerotic synthesis.[750,766] Glutamate was found to inhibit pyruvate carboxylase from both rat and chicken liver, wherease α-ketoglutarate inhibited only the enzyme from chicken liver.[766] The activity of purified pyruvate carboxylase was also found to be inhibited by citrate. The results of studies of the mechanism of this effect indicate that the tricarboxylic acid does not act primarily by chelating Mg^{2+}.[766]

While the apparent K_i values for the inhibition of pyruvate carboxylase by glutamate, α-ketoglutarate, and citrate are relatively high, the estimated concentration of these metabolites in the mitochondrial matrix is also high (Table 3). Thus, the available data are consistent with a physiological function for these carboxylic acids in the regulation of pyruvate carboxylase activity. Further support for a role for glutamate comes from the observation that the action of glucagon on isolated hepatocytes induces a five-fold decrease in the intramitochondrial concentration of this metabolite.[785] It is proposed that this change, together with the decrease in succinyl-CoA concentration[786] described above, is partly responsible for the stimulation of pyruvate carboxylation and gluconeogenesis by glucagon. It has also been proposed that an increase in mitochondrial glutamate may be responsible for the inhibition of pyruvate carboxylation (and gluconeogenesis) in isolated hepatocytes during the synthesis of urea from ammonia.[557]

On the basis of information presently available, it may be concluded that succinyl-CoA, methylmalonyl-CoA, propionyl-CoA, α-ketoglutarate, glutamate, and citrate all contribute to the regulation of pyruvate carboxylase activity under certain conditions, in addition to control exerted by acetyl-CoA, pyruvate, ATP, and ADP. The relationship of these metabolites to the pathways of gluconeogenesis, lipid synthesis, branched-chain amino acid degradation, and amino acid synthesis is shown in Figure 13.

5. Metal Ions

The activity of pyruvate carboxylase is influenced by a variety of metal ions, including Mg^{2+} (Chapter 4, Section IV), Mn^{2+},[545,755] Ca^{2+},[247,375,543,755,773,975,977] and K^+.[733] However, with the possible exception of Ca^{2+} and Mn^{2+} there is no compelling evidence to indicate that metals play an important part in the physiological regulation of enzyme activity.

The effects of Ca^{2+} on pyruvate carboxylase would appear to be exerted at the active site in competition with free Mg^{2+} and/or through the formation of $CaATP^{2-}$.[755,773] In this respect, the mechanism by which Ca^{2+} inhibits pyruvate carboxylase is similar to that by which this divalent cation inhibits α-ketoglutarate dehydrogenase[547] and isocitrate dehydrogenase.[226] However, the actions of Ca^{2+} on these enzymes differ from those on the pyruvate dehydrogenase complex, which are exerted through activation by Ca^{2+} of pyruvate dehydrogenase phosphate phosphatase.[225,646]

The concentration of total Ca^{2+} which gives 50% inhibition of isolated or purified pyruvate carboxylases is in the range of 0.4 to 1.0 m*M*.[247,375,543,975,977] This can be compared with estimates of about 2.5 m*M* (calculated from the data of Claret-Berthon et al.,[185] assuming that the mitochondrial volume is 1 μl/mg protein[376]) and 1 to 60 μ*M*[200,224,548,969] for the concentration of total exchangeable and free intramitochondrial Ca^{2+}, respectively. From this comparison alone, it is not possible to reach a conclusion about the possible physiological role of Ca^{2+} in the regulation of pyruvate carboxylase activity because the concentration of free Ca^{2+} in experiments conducted with the iso-

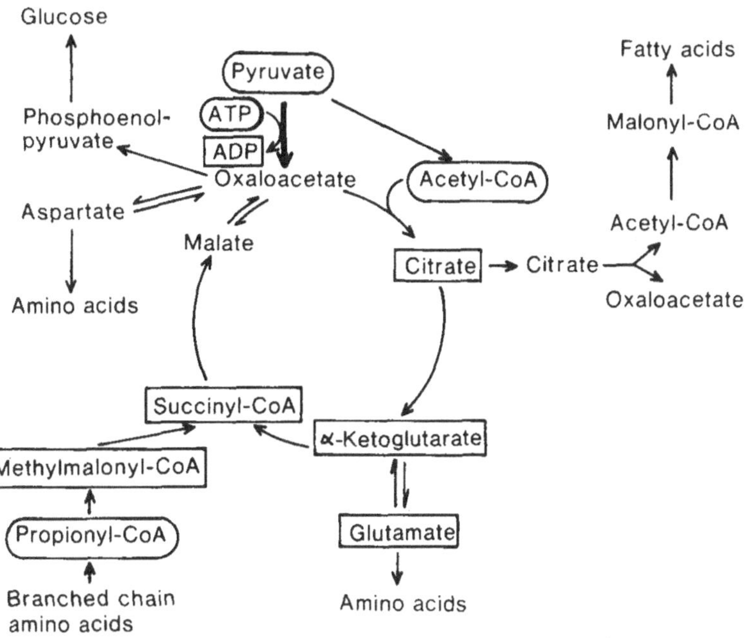

FIGURE 13. A summary of proposed activators and inhibitors of pyruvate carboxylase activity in the mammalian liver in relation to the metabolic pathways in which they are involved. Substrates or activators (⬭) and inhibitors (▭) which appear to have a physiological role in the regulation of enzyme activity (see text) are shown. With the exception of the transport of citrate from the mitochondria to the cytoplasm, the location of a given metabolite (mitochondria or cytoplasm) is not indicated.

lated enzyme has not been calculated. Moreover, little is known of the distribution of bound Ca^{2+} between proteins, lipids, and anions, including ATP and phosphate ions in the mitochondrial matrix.

In an alternative approach to the question of the role of Ca^{2+} in the regulation of pyruvate carboxylase activity in vivo, the effects of Ca^{2+} on the rate of pyruvate carboxylation in isolated mitochondria have been investigated. Early studies showed that the addition of Ca^{2+} to isolated rat liver mitochondria incubated in the presence of sucrose inhibits the rate of pyruvate carboxylation.[443] However, Morikofer-Zwez et al.[575] demonstrated that high concentrations of extramitochondrial sucrose, which depressed the basal rate of pyruvate carboxylation, were required to observe maximal inhibition. Furthermore, it was found that when mitochondria were preloaded with high concentrations of Ca^{2+}, and incubated in the presence of sucrose and mannitol, the intramitochondrial ratio of ATP:ADP concentrations was depressed.

More recent studies in which KCl was used to adjust the osmolarity of the incubation medium showed that low concentrations of Ca^{2+} inhibited the rate of pyruvate carboxylation in mitochondria isolated from the livers of starved rats.[282] This inhibition was shown to immediately follow the uptake of the added Ca^{2+} by the mitochondria and was completely reversed by the subsequent addition of Mn^{2+} (Figure 14), which is known to stimulate pyruvate carboxylase.[282,375,755,977] These changes were not associated with significant alterations in the concentrations of ATP, ADP, and acetyl-CoA.[282] Walter et al.[934] repeated these experiments and confirmed the inhibition by Ca^{2+} in mitochondria isolated from the livers of starved rats, but observed no inhibition in mitochondria isolated from fed rats. These authors also observed a decrease in the ratio of ATP:ADP in the presence of Ca^{2+} and suggested that this was due to the utilization of energy for mitochondrial Ca^{2+} uptake, although under the conditions

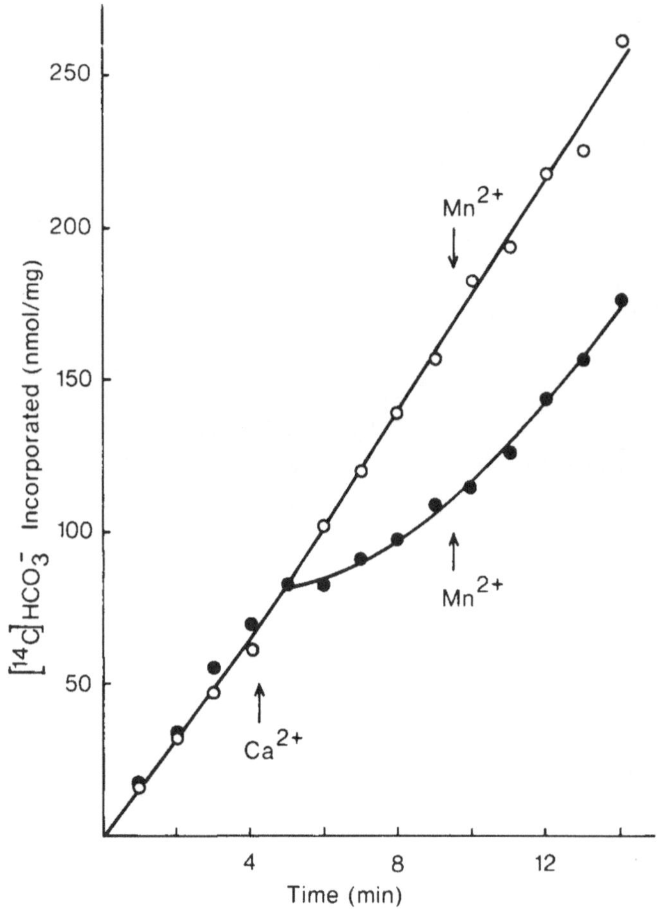

FIGURE 14. Inhibition by exogenous Ca^{2+} of the rate of pyruvate carboxylation in isolated rat liver mitochondria and its reversal by Mn^{2+}. The rate of $H^{14}CO_3^-$ incorporation into acid-stable intermediates was measured in the absence of added $CaCl_2$ (O) and in the presence of 50 nmol/mg protein of $CaCl_2$ (●) added at 4.5 min (indicated by the arrow). $MnCl_2$ (125 nmol of Mn^{2+}/mg of protein) was added at 9.5 min after the initiation of $H^{14}CO_3^-$ incorporation. (From Foldes, M. and Barritt, G. J., *J. Biol. Chem.*, 252, 5372, 1977. With permission.)

employed most of the Ca^{2+} is taken up by the mitochondria within 1 min after its addition,[282] wherease the deceased ratio of ATP:ADP was measured at 3 min.[934]

Ca^{2+} was also shown to inhibit adenine nucleotide exchange[282,320] and pyruvate transport[282,905] in isolated mitochondria. Thus it appears that the inhibition by Ca^{2+} of pyruvate carboxylation in intact mitochondria is likely to be due, at least in part, to a decrease in the intramitochondrial concentrations of $MgATP^{2-}$ and pyruvate[282] and in the ratio of ATP:ADP.[575,934]

Evidence which indicates that the activity of some other mitochondrial enzymes may be controlled, in part, by changes in the concentration of free Ca^{2+} in the mitochondrial matrix has recently been obtained.[224,373,548,558] However, further experiments are required in order to fully test whether this role of Ca^{2+} extends to the control of pyruvate carboxylase activity.

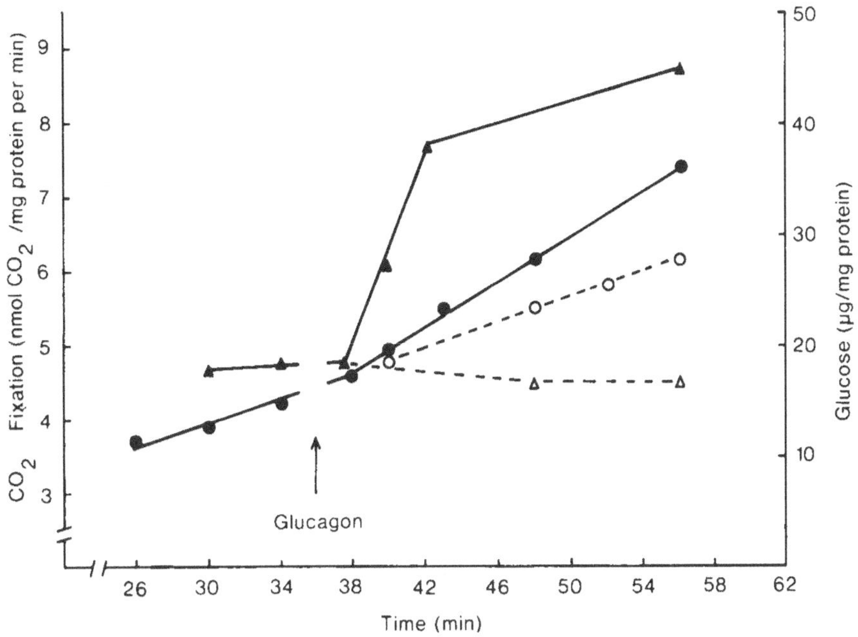

FIGURE 15. Effect of the incubation of hepatocytes isolated from the livers of starved rats with glucagon on the rate of CO_2 fixation catalyzed by mitochondria subsequently isolated from the cells (▲,△) and the synthesis of glucose from 20 m*M* lactate (●,○). Hepatocytes were incubated in the presence (closed symbols, solid lines) or absence (open symbols, broken lines) of 10 μ*M* glucagon added at 36 min. (From Garrison, J. C. and Haynes, R. C., Jr., *J. Biol. Chem.*, 250, 2769, 1975. With permission.)

6. Hormonal Control of Enzyme Activity in Liver Mitochondria

a. Effects of Glucagon

The treatment of liver cells or tissue with glucagon,[4] adrenaline,[306] glucocorticoids,[4,924] or thyroid hormones[4,924] has been shown to alter the rate of pyruvate carboxylation in mitochondria subsequently isolated from the cells or tissue without apparently changing the total amount of enzyme protein. This effect may be distinguished from the actions of hormones, which induce a change in the total amount of pyruvate carboxylase protein (Section I.B), but like changes in the rate of synthesis of the enzyme is proposed to be part of the mechanism by which these hormones stimulate hepatic gluconeogenesis.[4,306,924] While the effects of hormones on pyruvate carboxylation in liver mitochondria have been studied extensively, little is known of the actions of these hormones (if any) on pyruvate carboxylation in kidney or adipose tissue.

Adam and Haynes[4] first reported that mitochondria isolated from the livers of rats treated with glucagon, adrenaline, or cortisol exhibit an enhanced rate of incorporation of HCO_3 into malate and citrate. It was concluded that each of these hormones increases the rate of the reaction catalyzed by pyruvate carboxylase.[4] In subsequent investigations,[4,166,308,386-388,879,924,989] rates of mitochondrial pyruvate carboxylation have chiefly been estimated by measuring either the rate of incorporation of [^{14}C] from [^{14}C]HCO_3^- to acid-stable intermediates, or the rate of increase in malate and citrate in the presence of HCO_3^-.

The effect of glucagon is detected within 6 min (Figure 15)[308] and is maximal within 10 min of exposure of the liver or hepatocytes to the hormone.[308,988,990] No stimulation of pyruvate carboxylation[4] and no change in the total amount of enzyme activity[4] was detected in disrupted mitochondria prepared from control and glucagon-treated livers.

Although the latter observation indicates that there is no change in the total amount of enzyme protein, this conclusion has not yet been confirmed directly, e.g., using immunoprecipitation techniques.

Since the effect of glucagon is detected in isolated mitochondria, it is likely to represent a change in either the properties of the mitochondrial inner membrane or matrix,[26,361,362] or the intramitochondrial concentration of a substrate, activator, or inhibitor of pyruvate carboxylase.[386,878] This change appears to be relatively stable during the period required for isolation of the mitochondria.

Some authors have suggested that the observed effect of glucagon on mitochondrial pyruvate carboxylation is an artifact of the isolated mitochondrial system.[787] However, measurement of the effects of glucagon on (1) the distribution of metabolites (including pyruvate) in intact cells,[129,631,632,785,786] (2) pyruvate carboxylase activity measured in permeable hepatocytes,[8] and (3) pyruvate carboxylase activity measured in mitochondria isolated under a variety of conditions[428] indicate that the results obtained with isolated mitochondria do reflect changes in pyruvate carboxylase activity in intact cells.[8,26,428]

b. Possible Mechanisms of Glucagon Action

The process by which glucagon increases the rate of mitochondrial pyruvate carboxylation has not been elucidated. While there are some reports that the stimulation of pyruvate carboxylation is associated with an increase in the mitochondrial concentration of acetyl-CoA,[4,87,785] it is unlikely that an increase in the concentration of this activator catalyzes the stimulation by glucagon of pyruvate carboxylation.[129,964]

It has been suggested[166] that glucagon stimulates pyruvate carboxylation by decreasing the intramitochondrial concentration of Ca^{2+}, an inhibitor of the enzyme.[247,375,543,975,977] While glucagon has been shown to decrease mitochondrial Ca^{2+} in intact liver cells under certain experimental conditions,[44,106,166,293,863] the effect is relatively small at physiological concentrations of Ca^{2+} and the hormone.[44,106] Moreover, experiments with isolated mitochondria incubated under ionic conditions which approach those present in the cell cytoplasm provide no evidence to indicate that glucagon decrease mitochondrial Ca^{2+} under physiological conditions.[44,122]

The results of a variety of experimental approaches indicate that the stimulation of mitochondrial pyruvate carboxylation by glucagon is chiefly due to increases in the rates of ATP synthesis[20,137,166,665,877,878,888,913,988,989] and pyruvate transport across the inner mitochondrial membrane.[4,308,360,387,785,873,876] (but see References 166 and 876). Other factors which may contribute to the stimulation by glucagon of pyruvate carboxylation include changes in the concentrations of succinyl-CoA[786] and glutamate,[785] although the possible role of these metabolites remains to be fully elucidated.

The increased rates of pyruvate carboxylation, pyruvate transport and ATP synthesis observed in mitochondria isolated from livers or hepatocytes treated with glucagon are associated with changes in a large number of other properties of the mitochondria (reviewed in References 428 and 8). It is likely that these changes are all due to the same primary event which is caused by phosphorylation of a protein or proteins catalyzed by a cyclic AMP-dependent protein kinase. However, the nature and location of this protein (whether in the mitochondria or cytoplasm) are not known. Reasonable evidence that this particular effect of glucagon is mediated by cyclic AMP has been presented,[308] although the criteria defined by Robinson et al.[689] and Krebs and Beavo[474] for establishing roles for cyclic AMP and a cyclic AMP-dependent protein kinase in a given process have not yet been fully satisfied.

The nature of the postulated primary change in mitochondria induced by glucagon is not well understood. Two proposals are that the action of cyclic AMP-dependent protein kinase leads to a rapid net increase in intramitochondrial ATP and ADP[878,879] (compare with the rapid increase in total adenine nucleotide concentrations observed in neonatal rat livers at birth[18,19,21,593,711,844]) and a change in the lipid environment of enzymes of the respiratory chain.[362,787]

The results of experiments designed to test whether glucagon induces a covalent modification by phosphorylation of pyruvate carboxylase itself have not been reported. However, it seems unlikely that this regulatory mechanism accounts for the effects of glucagon since it has been shown that pyruvate carboxylase from rat liver is not modified covalently by a phosphorylation-dephosphorylation cycle under conditions where the activity of pyruvate dehydrogenase is altered by this mechanism[496]

c. Other Hormones

Adrenaline and other α-adrenergic agonists have also been shown to stimulate pyruvate carboxylation by isolated liver mitochondria[306] and induce changes in a number of other properties of the mitochondria[863,877] similar to those induced by glucagon. The action of α-adrenergic agonists on rat liver is mediated by an increase in the cytoplasmic Ca^{2+} concentration.[306,863,877] Thus the primary change in liver mitochondria which is induced by an increase in cytoplasmic cyclic AMP can also be induced by an increase in the concentration of cytoplasmic Ca^{2+}, which may act on a Ca^{2+}-dependent protein kinase.[307]

Mitochondrial pyruvate carboxylation is also stimulated by the administration of glucocorticoids of triiodothyronine to rats.[4,924] The onset of the effects of these hormones is much slower than that for the effects of glucagon. Moreover, the changes induced by glucocorticoids are blocked by cycloheximide, suggesting that nuclear DNA-directed protein synthesis is required.[924] As discussed above (Section I.B.2.d), there appears to be no change in the total amount of hepatic pyruvate carboxylase activity following the administration of glucocorticoids to fed rats under normal conditions,[643,829] although this point has not been completely established. Glucocorticoids also induce changes in the properties of isolated mitochondria similar to a number of changes induced by glucagon.[924] These observations suggest the existence of a slow-onset mechanism for the regulation of liver mitochondrial pyruvate carboxylase activity which does not involve a change in the total amount of pyruvate carboxylase protein, but may involve a change in the structure of the mitochondrial inner membrane.

D. Control of Pyruvate Carboxylation in Relation to Carbon Cycling and Pyruvate Oxidation

1. Cycling of Carbon Between Pyruvate, Oxaloacetate, and Phosphoenolpyruvate

When the enzymes pyruvate carboxylase, phosphoenolpyruvate (PEP) carboxykinase, and pyruvate kinase are present in the same cell, as is likely in the liver and kidney, a cycling of carbon can occur (Figure 1).[599,762] Experimental measurement of the rate of this cycling has been difficult. Nevertheless, it has been shown that flux through the cycle depends on the physiological state of the cells and can be up to twice the net rate of glucose formation from pyruvate.[292,695,697,805,967] Since the net reaction catalyzed by the combined reactions of the cycle is the hydrolysis of ATP, this and other similar cycles have been described as "futile" cycles.[599,762] However, it has also been proposed that the presence of these cycles may provide a more sensitive mechanism for the control of flux through the gluconeogenic and glycolytic pathways.[599] Furthermore, Rognstad and Katz[697] have proposed that the activity of the pyruvate-oxaloacetate-PEP cycle contributes to the energy-dependent transfer of reducing equivalents from the mitochondrial matrix to the cytoplasm during gluconeogenesis.

The intramitochondrial concentration of acetyl-CoA may play an important part in controlling the rate of carbon cycling. Thus, the activity of pyruvate carboxylase, and hence the cycle, would be increased only when the concentration of acetyl-CoA is elevated.

Scrutton and Utter[762] have pointed out that the maximum activity of pyruvate kinase is about ten times that of pyruvate carboxylase and PEP-carboxykinase.[600,762] It was argued that the activity of pyruvate kinase, and hence the rate of carbon cycling, must be inhibited when the rate of gluconeogenesis is enhanced.[762] In the short term this inhibition is chiefly achieved by the actions of the hormones glucagon and adrenaline. The stimulation by glucagon of gluconeogenesis in the liver involves, among other changes, phosphorylation and inhibition of pyruvate kinase. This is catalyzed by a cyclic AMP-dependent protein kinase.[427,684] As discussed above (Section I.C.6), it is proposed that this same protein kinase stimulates pyruvate carboxylation by altering the properties of the mitochondria so that the rate of pyruvate transport and the steady-state ratio of matrix ATP:ADP is increased. In addition to these changes, which are rapid in onset, the total amount of hepatic pyruvate kinase activity decreases in starvation[908,909] and increases on feeding,[498] whereas the activities of pyruvate carboxylase (Section I.B.2.b) and PEP-carboxykinase (reviewed by Tilghman et al.[874]) increase in starvation and decrease on refeeding.

2. Coordinate Control of Pyruvate Carboxylation and Oxidation

It has been argued that in the liver and kidney there is likely to be an inverse relationship between flux through pyruvate carboxylase and pyruvate dehydrogenase when changes from glycolysis to gluconeogenesis or vice versa are considered.[215,675] There is some experimental evidence to support this argument.[675]

The regulation of pyruvate carboxylase by the concentrations of substrates and modifiers, and covalent modification and product inhibition of pyruvate dehydrogenase appear to constitute the main mechanism which cause the activities of these enzymes to vary inversely. Thus, high ratios of acetyl-CoA:CoA-SH and ATP:ADP stimulate pyruvate carboxylation (Sections I.C.3 and I.C.2), but inhibit pyruvate dehydrogenase (reviewed by Denton and Halestrap[223]). Ca^{2+} may also play a role in preferentially directing carbon flow through one of these pathways.[223]

In contrast to the roles of pyruvate carboxylase and pyruvate dehydrogenase in gluconeogenesis, rates of carbon flow through both reactions are likely to be similar during stimulated lipogenesis (cf. References 51, 341, 459, 532, 696, 904, and 940). Flux through pyruvate dehydrogenase is presumably equal to that through pyruvate carboxylase plus flux through the citric acid cycle (these pathways are shown schematically in Figure 2). The mechanism by which this coordination is achieved has not been clearly defined. It has been proposed that in adipose tissue the activation of pyruvate dehydrogenase by insulin leads to an increase in the steady-state concentration of intramitochondrial acetyl-CoA which, in turn, increases the activity of pyruvate carboxylase.[223]

II. OTHER TISSUES OF VERTEBRATES

The chief function of pyruvate carboxylase in cells of the liver, kidney, and adipose tissue appears to be the synthesis of oxaloacetate in an anaplerotic reaction which provides intermediates of the citric acid cycle for a number of biosynthetic pathways[460] (Figure 16 and Chapter 2, Section I). It is likely that pyruvate carboxylase also has an anaplerotic function in the liver, kidney, and adipose tissue[79,750] in addition to its major roles in gluconeogenesis and lipogenesis.

Scrutton[746] has defined the anaplerotic function of pyruvate carboxylase described above (Figure 16) as an "external" one since intermediates of the citric acid cycle are used in biosynthetic pathways in which these metabolites are substrates. This has been

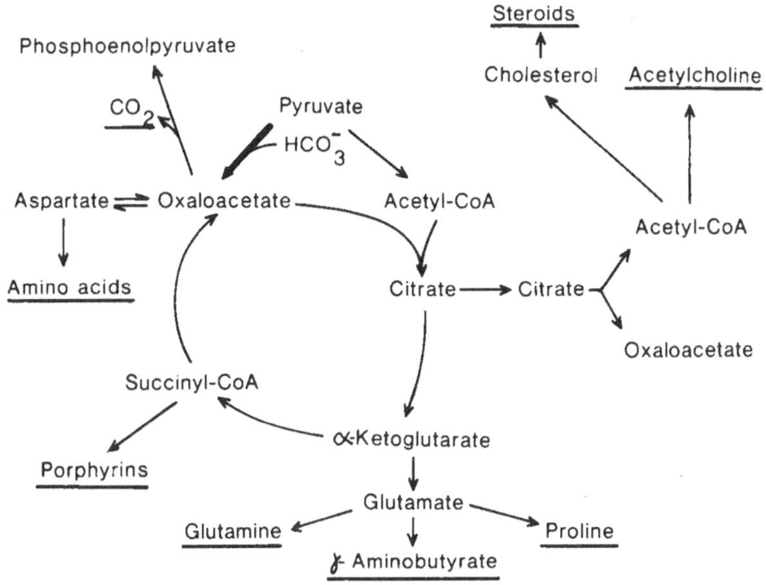

FIGURE 16. A schematic representation of the anaplerotic function of pyruvate carboxylase in vertebrate tissues in relation to the proposed biosynthetic pathways in which the oxaloacetate is utilized (Table 5). Some specialized functions of the enzyme have been omitted. The release of CO_2 in the conversion of oxaloacetate to PEP schematically represents the proposed role of pyruvate carboxylase in HCO_3^- resorption in renal tubular cells.[143] With the exception of the transport of citrate from the mitochondria to cytoplasm, the intracellular location of a given metabolite is not indicated.

distinguished from an "internal" anaplerotic function in which it is proposed that the oxaloacetate is used solely to increase the steady-state concentration of all intermediates of the citric acid cycle and hence increase the rate of oxidation of acetyl-CoA and synthesis of ATP.[746] In brown adipose tissue pyruvate carboxylase may play such a role when respiration is stimulated by noradrenaline[149] (compare also with the function of pyruvate carboxylase in insect flight muscle [Section III.A below]).

In comparison with knowledge of the regulation of pyruvate carboxylase in the liver and kidney, very little is known about the regulation of either the intracellular concentration of the enzyme or its activity in other vertebrate tissues. This is partly due to the relatively small quantities of enzyme present in these tissues.

Where the response to acetyl-CoA has been tested, pyruvate carboxylase isolated from vertebrate tissues other than liver and kidney, including brain[275,525] and avian red blood cells,[72] have been shown to be markedly stimulated by this activator. Spydevold et al.[811] have shown that the rate of pyruvate carboxylation in isolated skeletal muscle mitochondria is stimulated by acetylcarnitine, suggesting that skeletal muscle pyruvate carboxylase is also activated by acetyl-CoA. As pointed out previously (Section I.C.3) it is proposed that one of the functions of the activation of pyruvate carboxylase by acetyl-CoA is to ensure the provision of acetyl-CoA and oxaloacetate to the citric acid cycle at equal rates.[460]

The results of experiments conducted with isolated brain mitochondria indicate that in addition to activation by acetyl-CoA the activity of pyruvate carboxylase in this tissue may also be controlled by the ratio of ATP:ADP concentrations, the intramitochondrial concentrations of pyruvate, and the steady-state concentration of oxaloacetate.[634,636] Measurement of pyruvate carboxylase present in brain tissue during development has shown that the total amount of pyruvate carboxylase activity increases slowly during a period of 40 days after birth.[958]

III. INVERTEBRATES

A. Insect Flight Muscle

The results of studies of the properties of pyruvate carboxylase isolated from a wide variety of invertebrate species have led to a better understanding of the mechanisms which regulate the activity of this enzyme under physiological conditions in both invertebrate and vertebrate tissues.[746,747,750,768,904] The value of this comparative approach has been emphasized by Scrutton and colleagues.[746,747,750] Although knowledge is still far from complete, in the higher invertebrates information on the regulatory properties of pyruvate carboxylase has been obtained for the enzyme from insect flight muscle,[211,708] and for the lower invertebrates for the enzymes from a variety of fungi, including yeast and many species of bacteria (Table 4).

While pyruvate carboxylase activity has been detected in tissues of a variety of higher invertebrates (summarized in Table 3, Chapter 2), with the exception of pyruvate carboxylase from insect flight muscle,[211,708] little is known of the mechanisms of regulation of these enzymes. Evidence has been presented which indicates that, in insect flight muscle, as in cells of vertebrate tissues (Chapter 2, Section V.B) pyruvate carboxylase is located in the mitochondria.[211] is activated by acetyl-CoA (the apparent K_a value is 30 μM at pH 7.4), is insensitive to inhibition by di- and tricarboxylic acids,[211,708] and has an apparent K_m for HCO_3^- of 16 mM.[708] This is an order of magnitude greater than the apparent K_m of the enzyme from vertebrate tissues for HCO_3^-.[708]

On the basis of the observation that pyruvate carboxylase is located in flight muscles but not other muscles of insects,[211] and evidence that in flight muscles flux through the citric acid cycle increases about 50-fold at the onset of flight,[715] it has been proposed that the role of pyruvate carboxylase in this tissue is solely an internal anaplerotic one.[211,708] Thus, it is envisaged that the enzyme provides a rapid increase in the steady-state intramitochondrial concentration of oxaloacetate which, in conjunction with the stimulation of pyruvate dehydrogenase, permits rapid activation of the citric acid cycle.[211,708] Coordination of the activities of pyruvate carboxylase and pyruvate dehydrogenase may be primarily achieved by an increase in the concentrations of pyruvate and HCO_3^-,[708] and activation of pyruvate carboxylase by acetyl-CoA,[211] although Rowan et al.[708] have reported that the tissue concentration of this metabolite does not change significantly with the onset of flight.

B. Fungi

The role of pyruvate carboxylase in the provision of four-carbon precursors for a variety of biosynthetic pathways (Figure 17) in microorganisms is discussed in Chapter 2. In order to identify and evaluate the major mechanism for the regulation of pyruvate carboxylase activity in microorganisms, Scrutton[747,750] has proposed a classification of pyruvate carboxylase enzymes on the basis of the regulatory properties, intracellular location, and proposed biosynthetic function of each enzyme. The major subgroups of pyruvate carboxylase enzymes and the proposed mechanisms of regulation are summarized in Table 4. Only those enzymes for which the regulatory properties have been studied in some detail are included in the table. A complete list of the distribution of pyruvate carboxylase in species of microorganisms is presented in Tables 1 and 2 (Chapter 2).

For the first two groups of pyruvate carboxylase enzymes listed in Table 4, modification of enzyme activity by metabolic effectors (aspartate, α-ketoglutarate, and acyl-CoA compounds) constitutes one of the main, or possibly the sole, mechanism of regulation. In contrast to knowledge of the regulation of pyruvate carboxylase activity in liver and kidney, with a few exceptions.[141,680] relatively little attention has been paid to the question of how the intracellular concentrations of the effectors and substrates

Table 4
SOME REGULATORY PROPERTIES OF PYRUVATE CARBOXYLASE ENZYMES FROM MICROORGANISMS[a]

Classification of organisms	Feature of regulation: Modifiers of pyruvate carboxylase activity	Feature of regulation: Changes in amount of enzyme
Fungi	Inhibited by aspartate and α-ketoglutarate, but not glutamate Species investigated include: *S. cerevisiae*,[158,503,627] *A. niger*,[269] *A. nidulans*,[615] *P. camemberti* var. *candidum*,[818,819] *V. albo-atrum*,[379] and *Phycomyces blakesleeanaus*.[723]	Evidence for change in total amount of enzyme activity and a decrease in activity when grown in the presence of aspartate and other 4-carbon compounds Species investigated include: *S. cerevisiae*,[156,355-357,619] *C. boidinii*,[147] *R. glutinis*,[710] and *R. nigricans*.[620]
	Activated by acyl-CoA[b] although under special conditions in some cases[c] Species investigated include: *S. cerevisiae*,[158,203,204,304,710,881] *A. nidulans*,[615,617] *C. methylica*,[509] *P. camemberti*,[818] *V. albo-atrum*,[379] *N. vasinfecta*,[141] *P. blakesleeanaus*,[723] *R. nigricans*,[620] and *A. bisporus*.[672]	
Bacteria: pyruvate carboxylase sensitive to activation by acetyl-CoA	Inhibited by aspartate and α-ketoglutarate[304,509,617] Species investigated include: *A. globiformis*,[350] *B. stearothermophilus*,[163,503] *B. lichenformis*,[680] *B. megaterium*,[10] *S. faecalis* var. *liquefaciens*,[377,483] *S. lactis*.[405]	No evidence for marked changes in total amount of enzyme activity Species investigated include: *B. licheniformis*,[680,838] *B. coagulans*,[838] *B. subtilis*,[237] and *R. spheroides*[639]
	Activated by acyl-CoA[d] Species investigated include: *A. globiformis*,[128,350] *B. stearothermophilus*,[163,503] *B. subtilis*,[237] *B. coagulans*,[838] *B. licheniformis*,[680,838] *B. megaterium*,[10] *R. spheroides*,[639] *R. trifolii*,[701] *A. tumefaciens*,[181] *T. A2*,[178] *S. faecalis* var. *liquefaciens*,[377,483] *S. lactis*,[405] *C. erificans*,[499] *C. aquaticum*,[84] *N. erythropolis*,[84] and *B. ammoniagenes*.[84]	
Bacteria: pyruvate carboxylase apparently insensitive to effects of acetyl-CoA and dicarboxylic acids	Acyl-CoA, aspartate, and α-ketoglutarate appear to have no effect Species investigated include: *P. citronellolis*,[772,856] *P. fluorescens*,[404] *A. vinelandii*,[758]	Evidence for change in total amount of enzyme activity and a decrease in activity when grown in the presence of aspartate and other 4-carbon compounds Species investigated include: *P. citronellolis*,[858] *P. fluorescens*,[404] and *A. vinelandii*.[758]

[a] The classification has been made on the basis of current knowledge, following the proposals of Scrutton[747] and Scrutton and Griffiths.[750]

[b] For pyruvate carboxylase isolated from *S. cerevisiae*[881] and *A. nidulans*,[615] long-chain alkylacyl-CoA compounds are more effective activators than acetyl-CoA.

[c] The results of early investigations suggested that pyruvate carboxylase enzymes from *Aspergillus* species are not activated by acyl-CoA metabolites,[111,269] or are activated only in the presence of an inhibitor, aspartate, or α-ketoglutarate.[615] Subsequent studies have shown that an activation of pyruvate carboxylase from *A. nidulans* by acetyl-CoA is observed when the pH of the assay medium is between 7.4 and 7.8 and nonsaturating concentrations of pyruvate and $MgATP^{2-}$ are employed.

[d] For pyruvate carboxylase isolated from *B. stearothermophilus*, acetyl-CoA is a more effective activator than long-chain alkylacyl-CoA.

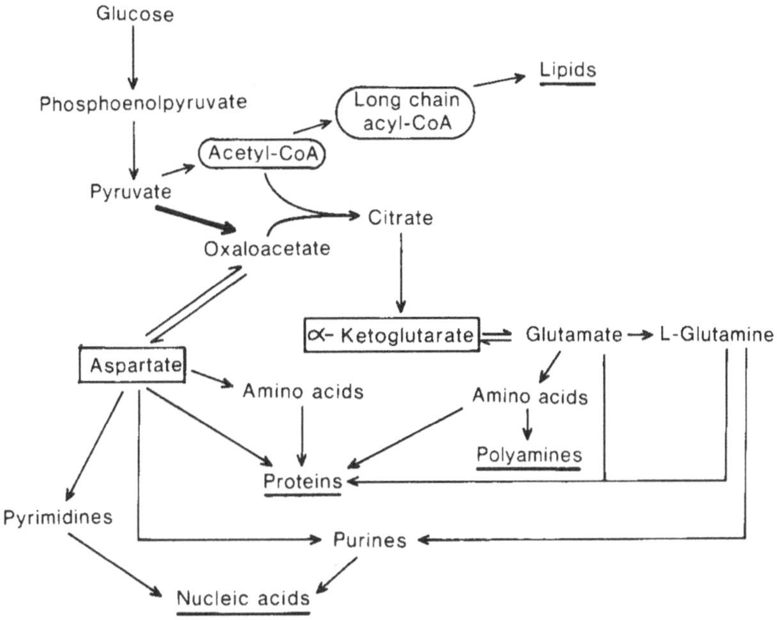

FIGURE 17. A schematic representation of the proposed anaplerotic function and regulation of pyruvate carboxylase activity in *A. nidulands* based on the proposal of Scrutton and Griffiths,[750] which empasizes coordination of the rate of provision of 4-carbon precursors for the synthesis of lipids, proteins, and nucleic acids on the one hand with the synthesis of fatty acids (for lipid synthesis) on the other. Activators (⬭) and inhibitors (▭) of pyruvate carboxylase activity are indicated. Similar regulatory mechanisms may be present in *S. cerevisiae.* A distinction between the cytoplasmic and mitochondrial location of different metabolites has not been made, although it has been shown that pyruvate carboxylase is located in the cytoplasm in *A. nidulands* and *S. cerevisiae.*[358,618] The amino acids formed from glutamate are lysine, proline, and arginine, while asparagine, methionine, threonine, and isoleucine are from formed aspartate.

(in particular pyruvate and ATP) compare with the apparent K_i (or K_a) and K_m values of the enzyme for a given effector or substrate, respectively. In the case of the major feedback inhibitor, aspartate, the intracellular concentration of this amino acid in *Saccharomyces cerevisiae* and a number of species of bacteria is estimated to be between 1 and 10 m*M*[411] compared with an apparent K_i value of pyruvate carboxylase for aspartate of 0.5 to 5 m*M* (Tables 1 and 2 of Chapter 2). These values are consistent with the conclusion that for pyruvate carboxylase enzymes which are susceptible to inhibition by aspartate this amino acid acts as a physiological inhibitor in vivo.

Subcellular fractionation studies (discussed in Chapter 2, Section III.A.2) have provided evidence to indicate that in cells of *S. cerevisiae,*[358] *A. nidulans,*[618] and possibly some other fungi pyruvate carboxylase is located in the cytoplasm. This is in contrast to the mitochondrial location of the enzyme in all other eukariotic cell types so far examined. Since pyruvate carboxylase is not present in the same intracellular compartment as enzymes of the citric acid cycle in these fungi, it has been pointed out that the mechanisms which achieve close coordination between the rates of synthesis of oxaloacetate and acetyl-CoA in other cell types are unlikely to operate in these microorganisms.[618,747,750]

Data presently available indicate that the two principal mechanisms for the control of pyruvate carboxylase activity in fungi are (1) the action of the metabolic effectors aspartate, α-ketoglutarate, and acyl-CoA compounds and (2) changes in the total amount of enzyme activity (Table 4). For *S. cerevisiae* and *A. nidulans* it has been proposed that feedback inhibition of pyruvate carboxylase activity by aspartate and α-ketoglutarate is supplemented by activation of the enzyme by long-chain acyl-CoA compounds.[618,747,750] The latter mechanism may provide feed-forward activation which coordinates the rate of fatty acid synthesis with the synthesis of other components of lipids, as well as the synthesis of amino acids, proteins, purines, and pyrimidines during cell growth (Figure 17).[618,747,750] It has been argued that aspartate and α-ketoglutarate play a key role in feedback inhibition of pyruvate carboxylase activity in many microorganisms because these metabolites can be considered as the initial precursors of a number of other amino acids (Figure 17).[747,750] While it is likely that similar regulatory mechanisms are present in the cells of other fungi, further information is required to verify this proposal.

Pyruvate carboxylase enzymes isolated from *S. cerevisiae* and *Penicillium camemberti* var. *candidum*[819] are inhibited by ADP, and evidence has been presented to indicate that changes in intracellular energy charge {[(ATP) + 0.5 (ADP) / [(AMP) + (ADP) + (ATP)]}[36] contribute to the regulation of enzyme activity[569,818,819] through a mechanism which alters the maximum velocity of the reaction.[569] Other proposed regulators of pyruvate carboxylase from *S. cerevisiae* include the pH of the incubation medium and the concentration of intracellular NADH,[159] although Scrutton and Griffiths[750] have noted that the effect of NADH has not been confirmed. Pyruvate carboxylase from *Verticillium albo-atrum* has been reported to be inhibited by $NADP^+$ and 2,3-diphosphoglycerate.[379] However, a high concentration of these compounds was required (about 5 m*M* for 50% inhibition), and the physiological significance of these observations has not been evaluated. Further experiments are required in order to fully assess the roles of energy charge, NADH, NADP, 2,3-diphosphoglycerate, and H^+ in the regulation of pyruvate carboxylase in vivo.

The second principal mechanism for the control of pyruvate carboxylase activity in fungi is regulation of the total amount of enzyme present in the cell (Table 4). The magnitude of these changes and evidence that aspartate or its metabolites repress *de novo* synthesis of the enzyme is presented in Chapter 2 (Section III.A.1).

C. Bacterial Pyruvate Carboxylase Enzymes Sensitive to Activation by Acetyl-CoA

The second group of pyruvate carboxylase enzymes from microorganisms is characterized by a sensitivity to metabolic effectors but, on the basis of present data, no marked change in the total amount of enzyme present (Table 4). While the effects of all possible metabolite effectors has only been fully investigated for pyruvate carboxylase enzymes from a few species, it appears that the principal mechanism for the regulation of pyruvate carboxylase activity in this group of microorgaisms in inhibition by asparate and α-ketoglutatate and activation by acetyl-CoA (Table 4). In contrast to the proposed mechanisms for the control of pyruvate carboxylases in fungi, activation by acetyl-CoA may provide a mechanism for coordination of the rates of synthesis of oxaloacetate and acetyl-CoA.

In *Bacillus stearothermophilis,* aspartate and acetyl-CoA may also control the activity of pyruvate carboxylase by inhibiting and activating, respectively, the addition of biotin to the apoenzyme to form the holoenzyme in the reaction catalyzed by holoenzyme synthetase.[161,162,838] This mode of regulation, which does not appear to have been observed in other bacterial species, is discussed in more detail in Chapter 2 (Section III.A.1).

D. Bacterial Pyruvate Carboxylase Enzymes Apparently Insensitive to Regulation by Metabolites

In *Pseudomonas citronellolis, Ps. fluorescens,* and *Azotobacter vinelandii,* flux from pyruvate to oxaloacetate appears to be controlled solely by changes in the total

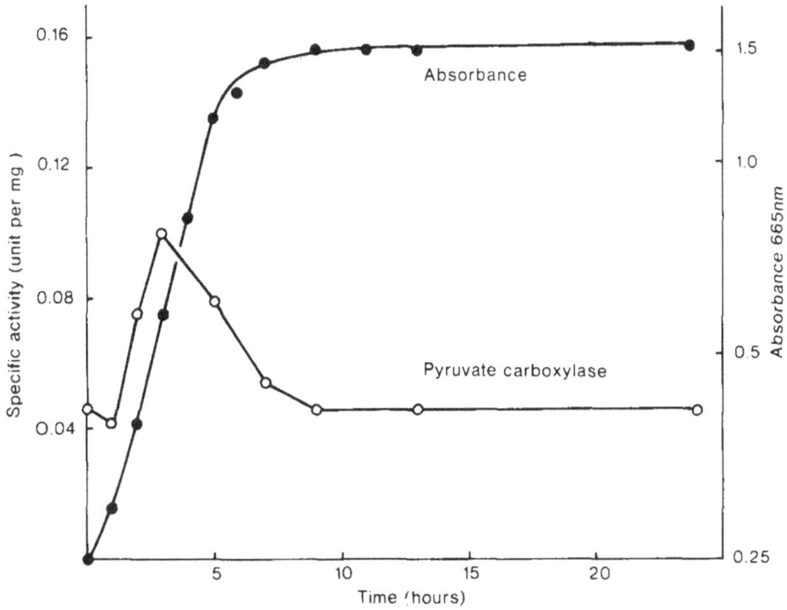

FIGURE 18. Changes in the total amount of pyruvate carboxylase activity (O) during growth monitored by absorbance (660 nm) of the growth medium (●) of *P. citronellolis* on lactate as the sole carbon source. (From Taylor, B. L., Routman, S., and Utter, M. F., *J. Biol. Chem.*, 250, 2376, 1975. With permission.)

amount of pyruvate carboxylase activity (Table 4). As discussed in Chapter 2 (Section II.B), maximum levels of pyruvate carboxylase activity correspond with maximum rates of growth (Figure 18) when the bacteria are grown on glucose, pyruvate, or lactate as sole carbon substrate. It has been proposed that the agent which is responsible for enhancing the rate of synthesis of the enzyme during induction is pyruvate or a metabolite of pyruvate.[758,858] Low or negligible amounts of pyruvate carboxylase are present in cells grown on succinate, glutamate, or malate.[758,858] Although the activity of pyruvate carboxylase from *P. citronellolis* is not inhibited by aspartate,[856] it has been shown that this amino acid does inhibit synthesis of the enzyme,[858] and is presumably responsible for inhibiting the synthesis of pyruvate carboxylase when the bacteria are grown in the presence of other four-carbon substrates.

The fact the pyruvate carboxylase is one of five enzymes involved in the interconversion of pyruvate, oxaloacetate, and PEP in *P. citronellolis* is discussed in Chapter 2 (Section I). The enzymes involved are shown in Figure 19. O'Brien et al.[612] have drawn attention to the complexity of this system and the need for tight control of the activity of each enzyme in order to prevent the wasteful cycling of carbon. The result obtained from studies conducted in a number of laboratories (summarized by O'Brien et al.[612]) indicate that this control is achieved by (1) the action of metabolites on four of the five enzymes and (2) regulation of the total amount of pyruvate carboxylase activity (summarized in Figure 19).

IV. CONCLUSIONS

The two principal mechanisms for the regulation of the activity of pyruvate carboxylase in microorganisms appear to be feedback inhibition by aspartate and α-ketoglutarate, and changes in the rate of synthesis of the enzyme. The purpose of these controls is to regulate the provision of oxaloacetate for the biosynthesis of amino acids (via aspartate and α-ketoglutarate), proteins, and other molecules. In some cases these

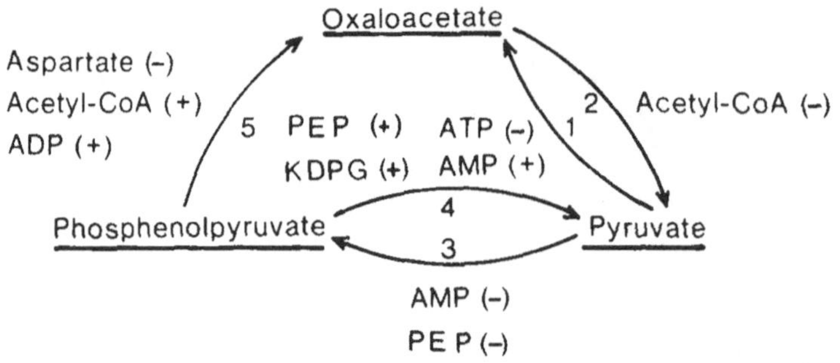

FIGURE 19. Enzymes involved in the interconversion of pyruvate, PEP, and oxaloacetate in *P. citronellolis*, showing schematically the presence of two possible alternative routes for the formation of oxaloacetate from pyruvate in this organism. (Taken from O'Brien *et al.*[612]). The enzymes indicated are (1) pyruvate carboxylase, (2) oxaloacetate decarboxylase, (3) PEP-synthetase, (4) pyruvate kinase, and (5) PEP-carboxykinase. It is proposed that regulation of the activity of these enzymes, and hence metabolic flux between pyruvate, PEP, and oxaloacetate is achieved by (1) the actions of the metabolic effectors indicated and (2) regulation of the total amount of pyruvate carboxylase (summarized by O'Brien et al.[612]). While the activity of pyruvate carboxylase in *P. citronellolis* is apparently not subject to control by metabolic effectors, induction of the synthesis of the enzyme is inhibited by aspartate.[111] The abbreviations are PEP, phosphoenolphyruvate, and KDPG, 2-keto, 3-deoxy, 6-phosphonogluconate.

controls are supplemented by activation of the enzyme by acetyl-CoA, permitting coordination of the rates of synthesis of oxaloacetate and acetyl-CoA, and in other cases by long-chain alkylacyl-CoA compounds, which may contribute to a coordination of the rates of lipid and protein synthesis.

Many of the mechanisms for the control of pyruvate carboxylase activity in microorganisms contribute to regulation of enzyme activity in muscles of invertebrates and vertebrates, and in the liver and kidney of vertebrates. In muscle cells, control of enzyme activity by acetyl-CoA would appear to be the predominant mechanism which provides coordination of the provision of oxaloacetate and acetyl-CoA for the formation of citrate and other intermediates of the citric acid cycle. However, it may be noted that less is known about the regulation of pyruvate carboxylase activity in muscle and other non-gluconeogenic and non-lipogenic tissues of the vertebrates than for the enzymes from other vertebrate and invertebrate sources.

In the liver and kidney the chief mechanisms for the control of enzyme activity are (1) activation by acetyl-CoA, which is proposed to provide a link between fatty acid oxidation and gluconeogenesis, (2) changes in the intramitochondrial concentration of pyruvate, adenine nucleotides, and possibly Ca^{2+} (short-term hormonal control), and (3) changes in the rate of synthesis of the enzyme (long-term hormonal control). Regulation by other acyl-CoA metabolites may contribute to the regulation of pyruvate carboxylase in relation to amino acid degradation and fatty acid synthesis in these tissues.

Chapter 6

DISEASED STATES IN MAN AND OTHER VERTEBRATES

Gregory J. Barritt

TABLE OF CONTENTS

I. INTRODUCTION

The diseases or diseased states which involve an alteration of pyruvate carboxylase activity in man and other vertebrates are summarized in Table 1. These have been divided into two categories: (1) those in which decreased pyruvate carboxylase activity is the primary cause, or one of the primary causes of the abnormalities, and (2) diseases, drugs, or other agents which cause a change in pyruvate carboxylase activity that is secondary to the primary cause of the disease or the action of the drug. In this chapter, the diseased states are grouped together according to the cause or proposed cause of the change in pyruvate carboxylase activity. Most emphasis will be placed on the inherited deficiency of pyruvate carboxylase and biotin-responsive multiple carboxylase deficiency (BRMCD) in children, and fatty liver and kidney syndrome (FLKS) in chickens. In order to provide a perspective for subsequent discussion, the main features of these diseases will be briefly summarized before each is discussed in detail.

As shown in Table 1, there are three diseases which are due solely or principally to a decrease in pyruvate carboxylase activity. These are: inherited abnormality in the pyruvate carboxylase protein in children,[42,687,694] BRMCD in children,[624,626] and FLKS in chickens.[68] Diseases which are due solely and directly to a dietary deficiency of biotin are rare. However, it has been suggested that in some situations impaired pyruvate carboxylase activity which results from decreased amounts of dietary biotin may lead to disease and death in man and some other mammals.[430,955]

The majority of children with an inherited abnormality in the pyruvate carboxylase protein have died.[42,687,694] The most critical pathological lesions which result from a severe decrease in activity of this enzyme in liver and brain are in the central nervous system.[42,687,694] The lethal nature of this inherited deficiency emphasizes the importance of the reaction catalyzed by pyruvate carboxylase in intermediary metabolism and anaplerotic synthesis.

The symptoms and signs exhibited by children with BRMCD arise from abnormaltites in a number of different metabolic pathways.[624,626] As its name implies, this disease can be effectively treated by the administration of large oral doses of biotin.[624,626] This is not the case for the inherited deficiency of pyruvate carboxylase.[42,687,694] Two causes of BRMCD have so far been postulated. These are: (1) an impairment in the activity of holocarboxylase synthetase[706,728,948] and (2) decreased activity of the enzyme which releases free biotin from biocytin in the plasma.[981]

FLKS in chickens is proposed to be due to a decrease in the amount of dietary biotin coupled with an increase in stress.[68] In untreated birds, death is thought to be due to hypoglycemia, which results from impaired gluconeogenesis.[68] Supplementation of the diet with biotin prevents onset of the disease.[68]

A number of other diseases in man, other mammals, and birds are also associated with alterations in either the total amount of enzyme protein or the activity of pyruvate carboxylase. These are summarized in Table 1. The changes in pyruvate carboxylase activity have chiefly been detected in the liver and kidney. In each of these diseased states, the change in pyruvate carboxylase appears to be a secondary consequence of the primary cause of the disease, although it may lead to some of the clinical symptoms. Exposure of tissues to alcohol, other drugs, antimetabolites, and heavy metals also leads to an alteration in the activity or total amount of pyruvate carboxylase (Table 1).

II. INHERITED DEFICIENCIES OF PYRUVATE CARBOXYLASE IN CHILDREN

A. Clinical, Biochemical, and Neuropathological Findings

About 30 patients in which there has been reported to be an inherited deficiency of

Table 1
DISEASED STATES WHICH INVOLVE ALTERED PYRUVATE CARBOXYLASE ACTIVITY

Disease	Comments	Refs.[a]
Altered Pyruvate Carboxylase Activity is Primary, or One of the Primary, Causes of Clinical Symptoms		
Inherited deficiency of pyruvate carboxylase in children	Rare; inactive or partially active enzyme protein is thought to be formed	42, 687, 694
BRMCD in man	Rare; decreased concentration of holopyruvate carboxylase	624, 626
FLKS in poultry	Rare but important in commercial poultry industry; decreased concentration of holopyruvate carboxylase	68
Dietary biotin deficiency	Very rare; may be important in special circumstances, e.g., SIDS	430, 955
Altered Pyruvate Carboxylase Activity is Secondary to the Primary Cause of the Disease but Contributes, or may Contribute to the Clinical Symptoms		
Diabetes mellitus	Increased amount of enzyme protein	945, 976
Neonatal hypoglycemia	Inadequate induction of enzyme synthesis and allosteric activation in liver	371, 700
Metabolic acidosis	Inhibition by H^+ of enzyme in the liver is proposed	425
Phenylketonuria	Inhibition by phenylpyruvate of enzyme in brain is proposed	635
Hepatomas	Total amount of enzyme activity (per milligram protein or gram wet weight) is decreased	173
Diseases which lead to mitochondrial damage e.g., Reye's syndrome	Total amount of hepatic activity is decreased	690
Toxic effects of drugs and antimetabolites	Activation or inhibition of enzyme activity by agent or a metabolite, or change in total amount of enzyme protein	6, 221, 521, 555, 824, 832, 997
Heavy metal toxicity	Change in concentration of substrate or change in total amount of enzyme protein	562, 564, 652, 823

[a] The references cited are articles in which the disease or diseased state is reviewed or summarized. More complete lists of references are given in Tables 2, 6, 9, 13, and 14, and in the text.

hepatic pyruvate carboxylase have been described. The majority of these are summarized in Table 2. A child with an apparent deficiency of pyruvate carboxylase was first reported in 1968 by Hommes et al.[416] There is reasonable evidence to indicate that many of the cases so far described (Table 2) represent a true deficiency or abnormality of the pyruvate carboxylase protein. Evidence which supports this conclusion is discussed below. The decrease in pyruvate carboxylase activity is not due to a deficiency of biotin,[230,344,726,910] and hence holopyruvate carboxylase, and is not a secondary con-

Table 2
SOME REPORTED CASES OF PRIMARY PCD IN CHILDREN[a]

Case (Ref.)	Pyruvate carboxylase: Tissue investigated	Pyruvate carboxylase: Activity (% of lowest control value)	Diagnosis of subacute necrotizing encephalomyelopathy (SNE) by autopsy
254, 416, 219, 415	Liver (biopsy)[b]	<1	Yes
			Autopsy-proven SNE also in sibling
415	Liver (biopsy), 4 cases	7, 0.7, 6, 9	No autopsy reported
344, 853, 995	Liver (biopsy)	28	No autopsy reported
	Liver (biopsy)[b]	28	No autopsy reported
	Liver (biopsy)[b]	311	Yes
	Liver (autopsy)[b]	10	Autopsy-proven SNE also in sibling
	Brain (autopsy)[b]	5	
855	Liver (biopsy)	0	No autopsy reported
			Autopsy-proven SNE found in sibling
134	Liver (biopsy)	Change in apparent K_m for pyruvate	No autopsy reported
851	Liver (biopsy)	45	No autopsy reported
345	Liver (biopsy); case 1[a]	100	Not found (autopsy performed)
	Kidney (autopsy), case 2[b]	3	Yes
603	Liver (autopsy)	Deficient	Yes
5	Not Specified	Deficient	No autopsy reported
343	Liver (biopsy)	About 3% (control values not reported)	No autopsy reported
149	Not Specified	Deficient	No autopsy reported
726	Liver (autospy), case 1	0.3	Brain not examined
	Kidney (autopsy), case 2	3	Not found (autopsy performed)
523	Liver (biopsy)	25	No autopsy reported
910	Liver (biopsy)	<1	No autopsy reported
230, 272	Liver (biopsy)	9	No autopsy reported
	Cultured skin fibroblasts	<3	
864	Liver (biopsy)	Greatly reduced	No autopsy reported

852	Liver (biopsy), cases 1 and 2	23, 35	No autopsy reported
911	Liver (autopsy)[b]	30—50	Yes
34, 35, 272	Liver (biopsy)	20	Not found (autopsy preformed)
	Liver (autopsy)	8	
	Kidney (autopsy)	0.2	
	Brain (autopsy)	<1	
	Leukocytes	<1	
	Cultured skin fibroblasts	2	
693, 694	Fibroblasts (5 cases)	<10%	No evidence for SNE
42	Liver (biopsy)	<0.5%	No autopsy reported
	Leukocytes	100	
	Fibroblasts	100	
718	Liver (biopsy) cases 1 and 2	1, 50	Not found (autopsy performed)
	Liver (autopsy) case 2	1	
372	Fibroblasts	<10%	No autopsy reported

[a] Available data indicate that the observed decreases in pyruvate carboxylase activity are not due to secondary deficiencies, including the availability of biotin.

[b] As discussed in the text, these cases may not represent true pyruvate carboxylase deficiencies.

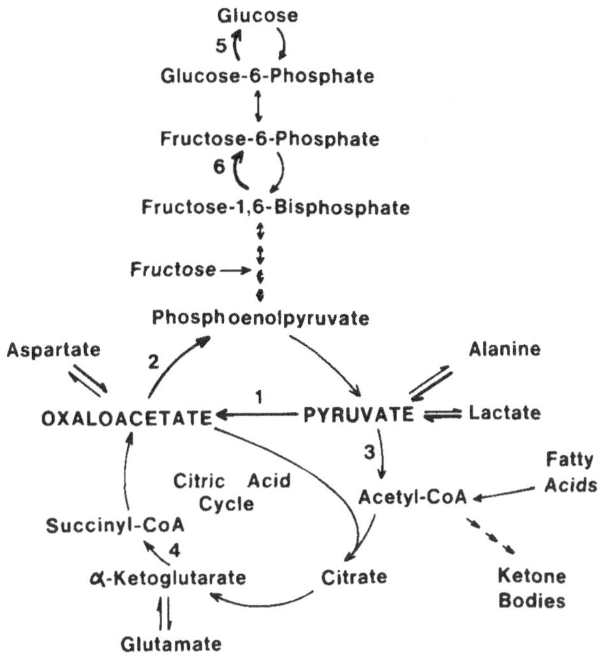

FIGURE 1. Relationship between the reaction catalyzed by pyruvate carboxylase and the interconversion of pyruvate and lactate, the gluconeogenic pathway, the citric acid cycle, and the pathways of fatty acid oxidation, ketone body synthesis, and glutamate metabolism in the liver. In the gluconeogenic pathway, the citric acid cycle, and the reactions of pyruvate oxidation, inherited deficiencies have been observed in the enzymes pyruvate carboxylase (1), PEP carboxykinase (2), the pyruvate dehydrogenase complex (3), the α-ketoglutamate dehydrogenase complex (4), glucose-6-phosphatase (5), and fructose-1,6-bisphosphatase (6).[109] (From Atkin, B. M., Buist, M. R. M., Utter, M. F., Leiter, A. B. and Banker, B. Q., *Pediatr. Res.* 13, 109, 1979. With permission.)

quence of general damage to the liver, which would also affect other mitochondrial enzymes.[34,694]

Pyruvate carboxylase is one of a number of enzymes in the pathways of pyruvate metabolism and gluconeogenesis in which inherited deficiencies have been reported.[109,694] These enzymes include phosphoenolpyruvate (PEP) carboxykinase (E.C. 4.1.1.32), glucose-6-phosphatase (E.C. 3.1.3.9), fructose-1,6-*bis*-phosphatase (E.C. 3.1.3.11), components of the pyruvate dehydrogenase complex, pyruvate dehydrogenase phosphate phosphatase (E.C. 3.1.3.43), and the α-ketogluturate dehydrogenase complex.[109,688,694] The reactions catalyzed by these enzymes are summarized in Figure 1.

Two main clinical features of patients with inherited pyruvate carboxylase deficiency are congenital lactic acidosis and deterioration of the central nervous system (CNS) (summarized in References 34, 42, 687, and 694). Other symptoms and signs are listed in Table 3. In many cases the lactic acidosis is severe and contributes to neonatal death, while in some cases lactic acidosis is mild, late in onset, and is associated with less severe lesions in the CNS (summarized in Reference 34). Although the activity of pyruvate carboxylase in patients with a deficiency of this enzyme is quite low, as discussed below, chronic hypoglycemia and episodes of pronounced acute hypoglycemia are not generally observed.[35,230] Proximal renal tubular acidosis has been reported in three cases.[34,344,345]

Table 3
CLINICAL, BIOCHEMICAL, AND NEUROPATHOLOGICAL FINDINGS IN 15 PATIENTS WITH REPORTED HEPATIC PCD[a]

Clinical findings	Number of patients[b]
Onset	
0—3 months	7
0—6 months	9
0—12 months	12
Course	
Rapidly progressive	4
Progressive	11
Progressive with episodic exacerbations	2
Clinical symptoms and signs	
Mental retardation	13
Hypotonia	8
Convulsions	7
Respiratory distress	3
Hypoflexia/areflexia	3
Eye-movement disorder	3
Microcephaly	3
Dystonia	1
Epileptic electroencephalogram	10
Enlargement lateral ventricle	2
Biochemistry	
Serum lactate raised	15(15)
Serum pyruvate raised	11(12)
Serum alanine raised	10(12)
Hypoglycemia	3(8)
Cerebrospinal fluid lactate raised	4(4)
Cerebrospinal fluid pyruvate raised	3(3)
Cerebrospinal fluid alanine raised	1(1)
Autopsy	
Poliodystrophy	2(4)
Hypomyelinisation	3(4)
Leigh's syndrome	1(4)

[a] All 15 cases reported here are included in the summarized data in Table 1. As discussed in the text, these cases include some in which it is doubtful whether a true deficiency of hepatic pyruvate carboxylase was present.

[b] Where a given parameter was not measured in all 15 cases, the number in parenthesis represents the total number of cases investigated.

From Baal, E. G., Gabreels, F. J., Renter, W. O., Hommes, F. A., Gijsberts, T. H., Lamers, K. J., and Kok, J. C., *Dev. Med. Child Neurol.*, 23, 521, 1981. With permission.

It has been suggested that patients with an inherited deficiency of pyruvate carboxylase can be classified in two groups. These are patients whose symptoms present in the 1st year of life, and patients whose symptoms present at between 1 and 5 years of age.[694] About 80% of the reported cases fall into the first group.[42]

Values for hepatic pyruvate carboxylase activity (expressed as a percentage of the amount in control livers) in patients with reported pyruvate carboxylase deficiency are summarized in Table 2. Early investigators were faced with two main problems in the measurement of hepatic pyruvate carboxylase activity. These were: (1) to obtain appropriate tissue samples for estimation of the normal activity and (2) the instability of the enzyme in post-mortem tissue samples.

In retrospect, it seems likely that pyruvate carboxylase activity was assayed under less than optimal conditions in some early investigations.[34,344,345,371,589,726,736] It has been suggested that low values for the activity of pyruvate carboxylase in extracts of the livers of patients may have been due to the use of hypotonic homogenization media[34,589] in the studies of Hommes et al.,[416] Hommes,[415] and Grobe et al.,[343] or to the loss of enzyme activity in samples of tissues taken at autopsy[34,344,345,371,726] in the studies of Gruskin et al.,[345] Grover et al.,[344] and Van Bierviet et al.[911] Furthermore, in studies of Hommes et al.[416] and Tada et al.[853] the values for enzyme activity for samples obtained from the livers of normal subjects are also very low compared with values obtained in other laboratories.[344] In a re-examination of the patient reported by Hommes et al.,[416] Schrijver[736] found much higher amounts of hepatic pyruvate carboxylase than those observed in the orignal study.

It is doubtful, therefore, whether a number of the cases listed in Table 2 (indicated by footnote "b") represent true pyruvate carboxylase deficiency (PCD). However, this conclusion cannot be fully substantiated without further data for the activity of hepatic pyruvate carboxylase measured under more appropriate assay conditions. In most cases, these data are not available.

In those cases in which there is reasonable evidence for a pyruvate carboxylase deficiency,[5,34,35,42,134,230,523,603,693,694,718,726,851,852,855,864,910,995] the activity of hepatic pyruvate carboxylase is decreased to between about 3 and 25% of the amount of enzyme present in control livers (Tables 2 and 4).

In contrast to studies of the total activity of pyruvate carboxylase in inherited PCD, there is presently less information about changes in the total amount of cellular pyruvate carboxylase protein in this disease.[109] Robinson et al.[693] have used [8,9-^{3}H] biotin to label biotin-dependent carboxylase enzymes in biotin-depleted cultured fibroblasts obtained from five patients[694] with an inherited PCD. In four of these cell lines, the labeling of fibroblasts with [^{3}H] biotin and subsequent separation of proteins by sodium dodecyl sulfate - polyacrylamide gel electrophoresis (SDS-PAGE) yielded a distinct band of radioactivity which corresponded to pyruvate carboxylase (Figure 2). This band was also observed if the fibroblast extracts were first treated with anti-pyruvate carboxylase antiserum before electrophoresis of the immunoprecipitate. In the fifth case no band of [^{3}H] biotin-labeled pyruvate carboxylase could be detected.

These results are consistent with the synthesis of inactive apopyruvate carboxylase protein to which biotin can link to form an inactive holoenzyme in patients with inherited PCD. Since the intensity of the [^{3}H] biotin-labeled bands was not measured quantitatively, the amount of inactive holoenzyme is not yet known. Preliminary reports of Feldman and Wolf[273] and Atkin et al.[33] are also consistent with the formation of enzymatically inactive but immunologically active pyruvate carboxylase protein in patients with inherited PCD. However, the results of Robinson et al.[693] (Figure 2) indicate that there may be other cases of PCD in which no apopyruvate carboxylase is formed or in which the apoenzyme cannot bind biotin. This would suggest that there is more than one type of mutation which leads to inherited PCD, although other explanations for the data are possible.

Decreased amounts of pyruvate carboxylase have also been observed in the kidney,[34,345,726] brain,[34,344] cultured skin fibroblasts,[35,272] and leukocytes[35] in patients with hepatic PCD (Tables 2 and 4). Baal et al.[42] did not detect a decrease in enzyme activity in leukocytes and fibroblasts (Table 2). This may be due to difficulties encountered in measurement of the activity of the enzyme in fibroblasts and leukocytes since the amount of enzyme present in these tissues is very much lower than that present in the

Table 4
LEVELS OF PYRUVATE CARBOXYLASE AND CITRATE SYNTHASE IN LIVER, KIDNEY, AND BRAIN OF A PATIENT WITH PCD AND IN CONTROLS[a]

Tissue		Pyruvate carboxylase (units/g fresh wt)	Citrate synthase (units/g fresh wt)
Liver			
Patient	(biopsy)	1.81 ± 0.20 (3)	20.8 ± 2.45
Patient	(autopsy)	0.75 ± 0.07 (4)	
Control	Anencephalic child (autopsy)	10.9 ± 0.3 (9)	12.3 ± 1.4
Control	Adult with carcinoma of pancreas (biopsy)	11.3	
Control	Adult; death from brain hemorrhage (autopsy)	9.5 ± 1.2 (3)	10.2 ± 0.45
Kidney			
Patient	(autopsy)	0.008 ± 0.004 (5)	23.0 ± 2.45 (3)
Control	Normal 16-year-old; death by auto accident	5.05 ± 0.24 (2)	13.3 ± 0.8 (2)
Brain			
Patient	(autopsy)	0.018	19.8 ± 0.30 (2)
Control	14-year-old with Down's syndrome (autopsy); death from congestive heart failure	0.44 ± 0.027 (3)	12.8 ± 1.1 (2)
Control	Normal 2 month old rhesus monkey (autopsy)	0.53 ± 0.015 (2)	20.6 ± 0.2
Control	Normal adult rhesus monkey (autopsy)	0.695 ± 0.02	

[a] Units of enzyme activity (measured at 37°C) are: for pyruvate carboxylase, μmole HCO_3^- fixed/min; and for citrate synthetase, μmole CoA-SH formed/min.

From Atkin, B. M., Buist, N. R. M., Utter, M. F., Leiter, A. B., and Banker, B. Q., *Pediatr. Res.*, 13, 109, 1979. With permission.

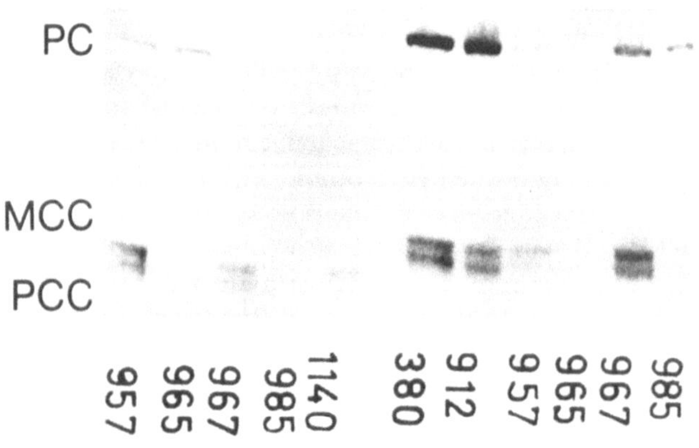

FIGURE 2. SDS-PAGE patterns of [^{3}H]biotin-labeled proteins in extracts of cultured skin fibroblasts from patients with inherited PCD (cell lines 957, 967, 1140, 380 and 912) and BRMCD due to a postulated defect in holocarboxylase synthetase (965 and 985). PC, the subunit of pyruvate carboxylase; PCC, the α-subunit of propionyl-CoA carboxylase; MCC, the α-subunit of methylcrotonyl-CoA carboxylase. (From Robinson, B. H., Oei, J., Saunders, M., and Gravel, R., *J. Biol. Chem.* 258, 6660, 1983. With permission.)

liver.[35] The observation that hepatic PCD is associated with a deficiency of the enzyme in a number of other tissues suggests that the same form of the enzyme is present in all tissues.[35] It also suggests that the activity of the enzyme is decreased in all tissues of patients with PCD.

B. Possible Relationship Between Pyruvate Carboxylase Deficiency and Subacute Necrotizing Encephalomyelopathy (Leigh's Disease)

The early investigations of hepatic PCD included a number of patients with symptoms which suggested subacute necrotizing encephalomyelopathy (SNE). Some of these patients later died and SNE was confirmed by post-mortem examination.[344,345,415,603,911] The symptoms of SNE, which were first described by Leigh,[494] include ataxia, seizures, tiredness, slow development, hypotonia, congenital lactic acidosis, and elevated concentrations of blood pyruvate and alanine.[110,494] The disease can only be diagnosed positively after death by the finding of a characteristic pattern of necrotizing lesions within the brainstem.[333,653] The apparent correlation between pyruvate carboxylase deficiency and SNE led to the proposal by some investigators that hepatic PCD is the underlying metabolic defect in SNE.[218,219,416,654,853]

Two lines of evidence suggest that (PCD) is unlikely to be the cause of SNE and is not directly associated with this disease. Firstly, there is presently no compelling evidence to indicate that there have been any cases of inherited PCD with SNE. Thus, only about five of the reported cases of PCD (Table 2) have been positively diagnosed as SNE by post-mortem examination.[34,371] These are the patients described by Hommes et al.,[219,415,416] Grover et al.,[344] Gruskin et al.,[345] Moosa and Hughes,[603] and Van Biervliet et al.[911] As discussed above, in four of these there is some doubt about the validity of the assay of pyruvate carboxylase activity, and in the fifth case (that of Mossa and Hughes[603]), details of the assay procedure were not reported. Futhermore, a recent thorough study of a case of PCD provided good evidence for a deficiency of the enzyme in the liver and in a number of other tissues. However, an intensive post-mortem examination revealed no evidence of SNE.[34] These observations do not exclude the possibility that other cases of inherited PCD listed in Table 2 had or have SNE, but in the absence of further information this appears to be unlikely.

Secondly, other studies have shown that a number of cases of SNE, diagnosed by post-mortem examination, did not have PCD.[371,588,589,886,890] Thus, it has been concluded that PCD is not the cause of SNE, but is a separate disease.[34,109,343,344,371,589] Since the association of some cases of PCD (Table 2) with SNE is not excluded (e.g., Tang et al.,[855] in view of a positive diagnosis of SNE in a sibling of the patient; Grover et al.,[344] in view of the observation of renal tubular acidosis), the possibility that PCD may, in some circumstances, contribute to abnormalites in patients with SNE cannot be excluded.

The primary biochemical defects in SNE have not been established. However, they may include an abnormality in the metabolism of thiamine triphosphate, and/or the reactions catalyzed by transketolase and the pyruvate dehydrogenase complex (summarized by Toshima et al.[886] and Hansen et al.[371]).

C. Detection and Mechanism of Inheritance

In order to detect PCD in children with congenital lactic acidosis and in the families of patients with a known deficiency of the enzyme, methods for the measurement of pyruvate carboxylase activity in cultured skin fibroblasts,[32,35,93,230,272,668,694] leukocytes,[35] and circulating peripheral lymphocytes[32] have been developed. These are based on the procedures of Utter and Keech[899] and Ballard and Hanson.[52] Since the amount of pyruvate carboxylase present in these cell types is much lower than that present in liver (Table 5 compared with Table 4), considerable care is required for the accurate estimation of pyruvate carboxylase activity.[32,35]

Table 5
PYRUVATE CARBOXYLASE ACTIVITES IN LEUKOCYTES AND FIBROBLASTS OF NORMAL SUBJECTS AND A PATIENT WITH PCD[a]

Subject	Pyruvate carboxylase (mU/mg protein)
Mixed Peripheral Leukocytes	
Patient	0.009 (2)
Controls: children	0.089 ± 0.005 (n = 2)
Controls: adults	0.095 ± 0.014 (n = 3)
Fibroblasts	
Patient: TC 5, 12 month	0.018 ± 0.001 (4)
Control: TC 16, 15 months, normal	1.05 ± 0.078 (5)
Control: TC 48, <3 days, normal	1.95 ± 0.096 (4)
Control: TC 65, <7 months, lactic acidosis	1.64 ± 0.105 (2)
Control: ST 12, adult normal	0.95 ± 0.069

[a] Units of pyruvate carboxylase activity are expressed as milliunits (mU/mg protein. A milliunit represents 1 nmole HCO_3^- fixed per min at 37°C.

From Atkin, B. M., Utter, M. F., and Weinberg, M. B., *Pediatr. Res.* 13, 38, 1979. With permission.

As discussed previously, decreased activity of pyruvate carboxylase (generally expressed as U/mg protein) is observed in cultured skin fibroblasts[32,35,272] (but see Baal et al.[42]), leukocytes[35] and lymphocytes[32] taken from patients shown to have decreased amounts of hepatic pyruvate carboxylase in liver biopsy samples. Takada et al.[854] have also shown that the activity of pyruvate carboxylase in lymphoblasts (phytohemaglutinin-transformed lymphocytes) from a patient with BRMCD is decreased. Thus, it has been concluded that the activity of pyruvate carboxylase in cultured skin fibroblasts and in lymphocytes, leukocytes, and lymphoblasts can be used as an indicator of hepatic pyruvate carboxylase activity.[32,35,272,694,854]

Measurement of pyruvate carboxylase activity in cultured fibroblasts has been used to scan cases of congenital lactic acidosis of previously unknown cause. In a group of 40 such cases, the activity of pyruvate carboxylase in five patients was found to be less than 10% of the normal value,[694] indicating a deficiency of hepatic carboxylase.

Pyruvate carboxylase activity has also been detected and measured in cells taken from amniotic fluid in normal subjects,[272,367-370,524] with a view to the use of these measurements in prenatal diagnosis of deficiency of the enzyme.[272,367-370] This approach has been used successfully in the prenatal detection and treatment of a case of biotin-responsive multiple carboxylase deficiency.[623]

The mechanism of inheritance of PCD has been investigated by measuring the activity of pyruvate carboxylase in cultured fibroblasts, lymphocytes, and leukocytes obtained from members of the families of patients with a deficiency of the enzyme. For two patients, PC (Portland) and PC (St. Louis), the activity of pyruvate carboxylase in cells obtained from the parents and a sister was decreased (Figure 3).[32,230] The results were found to be consistent with inheritance of the defect in an autosomal recessive manner (Figure 3).[32] The mutation which leads to inherited PCD has been designated *pyc*.[273] Genetic complementation studies conducted with fibroblast cell lines derived from patients deficient in pyruvate carboxylase and from two patients with biotin-responsive multiple carboxylase deficiency have indicated that the locus *pyc* is separate

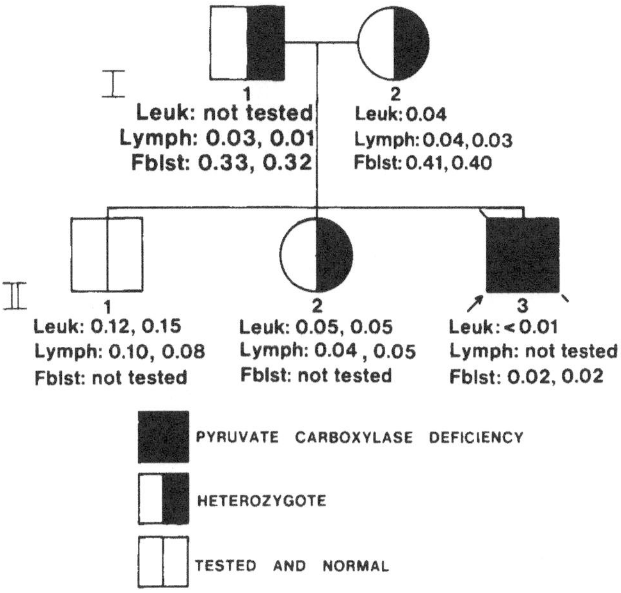

FIGURE 3. Pedigree of family with PC (Portland) deficiency. The propositus (II, 3) died at 46 months of age. The values represent nmoles HCO_3^- fixed per minute per milligram protein at 37°C. Fblst: activity determined in fibroblasts; Lymph: activity determined in lymphocytes; Leuk: activity determined in a mixed leukocyte preparation. Normal values are: fibroblasts, 1.35 ± 0.173 (n = 5); lymphocytes, 0.156 ± 0.011 (n = 19); leukocytes, 0.089 ± 0.013 (n = 5) mU/mg protein. (From Atkin, B., *Pediatr. Res.*, 13, 1101, 1979. With permission.)

from the loci of mutations for biotin-responsive multiple carboxylase deficiency (*bio*).[273]

D. Immediate Metabolic Consequences of Pyruvate Carboxylase Deficiency

The experimental evidence reviewed above indicates that a group of patients has been described in which there is a mutation which leads to PCD in all tissues. The enzymatic defect is likely to be the sole and primary cause of the clinical symptoms and signs exhibited by these patients, although this has not been shown conclusively.

On the basis of the metabolic pathways involved (Figure 1), it can be predicted that hepatic PCD would lead to an increase in the steady-state intracellular concentrations of pyruvate and lactate and to a decrease in the concentration of oxaloacetate. This, in turn, would be expected to cause a decrease in the rate of gluconeogenesis from three-carbon precursors and in flux through citrate synthetase.[34,134,230,344,726] The high concentrations of plasma lactate and pyruvate which are generally observed, and low concentrations of plasma glucose which are sometimes observed in children with PCD (Table 3) are consistent with these predictions.[34,134,230,344,726] The absence of chronic hypoglycemia and frequent episodes of acute hypoglycemia[34,230] may be due to the ability of liver cells to synthesize some glucose from three-carbon precursors through the low residual activity of hepatic pyruvate carboxylase (Tables 2 and 4) and from substrates other than three-carbon precursors in reactions which do not require pyruvate carboxylase.[34,230] Other consequences of a diminished rate of synthesis of oxaloacetate in liver cells are likely to be decreased activity of those redox shuttles (described in Reference 893) which utilize oxaloacetate,[726] and an increased steady-state concentration of mitochondrial acetyl-CoA.[230] This would lead to an increased rate of ketone body synthesis in response to an i.v. glucose tolerance test or high-fat diet.[230]

In cells of the CNS, the decreased activity of the citric acid cycle, and hence rate of ATP synthesis which results from a deficiency of pyruvate carboxylase and oxaloacetate is one factor which may contribute to the development of neurological lesions.[852,855] Other factors which may be important are a decrease in the rate of formation of four-carbon precursors for anaplerotic synthesis and the effect of abnormal plasma concentrations of lactic acid.[34,344] The relative contribution of each of these processes to the onset and progression of the neurological lesions and whether other factors are involved is yet to be determined.

Proximal renal tubular acidosis has been observed in at least three cases of hepatic PCD.[34,344,345] In one, the activity of pyruvate carboxylase in the kidneys was measured at post-mortem and found to be about 10% of the normal value (Table 4).[34] On the basis of these observations, Atkin et al.[34] proposed that the reaction catalyzed by pyruvate carboxylase is part of the mechanism by which HCO_3^- is resorbed in the proximal tubule. Other data consistent with this proposal have been reported.[143] If renal pyruvate carboxylase is involved in HCO_3^- resorption, a deficiency of the enzyme would be expected to lead to impaired retention of HCO_3^-, and may contribute to the observed acidosis.[34,143]

A number of diets for the treatment of patients with PCD have been devised. De Vivo et al.[230] proposed that the best approach is to enhance the rate of synthesis of oxaloacetate. Some improvement has been reported in patients treated with aspartate[42,219,343,416] or glutamate[855] with or without one or more vitamins (including thiamine, lipoic acid, and pyridoxine), although, as discussed above (Table 2), some of these patients may not have had PCD. There is no evidence that treatment with biotin[230,344,911] or thiamine alone[134,718] leads to a remission of symptoms. Notwithstanding the observation that precursors of oxaloacetate, such as aspartate or glutamate (Figure 1), may cause some remission of symptoms, no effective treatment of the primary consequence of PCD has been so far devised. In order to prevent the onset of major neurological lesions it would be necessary to make an early diagnosis of the disease and begin treatment immediately after birth (compare with References 272 and 367 to 370).

III. DECREASED AVAILABILITY OF BIOTIN

A. Biotin-Responsive Multiple Carboxylase Deficiency in Children

1. Clinical and Biochemical Findings

In the inherited disease, biotin-responsive multiple carboxylase deficiency (BRMCD), the concentration of holopyruvate carboxylase, and hence the activity of pyruvate carboxylase is greatly reduced (summarized in References 81, 583, 624, 626, 868, and 982). Three other enzyme reactions in the mammalian liver are known to require biotin as a cofactor. These are propionyl-CoA carboxylase (EC 6.4.1.3) and 3-methylcrotonyl-CoA carboxylase (EC 6.4.1.4), which are both located in the mitochondria, and acetyl-CoA carboxylase (EC 6.4.1.2), which is located in the cytoplasm[576] (enzymes 2, 3, and 4, respectively, in Figure 4). The activities of these enzymes are also decreased in BRMCD, although changes in the activity of acetyl-CoA carboxylase appear to be small. In untreated patients, the consequences of deficiencies of these enzymes are severe. However, patients can be effectively treated with pharmacological doses of biotin which restore the activity of the biotin-dependent carboxylases to normal or near-normal values (summarized in References 81, 583, 624, 626, 868, and 982). As discussed below, the underlying enzymatic defect appears to be either an impairment in the reaction which catalyzes the addition of biotin to the apocarboxylases,[706,728,948] or a decrease in the rate of release of free biotin from biocytin.[981]

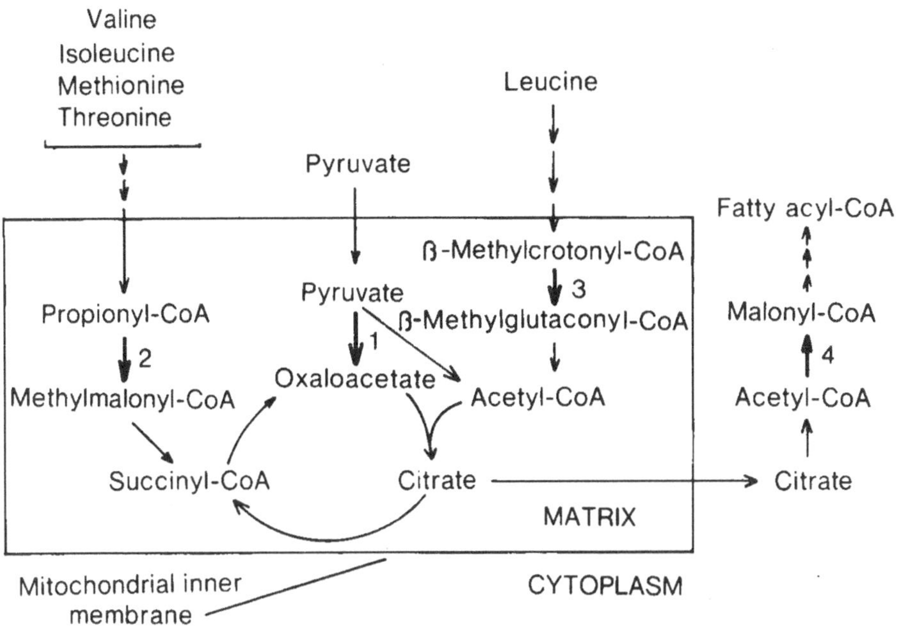

FIGURE 4. A schematic representation of the roles of the four known biotin-dependent carboxylase enzymes in intermediary metabolism in the liver cell.[576] The enzymes are (1) pyruvate carboxylase; (2) propionyl-CoA carboxylase (E.C. 6.4.1.4); (3) β-methylcrotonyl-CoA carboxylase (E.C. 6.4.1.3); (4) acetyl-CoA carboxylase (E.C. 6.4.1.2).

Two forms of BRMCD have been identified on the basis of differences in the age of onset and the clinical and biochemical symptoms. These have been designated "neonatal" and "infantile" (or "late onset").[583,622,626,846,868,979,981,982] In the future, it may prove more convenient to classify these diseases on the basis of the underlying biochemical defect and the nature of the major clinical symptoms rather than on the basis of the age of onset. Therefore, the terms "Type A" and "Type B" have been employed to describe the neonatal and infantile or late onset forms of the disease, respectively, in the present discussion. On the basis of information presently available, including the data summarized in Table 6, the cases described in References 82, 270, 321, 624, 625, 704, 706, and 982, and possibly those in References 705 and 493, have been classified as Type A (neonatal), and those described in References 88, 179, 208, 280, 581 to 586, 626, 721, 848, 868, 871, and 981 as Type B (infantile, late onset).

Both types of BRMCD are characterized by the excretion of abnormally high amounts of organic acids in the urine (e.g., those shown in Figure 5). These include lactate, 3-hydroxypropionate, methylcitrate, 3-hydroxyisovalerate, and 3-methylcrotonylglycine.[82,88,179,208,321,583,585,586,624,704-706,848,868,871,982] Symptoms and signs of patients with Type A (neonatal) BRMCD also include lethargy, vomiting, hypotonia, metabolic ketoacidosis and lactic acidosis, hyperammonemia, and coma.[82,624,705,982]

The symptoms and signs of patients with Type B (infantile or late-onset) disease include alopecia (Figure 6), skin rash, canidia dermatitis, hypertonia, lactic acidosis, ketoacidosis, ataxia, keratoconjuctivitis, motor retardation and developmental delay,[179,208,280,583-586,721,848,868,871] and a decreased concentration of plasma biotin (Table 6). There is also evidence for a deficiency of the immune system in the Type B disease.[208,280,581,721] For both types of BRMCD, the symptoms and signs are almost completely reversed by the administration of pharmacological doses of biotin (e.g., Figure 5).[82,88,179,208,280,321,583,585,586,624,704,706,721,778,846,848,868,871,982] The doses of biotin employed range from 5 to 40 (often 10) mg per child per day compared with the recommended

Table 6
CHANGES IN THE CONCENTRATIONS OF BIOTIN AND BIOTIN-DEPENDENT CARBOXYLASE ENZYMES IN CASES OF BRMCD IN COMPARISON WITH NORMAL VALUES[a]

Parameter	Type A (neonatal) Change[b]	Type A (neonatal) Ref.	Type B (infantile or late-onset) Change[b]	Type B (infantile or late-onset) Ref.
Plasma or urinary biotin concentration	Unchanged	624	Decreased	88, 280, 582, 583, 585, 626, 868
	Decreased	704		
Pyruvate carboxylase activity in:				
Liver (biopsy)[c]	Not reported		Decreased	
Cultured skin fibroblasts[c]	Decreased in presence of low or normal biotin in culture medium. Return control values in presence of high biotin in culture medium[d]	82, 270, 530, 622, 623, 624, 728, 729, 778, 982	Unchanged in presence of either low or high biotin in culture medium	88, 179, 280, 530, 581, 583, 584, 622
Peripheral blood leukocytes[c]	Decreased (return to control values after treatment of patient with biotin)	982	Decreased (return to control values after treatment of patient with biotin)	583, 626, 848
Lymphocytes or lymphoblasts[c]	Not reported		Decreases (return to control values after treatment of patient with biotin)	88, 854, 871
Acetyl-CoA carboxylase activity[e]	Decreased	274	Decreased (40—50% of control)	280, 581, 583, 585

[a] The classification of a given case of BRMCD as Type A or Type B has been made using available data which, in some cases, are reasonably limited.

[b] Compared with values for the control or normal state.

[c] The primary data for enzyme activity were expressed as micromoles oxaloacetate formed per minute per gram wet wt or as nanomoles per minute per milligram protein. Changes in the activities of propionyl-CoA carboxylase and methylcrotonyl-CoA carboxylase in liver, cultured fibroblast, leukocytes, and lymphocytes in type A[82,270,622,624,728,729,982] and Type B[88,179,280,581,583,584,622,626,848,854,871] MCD are similar to those for pyruvate carboxylase.

[d] In two cases the increase in pyruvate carboxylase activity which followed the inclusion of high concentrations of biotin was either very slow[270] or did not reach the normal (control) value.[982]

[e] Assayed in liver tissue[280,581,583,585] or cultured fibroblasts.[274]

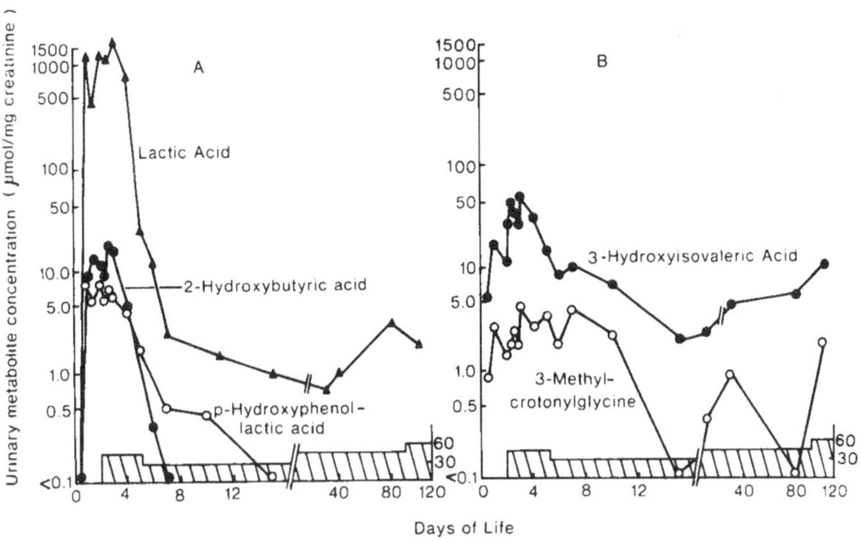

FIGURE 5. Effect of the administration of biotin on the concentration of organic acids in the urine excreted by a patient with Type A BRMCD. (A) Metabolism caused by deficiencies of pyruvate carboxylase and propionyl-CoA carboxylase activities and other secondary enzymes; (B) metabolites caused by deficiency of β-methylcrotonyl-CoA carboxylase activity. (From Wolf, B., Hsia, Y. E., Sweetman, L., Feldman, G., Boychuk, R. B., Bart, R. D., Crowell, D. H., DiMauri, R. M., and Nyhan, W. L., *Pediatrics,* 68, 113, 1981. With permission.

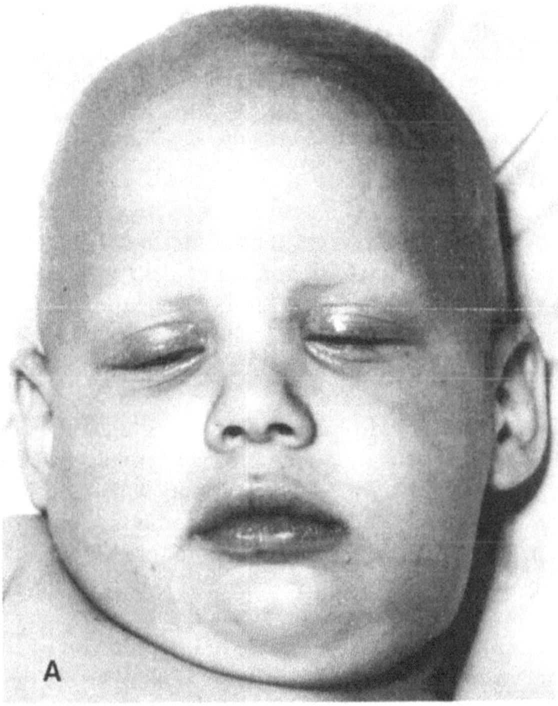

FIGURE 6. Effect of therapy with biotin on the appearence of a patient with Type B (infantile or late onset) BRMCD. (A) Shows the baldness (alopecia) of the patient at 2 years and 9 months of age before treatment with biotin was initiated; (B) shows the patient at 3 years and 1 month of age after 4 months of treatment with biotin (10 mg/day given orally). (From Thoene, J., Baker, H., Yoshino, M., and Sweetman, L., *New Engl. J. Med.,* 304, 817, 1981. With permission.

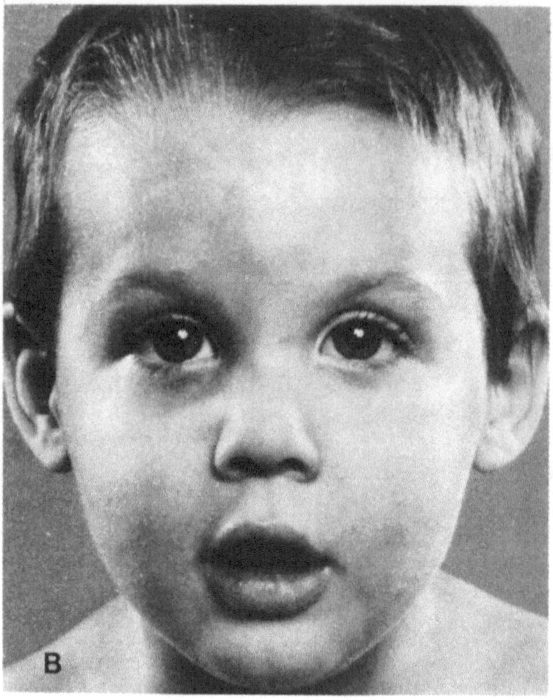

FIGURE 6B

daily allowance of biotin of about 0.1 to 1.0 mg per person per day.[115] In some cases a partial remission of symptoms has been observed following the inclusion of additional fatty acids in the diet in the absence of additional biotin.[584]

2. Pyruvate Carboxylase Activity

In Type A (neonatal) BRMCD, the activity of pyruvate carboxylase in skin fibroblasts cultured in the presence of low or normal concentrations of biotin, and in leukocytes, is decreased in comparison with control values (Tables 6 and 7). Treatment of patients with biotin, or the inclusion of high concentrations of biotin in the culture medium restores the activity of pyruvate carboxylase in leukocytes and fibroblasts, respectively, to normal values (Tables 6 and 7).

Decreased activity of pyruvate carboxylase is observed in the liver, leukocytes, and lymphocytes (Tables 6 and 8) and in lymphoblasts[854] of patients with Type B (infantile or late-onset) BRMCD. The activity of the enzyme in leukocytes and lymphocytes is restored to normal values after treatment of patients with biotin (Tables 6 and 8). In contrast, the activity of pyruvate carboxylase in fibroblasts derived from these patients is the same as that present in fibroblasts from normal subjects, even when the culture medium contains very low concentrations of biotin (Table 6). The reason for this is not clear,[179] although it has been suggested that it is due to insufficient removal of biotin from the culture medium.[143] It is also possible that fibroblasts derived from these patients have an enhanced ability to retain biotin or metabolize biocytin (see below) compared with that of leukocytes and lymphocytes.

Studies conducted with cultured skin fibroblasts taken from normal subjects have confirmed that the activity of pyruvate carboxylase in these cells normally depends on the concentration of biotin in the culture medium.[318,978] This depletion of biotin by prior treatment of the culture serum with avidin decreases the activity of pyruvate carboxylase by about 80%. The addition of high concentrations of biotin to cells in-

Table 7
EFFECT OF BIOTIN ON BIOTIN-DEPENDENT CARBOXYLASE ACTIVITIES IN CULTURED SKIN FIBROBLASTS FROM A PATIENT WITH TYPE A (NEONATAL) BRMCD

	Enzyme activities in fibroblasts (pmol/min/mg protein)		
Cultured medium conditions	Pyruvate carboxylase	Propionyl-CoA carboxylase	β-Methylcrotonyl-CoA carboxylase
Fibroblasts from patient; cells incubated without supplemented biotin (about 2 μg biotin per liter)	12	30	5
Fibroblasts from patient; cells incubated with supplemented biotin (10 mg biotin per liter)	380	390	71
Fibroblasts from normal subject; cells incubated without supplemented biotin	260—1150	660—1100	160—330

From Wolf, B., Hsia, Y. E., Sweetman, L., Feldman, G., Boychuk, R. B., Bart, R. D., Crowell, D. H., Di Mauri, R. M., and Nyhan, W. L., *Pediatrics*, 68, 113, 1981. With permission.

Table 8
EFFECT OF TREATMENT WITH BIOTIN ON THE ACTIVITY OF PYRUVATE CARBOXYLASE IN LEUKOCYTES FROM A PATIENT WITH TYPE B (INFANTILE OR LATE ONSET) BRMCD

Treatment of patient with biotin	Pyruvate carboxylase activity (pmol/min/mg protein)
Initial studies	
Patient: before treatment	0.045 ± 0.007 (2)
Patient: 1 day on biotin	0.48 ± 0.28 (2)
Controls	0.66 ± 0.63 (4)
Later studies	
Patient; days on biotin	
4	1.8
6	1.2
8	1.7
14	2.8
Controls	0.69 ± 0.40 (4)

From Packman, S., Sweetman, L., Yoshino, M., Baker, H., and Cowan, M., *J. Pediatr.*, 99, 421, 1981. With permission.

cubated in normal or biotin-deficient media increases the activity of the enzyme to about 200% of that present in cells incubated at normal concentrations of biotin, but does not increase the activities of propionyl- and methylcrotonyl-CoA carboxylase above the normal value.[318] This suggests that in fibroblasts incubated under normal conditions some pyruvate carboxylase enzyme is present as apocarboxylase. The administration of biotin (0.14 mg/kg/day) to normal subjects has also been shown to increase the specific activities of all three mitochondrial biotin-dependent carboxylase enzymes in leukocytes. These increases were not observed when leukocytes were incubated with biotin in vitro.[825]

The effects of BRMCD, and of the treatment of patients with this disease with biotin, on the activities of propionyl-CoA carboxylase and methylcrotonyl-CoA carboxylase are similar to those on pyruvate carboxylase (Table 6).[82,88,179,270,280,581,583,584,622,624,626,728,729,848,854,871,982] In contrast to the three mitochondrial biotin-dependent carboxylases, with one or two exceptions (e.g., Feldman and Wolf[274]), a much smaller decrease, or in some cases no decrease, in activity of acetyl-CoA carboxylase has been observed in patients with BRMCD (Table 6).[274,280,581,583,585] The reason for this difference may be because acetyl-CoA carboxylase is more difficult to assay, or that the activity of this enzyme does not change markedly with alterations in the availability of biotin.[825]

3. Proposed Underlying Metabolic Defects

The observation that plasma and urinary concentrations of biotin are within the normal range in patients with Type A (neonatal) BRMCD suggested that the absorption of biotin from the lumen of the intestine to the bloodstream is normal. Indirect evidence which suggests that this process is not impaired has been obtained.[624] The results of earlier investigations led to the proposal that the underlying defect in the Type A disease is a deficiency of holocarboxylase synthetase.[144,270,583,729,982] This enzyme catalyzes the addition of biotin to the apocarboxylase protein. The reaction, which is described in detail in Chapter 2 (Section II.B.4) is

$$\text{Apocarboxylase} + \text{ATP} + \text{Biotin} \rightarrow \text{Biotinyl} - \text{Carboxylase} + \text{AMP} + \text{PPi}$$

There is some evidence that cells contain two holocarboxylase synthetase enzymes, one located in the mitochondrial matrix and the other in the cytoplasm.[1,611]

Indirect evidence has been obtained which indicates that the activity of mitochondrial holocarboxylase synthetase is decreased in patients with Type A BRMCD.[530,706,727,728,948] Furthermore, direct experiments using (1) apopropionyl-CoA carboxylase dervied from biotin-deficient rats and lymphocyte homogenates or fibroblast sonicates[623,845-847] or (2) human fibroblast apopropionyl-CoA carboxylase and extracts of fibroblast mitochondria,[317] have shown that the apparent K_m for biotin of holocarboxylase synthetase is increased by about 200- to 400-fold in a number of patients with Type A disease (Figure 7). Packman et al.[623] also reported a decrease in the maximum velocity of holocarboxylase synthetase.

Further confirmation of the proposal that the central defect in Type A BRMCD is an increase in the apparent K_m of the mitochondrial holocarboxylase synthetase for biotin awaits direct measurement of the activity of this enzyme in partially purified extracts of the cells of other patients with both Type A and B BRMCD. Further experiments are also required to eliminate other possible explanations for decreased amounts of holopyruvate carboxylase in the cells of patients with Type A BRMCD. These include alterations in the rate of transport of biotin across the plasma membrane and between subcellular compartments.[271,624,982]

Robinson et al.[693] observed that the incubation of biotin-depleted cultured fibroblasts obtained from two patients with Type A BRMCD in the presence of [^{3}H] biotin leads to the incorporation of some [^{3}H] biotin into the band corresponding to the pyruvate carboxylase subunit (cell lines 965 and 985 in Figure 2). Some labeling of methylcrotonyl- and propionyl-CoA carboxylase was also observed, but this was weaker than that found for these enzymes in normal fibroblasts. These results suggest that in the cells of the two patients the addition of biotin to at least two biotin-dependent carboxylase enzymes is impaired. Quantitative measurements and comparison of rates of [^{3}H] biotin incorporation will be required to confirm this.

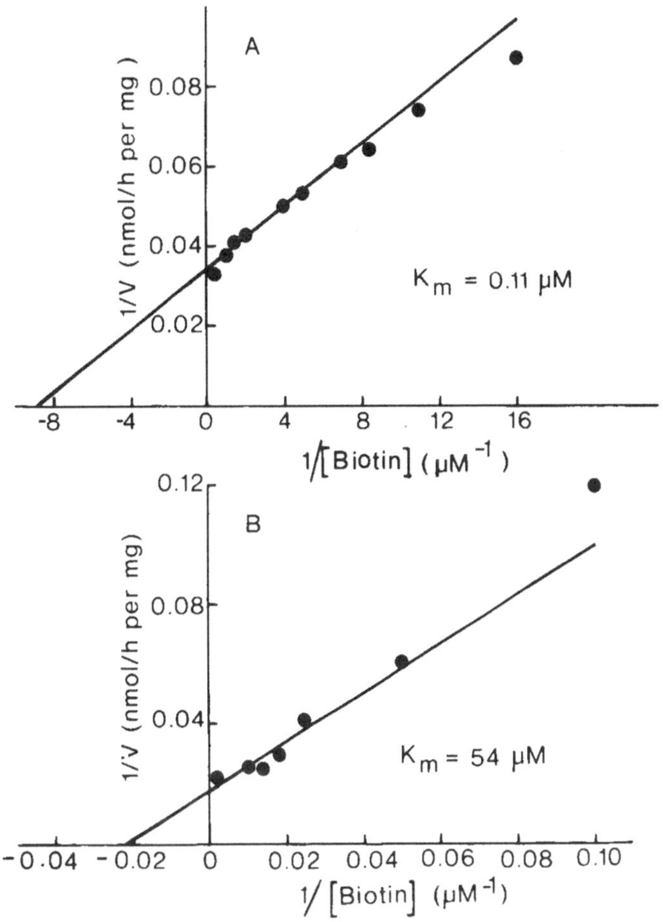

FIGURE 7. Estimation of the apparent K_m of holocarboxylase synthetase for biotin in mitochondrial homogenates prepared from cultured fibroblasts from a normal subject (A), and a patient with Type A BRMCD (B). The second substrate for holocarboxylase synthetase was apopropionyl-CoA carboxylase which was prepared from biotin-deficient fibroblasts. (From Gnheim, H. K. and Bartlett, K., *Lancet*, 1187, 1982. With permission.)

It was noted previously (Table 6) that in contrast to patients with the Type A disease, the concentration of biotin in the plasma of patients with Type B (infantile) BRMCD was lower than normal. This led to the proposal that an impairment in the absorption of biotin by epithelial cells of the intestine or in the transport of this cofactor in the plasma or across cell membranes was the primary defect in these patients.[179,582,583,585,626,728,729,848,868,869] However, recent studies have provided evidence that the primary enzymological defect may be a deficiency of the biotinidase.[311,872,980,981] This enzyme catalyzes the release of free biotin in the following reaction:

$$\text{Biotin} - \epsilon - \text{Lysine} + H_2O \rightarrow \text{Biotin} + \text{Lysine}$$

The activity of biotinidase in the plasma[311,980,981] and liver[311] was found to be 0 to 65% of the normal value in patients with Type B BRMCD. It has been proposed that decreased activity of this enzyme reduces the formation of free biotin from biotinyl-ε-lysine, including that derived from the diet, for holocarboxylase formation.[311,872,980,981] These results suggest that decreased activity of biotinidase may be the primary defect in patients with Type B BRMCD. Further experiments will be required to determine whether this postulated enzyme defect is present in all Type B patients, and to demonstrate an accumulation of biotin-ε-lysine in the plasma or tissues.

Patients with BRMCD have been placed in the *bio* complementation group. Complementation studies have shown that this differs from the *pcc* and *pyc* complementation groups for inherited deficiencies in the enzymes propionyl-CoA carboxylase and pyruvate carboxylase, respectively.[270,271,273,728,729,778] Thus, mutant fibroblasts from two patients with BRMCD, probably Type A, did not complement each other,[728,729] but were shown to complement fibroblasts carrying the *pcc*[728,729] or *pyc*[271,273] mutants. Other studies have provided evidence that the disease is inherited in a recessive manner.[728,729] Since, as discussed above, there appear to be at least two different primary genetic defects which cause BRMCD Types A and B, it is likely that the *bio* complementation group is heterogeneous.[270,273,982]

A decrease in the activity of either the holocarboxylase synthetase or biotinidase enzymes would be expected to cause a decrease in the amount, and hence activity, of each of the four biotin-dependent holocarboxylases (listed in Figure 4). This is clearly observed for the three mitochondrial enzymes and leads to the accumulation of lactate, pyruvate, and other organic acids in the plasma and urine, and to a decreased rate of gluconeogenesis from three-carbon precursors (summarized in References 624, 626, 688, and 982).

The consequences of decreased availability of intracelluar biotin on the amount of holoacetyl-CoA carboxylase present in the cytoplasm appear to be less clear. As mentioned, above not all investigators have observed a substantial decrease in the activity of this enzyme in the cells of patients with BRMCD. It has been suggested that this may be due to differences in the mitochondrial and cytoplasmic holocarboxylase synthetase enzymes,[583,584] or to the possibility that acetyl-CoA carboxylase holocarboxylase synthetase binds biotin more tightly than the mitochondrial biotin-dependent carboxylases.[61,70,956] Notwithstanding these considerations, there is both direct[274] and indirect[581,584] evidence that the activity of acetyl-CoA carboxylase is impaired in some patients with BRMCD. This leads to decreased synthesis of a variety of fatty acids from acetyl-CoA and may account for abnormalities in the immune sytem.[280,581,584]

B. Biotin Deficiency in Chickens

1. Dietary Biotin Deficiency

Fatty liver and kidney syndrome (FLKS) is a commercially important disease in poultry. An impairment in the activity of hepatic pyruvate carboxylase is a major factor in progression of the disease. The disease is caused by the combination of decreased availability of dietary biotin to the liver and at least one other factor, e.g., stress (reviewed in References 68 and 640). Since FLKS is closely related to the diseased state which results directly from a dietary deficiency of biotin[68,640] without the apparent involvement of other factors, the effects of dietary biotin deficiency in chickens will be briefly summarized before discussion of FLKS.

The maintenance of chickens on a purified diet deficient in biotin, or on a normal diet supplemented with avidin, leads to symptoms and signs which include poor growth and feather development, foot lesions, dermatitis, perosis, lethargy, and ultimately, death. Serum lactate concentrations are elevated and mild hypoglycemia is observed in the fasting state.[640,956] The specific activity of hepatic pyruvate carboxylase (expressed as micromoles HCO_3^- fixed per minute per milligram protein) is decreased to about 20% of the normal value.[24,40,295,522,641,953] Decreased in the specific activities of hepatic propionyl-CoA carboxylase[24] and acetyl-CoA carboxylase[24,534] have also been ob-

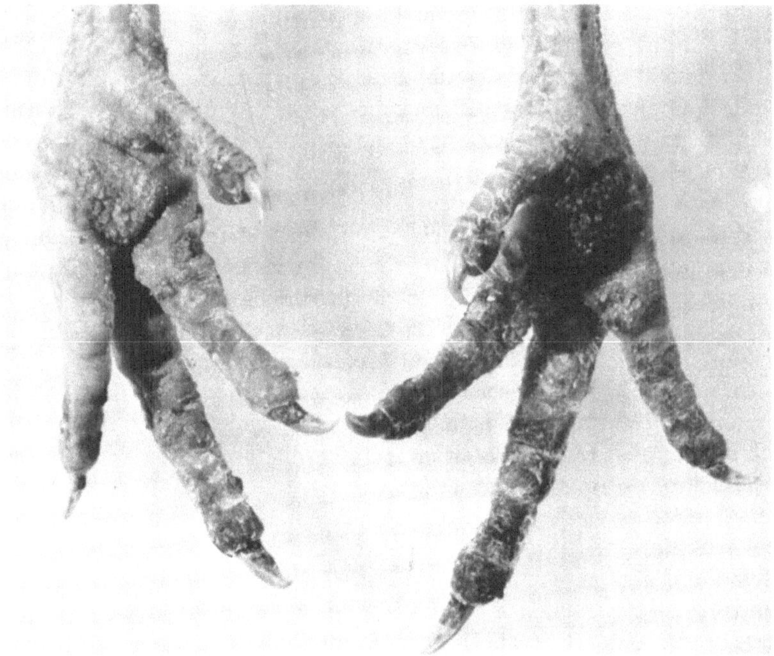

FIGURE 8. Dry brown crust on the plantar surface of the foot in a chicken with experimentally induced FLKS. (From Lohr, J. E., *N.Z. Vet. J.*, 23, 167, 1975. With permission.)

served. The rate of decline of acetyl-CoA carboxylase activity which follows the removal of biotin from the diet was found to be much slower than that for pyruvate carboxylase.[24]

The administration of biotin to biotin-deficient chickens increases the activity of pyruvate carboxylase from 15 to 60% of the activity of the enzyme present in control livers within a period of 4 hr.[522] This indicates that significant quantities of apopyruvate carboxylase are formed in the absence of adequate supplies of biotin.[522]

The effects of biotin deficiency on the growth and development of chickens may be due to a decrease in the rates of synthesis of proteins, purines, and other molecules caused by a decrease in the provision of four-carbon compounds for anaplerotic synthesis in the reaction catalyzed by pyruvate carboxylase.[2] A stimulation by biotin of protein synthesis in a number of tissues in biotin-deficient rats has been reported.[114] However, the explanation for reduced growth and development in biotin deficiency may be more complex and involve other reactions in a variety of tissues.[68,294]

2. Fatty Liver and Kidney Syndrome in Chickens

a. Signs and Symptoms

The main features of fatty liver and kidney syndrome (FLKS) in chickens are reviewed in References 66, 68, 492, 510, and 640. The disease affects immature chickens (3 to 4 weeks of age) in certain flocks. The number of birds which die is generally less than 6% but can be as high as 20%. The course of the disease follows two stages. In the preclinical stage the birds appear to be healthy but exhibit abnormal biochemistry. The main signs of the clinical (second) stage include lethargy, paralysis, and foot lesions (Figure 8) which are rapidly followed by death. Onset of the clinical stage is sudden and its course rapid; the birds generally die within 24 hr and can die within 6 to 10 hr.

In some cases, FLKS is associated with symptoms and signs of experimental biotin deficiency, but these are not always evident. Post-mortem examination reveals an enlarged liver with lipid (chiefly triglyceride) infiltration. Lipid infiltration of the kidneys and a number of other tissues is also observed. The biochemical findings in birds with FLKS include hypoglycemia, elevated concentrations of plasma Na^+, K^+, pyruvate, and uric acid, and decreased concentrations of biotin in the liver and plasma.

Composition of the diet is a major factor which determines the predisposition of chickens to FLKS (reviewed in References 66, 68, 492, 510, and 640). In addition to biotin, the quantities and nature of dietary lipid and protein are important. The syndrome can be induced experimentally by feeding birds diets which contain animal or fish protein but are low in total protein, low in fat and biotin, and high in carbohydrate (e.g., the diets listed in Table 9). There is some evidence that stress, possibly associated with starvation, is a necessary additional factor in precipitation of the onset of the disease. Mortality from the disease can be reduced by increasing the quantity of fat, protein, or amino acids, and the quantity of biotin in diets which are low in fat and protein and not supplemented with an increase in the quantity of biotin in the diet has also been shown to restore normoglycemia.

b. Enzyme Activities

In the livers of chickens in the clinical stage of FLKS, or of birds maintained on diets which increase the incidence of FLKS (presumably birds in the preclinical stage) the activity of pyruvate carboxylase is decreased to values between 5 and 70% of those present in control livers (Tables 9 and 10).[40,60-62,66,418,431,641,953,954] This decrease is associated with a decrease in the amount of biotin in the liver (Table 10).[418,641,953] The readdition of biotin to biotin-deficient diets, the supplementation of normal diets with biotin, or the incubation of liver slices with this cofactor restores the activity of pyruvate carboxylase to control values.[40,62,66,953,954] Indirect evidence that restoration of enzyme activity is rapid and does not involve the synthesis of new protein has been presented.[66]

Little or no decrease,[62,241,640,953] and in one case an increase[418] in the activity of acetyl-CoA carboxylase in birds with FLKS has been observed, suggesting that this enzyme is less susceptible to the lack of biotin than is pyruvate carboxylase.[61,70,956] Decreases in the amounts of some other hepatic gluconeogenic enzymes, glucose-6-phosphatase, fructose-1-6-bisphosphatase, and phosphoenolpyruvate (PEP) carboxykinase have also been observed.[66] The activity of these enzymes is not restored by supplementation of diet with biotin.[66] Little information on changes in the activities of the two other biotin-dependent enzymes, propionyl- and methylcrotonyl-CoA carboxylase in chickens with FLKS has been reported.

Pyruvate carboxylase activity in red blood cells is also decreased in chickens with FLKS (Table 10).[953] Furthermore, it has been shown that in apparently healthy birds there is a good correlation between the activity of the enzyme in the red blood cells and the amount of biotin present in the diet (Figure 9). Therefore, it has been proposed that the activity of pyruvate carboxylase in red blood cells can be used to monitor the biotin "status" of groups of chickens.[72,950-952] However, it has also been found that for individual birds the activity of the enzyme in red blood cells is not necessarily related to that in the liver.[949]

c. Metabolic Consequences of Pyruvate Carboxylase Deficiency

The main metabolic abnormalities in chickens with FLKS are in the pathways of carbohydrate and lipid metabolism.[68,640] Experiments conducted in vitro indicate that the rate of gluconeogenesis from three-[66,67,71] or four-carbon[69] precursors in liver slices from FLKS-affected birds is about 20 and 45%, respectively, of that for control liver slices (Figure 10). Addition of biotin to the diet or incubation medium restores the rate

Table 10
EFFECTS OF FLKS INDUCED BY BIOTIN-DEFICIENT DIETS, ON THE ACTIVITY OF PYRUVATE CARBOXYLASE AND THE CONCENTRATION OF BIOTIN IN THE LIVER AND BLOOD OF 4-WEEK-OLD CHICKENS

		Liver		Blood	
Dietary available biotin (μg/kg)	Severity of foot lesions[a]	Pyruvate carboxylase activity (μmol substrate converted per hr per 100 g body wt)	Biotin (ng/g)	Pyruvate carboxylase activity (μmol substrate converted per hr per g hemoglobin)	Biotin (ng/ℓ)
30	1.1 ± 0.2	1156 ± 129	802 ± 132	13 ± 2	1060 ± 250
70	0.2 ± 0.1	4079 ± 276	1940 ± 150	55 ± 7	1960 ± 330
130	0	7776 ± 856	2805 ± 310	116 ± 11	2560 ± 520
200	0.1	9770 ± 896	3054 ± 283	111 ± 4	4700 ± 460

[a] Assessed subjectively on a scale ranging from 0 (no lesions) to 5 (very severe). Compare with Figure 8.

From Whitehead, C. C. and Banister, D. W., *Br. Poult. Sci.*, 22, 467, 1981. With permission.

Table 9
EFFECTS OF FLKS INDUCED BY VARIOUS DIETS, ON HEPATIC PYRUVATE CARBOXYLASE ACTIVITY IN CHICKENS

Diet employed to induce FLKS[a]	Hepatic pyruvate carboxylase activity in birds with FLKS (% of control)[b]	Ref.
Low fat; post-mortem examination was used to define birds with FLKS.	3	66
Low protein, low biotin; biotin content of liver was used to define birds with FLKS.	16	418
Wheat-based, low-fat, low-protein, low-biotin; birds maintained on this diet for 4 weeks were defined as those with FLKS.	10	954
Wheat and meat meal diet, birds fasted 18 hr; pyruvate carboxylase measured in the livers of surviving birds.	15	60
Wheat and fish meal diet; birds subject to stress; controls were fed the same diet supplemented with biotin; no visible signs of FLKS reported.	73	62
Wheat, meat, and fish meal; controls were fed the same diet supplemented with biotin; visible signs used to monitor FLKS.	57	40
Wheat and meat meal diet; birds fasted for 18 hr; pyruvate carboxylase measured in livers of surviving birds.	47	61
Biotin deficiency induced by feeding avidin: birds starved for 18 hr; no visible signs of FLKS reported but marked decrease in liver biotin content.	9	641
Low-protein, biotin-deficient diet; FLKS assessed by visible signs and biotin content of liver and blood; controls are birds fed on the same diet supplemented with biotin.	12	953

[a] In many of the studies cited, the effects of a number of diets which induced FLKS were tested. Data for the most effective diet are reported here.

[b] Absolute values for pyruvate carboxylase activity were expressed in micromoles pyruvate carboxylated per minute per gram wet weight[40,66,418,954] per milligram protein,[60,62,641] and per 100 g body wt.[953]

of gluconeogenesis in liver slices from FLKS-affected birds to near control values (Figure 10).[66,67] The size of glucose pools and rates of glucose synthesis measured in vivo are also decreased in FLKS-affected birds.[63] Supplementation of the diet with biotin restores these parameters to normal.[63] In contrast to results obtained with the liver slices, rates of gluconeogenesis from alanine and four-carbon precursors in kidney slices from birds with FLKS have been found to be higher than those for controls.[69,71]

The oxidation of long-chain fatty acids in the livers of chickens affected by FLKS

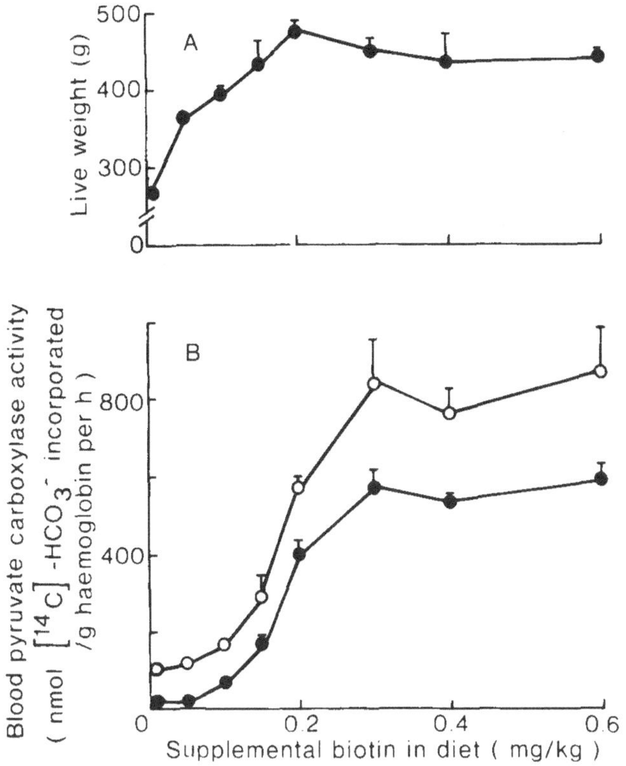

FIGURE 9. Relationship between the amount of biotin present as a supplement in the diet and the activity of pyruvate carboxylase in the blood at 2 (open circles) and 4 (closed circles) weeks of age (B) and the live weight at 4 weeks of age (A) in turkey poults. (From Whitehead, C. C. and Bannister, D. W., *Br. J. Nutr.*, 39, 547, 1978. With permission.)

does not appear to be reduced.[261] While an increase in the rate of hepatic lipid synthesis and in the activities of some lipogenic enzymes has been reported,[61,418] a number of other investigators have found no increase in this pathway or these enzymes.[62,261,640,641,954]

The mechanisms which cause the onset of the clinical stage of FLKS are not yet fully understood. It is currently proposed that in the preclinical phase, a deficiency of biotin leads to impaired pyruvate carboxylase activity, a decrease in the rate of gluconeogensis from three-carbon precursors, and an increase in the concentration of lactate in the plasma (reviewed in References 68 and 640). The absence of severe hypoglycemia at this stage may be due to the release of glucose from glycogen stores, the synthesis of glucose from other precursors which does not require the conversion of pyruvate to oxaloacetate,[68,954] and the possible contribution of renal gluconeogenesis.[68,939] (Compared with the proposed metabolic defects in patients with inherited pyruvate carboxylase deficiency in Section II.D.) It has been proposed that the absence of a decrease in the activity of acetyl-CoA carboxylase is due to the ability of cytosolic holocarboxylase to bind biotin more tightly than pyruvate carboxylase or propionyl-CoA carboxylase, or to a greater accessibility to biotin by this cytosolic enzyme.[61,70,956]

As mentioned earlier, onset of the clinical stage of FLKS appears to be precipitated by stress and/or starvation in birds already deficient in biotin and hence holopyruvate carboxylase.[68,954] Although the results of most studies indicate that FLKS occurs only in biotin-deficient birds (reviewed in References 68 and 640), there is at least one report

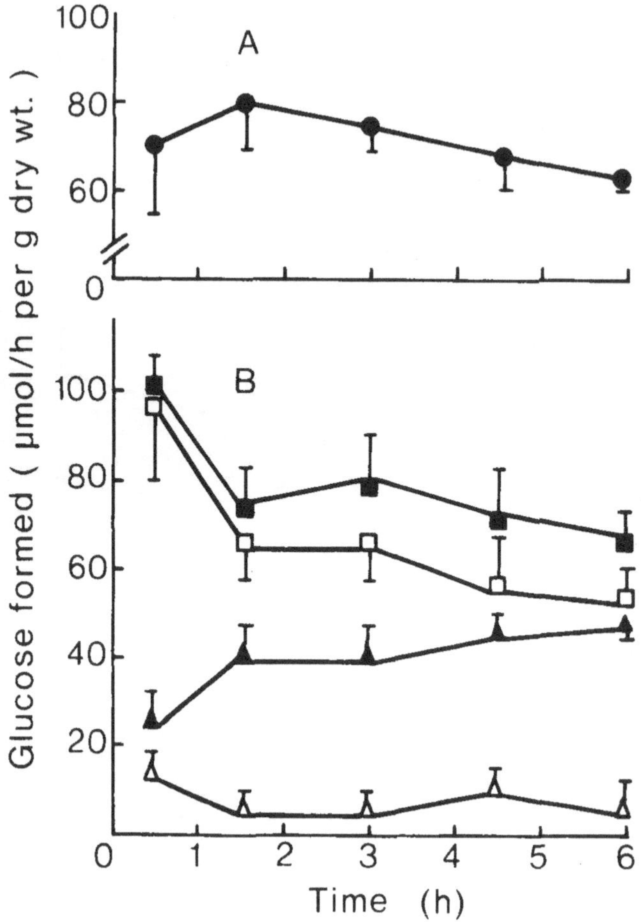

FIGURE 10. Effect of biotin on gluconeogenesis from lactate in liver slices from chickens fed on a normal diet (A) and a diet which induces FLKS (B). In B, the chickens were fed from day 1 on a diet with a low fat content for 3.5 to 4.5 weeks and starved (17 to 18 hr) before being killed. Two groups of chickens were identified. In one group, the diagnosis of FLKS was made on the basis of post-mortem examination (chickens with FLKS, triangles) and in the other group no signs of FLKS were observed (controls, squares). When present (closed symbols) the concentration of biotin in the incubation medium was 48 μM. (From Bannister, D. W., *Biochem. J.* 156, 167, 1976. With permission.)

which has questioned the role of biotin deficiency in onset of the disease.[820] It has been postulated that in the clinical stage of the disease severe hypoglycemia is caused by the depletion of glycogen reserves, and inhibition of pathways of gluconeogenesis which do not utilize pyruvate carboxylase as well as the inhibition of gluconeogenesis from three-carbon precursors.[68,954] There is probably no further decrease in the activity of pyruvate carboxylase at this stage.[68,954] These changes are associated with enhanced hydrolysis of lipids and the liberation of free fatty acids in peripheral tissues. The re-esterification of these fatty acids and possible enhanced hepatic lipogenesis are proposed to lead to an increase in triglyceride in this tissue.[61,418] However, hypoglycemia and lactic acidosis are the likely primary cause of death.[59,61,68]

Reduction in the incidence of FLKS, which is observed when the amount of fat or protein in FLKS-induced diets is increased, is proposed to be due to the release of

Table 11
BIOTIN CONTENT OF THE LIVERS OF INFANTS WHO DIED FROM SIDS AND OTHER CAUSES

Age at death and SIDS classification[a]	n	Liver biotin (ng/g)	95% Confidence limits of the median	
			Lower	Upper
Stillborn and less than 1 week	69	249	222	307
1—52 weeks; groups A and B	57	419	368	462
1—52 weeks; groups C and D (defined as SIDS)	35	336	277	356
>52 weeks	43	344	321	399

[a] The system of classification employed was: (A) children with gross congenital deformities; (B) children with definite and severe disease; (C) children with a mild disease which is of itself insufficient to explain death adequately; and (D) children with no definite diagnosable disease to account for death.

From Johnson, A. R., Hood, R. L., and Emery, J. L., *Nature (London)*, 285, 159, 1980.

biotin from acetyl-CoA carboxylase in response to a lower demand for the conversion of acetyl-CoA to fatty acid,[954] or in the case of elevated amounts of protein to the provision of four-carbon precursors for gluconeogenesis. In contrast to these predictions, Whitehead and Bannister[953] have reported that an increase in dietary protein leads to a greater severity of FLKS and a decrease in the concentration of plasma biotin. They suggested that under these conditions the provision of amino acids and synthesis of glucose from these precursors is enchanced and hence there is greater flux through pyruvate carboxylase, which is proposed to increase the requirement for biotin by holopyruvate carboxylase.

C. Dietary Biotin Deficiency in Man and Other Vertebrates

On the basis of the observation that a deficiency of biotin coupled with other factors can lead to hypoglycemia and death in chickens, Johnson et al.[430] have proposed that biotin deficiency may contribute to sudden infant death syndrome (SIDS) in children. In two studies, the concentration of biotin in the livers of infants who died from this disease was measured and found to be lower than that present in infants who had died from other causes (Table 11).[389,430] However in a separate study, Zielke et al.[1001] found no decrease in hepatic pyruvate carboxylase activity in infants who had died of SIDS.[1001] This result suggests that it is unlikely that a deficiency of biotin is a major cause of SIDS, but further experimental evidence is required before this possibility can be completely excluded.

There is some evidence that biotin deficiency leads to foot lesions and lameness in pigs fed a commercial diet.[132] The activity of pyruvate carboxylase in blood cells (white cells and reticulocytes) of pigs maintained on diets containing different amounts of biotin has been measured and shown to be related to the concentration of biotin in the blood.[955] It was proposed that, as with broiler chickens, blood pyruvate carboxylase activity can be used as an indicator of the concentration of biotin available to cells (the biotin "status") in the pig.[955]

A number of investigators have studied the consequences of dietary biotin deficiency on the activity of pyruvate carboxylase in rodents. In some cases, attempts have been made to use the biotin-deficient rat as an experimental model for pyruvate carboxylase deficiency (PCD) in man.[417,722,737,738] In rats or mice maintained on a biotin-deficient diet, the activity of pyruvate carboxylase in the liver and kidney is 5 to 20% of control values.[24,104,105,183,227,417,539,722,737,738,919] The activity of hepatic propionyl-CoA carboxylase is also decreased,[24,183,466] and a decrease in acetyl-CoA carboxylase activity has been observed in severe biotin deficiency.[104,105,228] No change in the hepatic activities of PEP carboxykinase, fructose-1,6-bisphosphatase, or glucose-6-phosphatase was observed.[227] Wagle[919] and colleagues[183,227] found a marked increase in hepatic pyruvate carboxylase activity following supplementation of a biotin-deficient diet with biotin, although Bitsch et al.[104] reported no increase in enzyme activity under these conditions.

Since inherited pyruvate carboxylase deficiency and BRMCD in children lead to disorders of the CNS the effect of dietary biotin deficiency on the activity of cerebral pyruvate carboxylase has been compared with that on the enzyme in the liver. Schrijver et al.[737,738] found that during the first 50 days in the life of rats maintained on a biotin-deficient diet the activity of cerebral pyruvate carboxylase (expressed as micromoles HCO_3^- fixed per minute per gram wet weight) was about 30% of control values. When the activity of the enzyme was measured in tissue dissected from the brain stem and cerebrum, a similar reduction in activity was observed.[737]

In brains of adult biotin-deficient rats the specific activity of pyruvate carboxylase (micromoles HCO_3^- fixed per minute per milligram protein) was found to be 50 to 70% of control values for tissue from the cerebral hemisphere, vermis, cerebral, and brain stem regions.[722] This decrease was considerably less than that observed for hepatic pyruvate carboxylase (3% of the control value) in the same animals.[722] Therefore, it was concluded that the activity of pyruvate carboxylase in cells of the CNS is not preferentially diminished by biotin deficiency. Rather, a greater proportion of the enzyme in this tissue is maintained in an active form compared with the activity of the enzyme in the liver.[722] These results are also consistent with the observation that dietary biotin deficiency in rats leads to a greater decrease in the concentration of biotin in the liver than in the brain.[99]

IV. PYRUVATE CARBOXYLASE IN OTHER DISEASED STATES

In man there are a number of diseases or abnormal states in which the total amount of pyruvate carboxylase activity is altered, but in which this alteration is not the primary cause of the disease (Table 1). These states include diabetes mellitus,[945,976] neonatal hypoglycemia,[371,700] the growth of hepatomas,[173] and some other inherited diseases.[690,842,882]

In animal models of diabetes mellitus the increased ratio of plasma glucagon to insulin concentrations is proposed to lead to the observed increase in the total amount of pyruvate carboxylase activity in the liver and kidney[945,976] (Chapter 5, Section I.B.2.c). This in turn contributes to the hyperglycemia associated with diabetes in these animals. It is likely that increased amounts of hepatic and renal pyruvate carboxylase contribute to hyperglycemia in patients with diabetes mellitus.

Studies on the development of guinea pigs have shown that retardation of growth of the fetus or neonate leads to a decrease of about 70 to 75% in the total activity of hepatic pyruvate carboxylase.[700] This may be caused by a low ratio of glucagon to insulin and a low concentration of thyroid hormone in the plasma.[700] On the basis of these observations it has been proposed that a decreased activity of hepatic pyruvate carboxylase is wholly or partially responsible for neonatal hypoglycemia.[371,700]

Decreased amounts of pyruvate carboxylase activity have been observed in a number of hepatomas (Table 12). These include Reuber and Morris hepatomas,[173] primary hepatomas taken from patients,[366,994] and cell lines derived from hepatomas in patients.[366] Chang and Morris[173] have shown that in a number of experimental hepatomas

Table 12
PYRUVATE CARBOXYLASE ACTIVITY, THE AMOUNT OF IMMUNOLOGICALLY ACTIVE PYRUVATE CARBOXYLASE PROTEIN, AND MITOCHONDRIAL PROTEIN CONTENT IN NORMAL RAT LIVER, HEPATOMAS, AND HOST LIVERS

		Pyruvate carboxylase		
Tissue	Growth rate	Activity (μmol HCO_3^- fixed/ min per g wet wt.)	Protein (mg/g wet wt.)	Mitochondrial protein (mg/g wet wt.)
Normal rat liver		4.71 ± 0.35	0.139 ± 0.006	21.0 ± 0.32
Regenerating rat liver	Fast	3.35 ± 0.24		17.6 ± 0.46
Hepatoma H35	Intermediate	0.41	0.004	7.5
Host liver		3.32	0.110	16.5
Hepatoma 9633	Slow	0.31 ± 0.04	0.003	10.4 ± 0.63
Host liver		4.30 ± 0.31	0.119 ± 0.007	19.5 ± 1.3
Hepatoma 9618A	Slow	4.77 ± 0.57	0.146 ± 0.007	16.6 ± 1.0
Host liver			0.156 ± 0.006	

From Chang, L. O. and Morris, H. P., *Cancer Res.*, 33, 2043, 1973. With permission.

the total amount of pyruvate carboxylase protein as well as the total amount of enzyme activity are reduced (Table 12). This reduction appears to be due to both a decrease in the number of mitochondria per cell and changes in the rate of synthesis and/or degradation of pyruvate carboxylase.[173] Evidence that the pyruvate carboxylase protein present in these hepatomas is identical with the present in normal liver cells was obtained using immunological techniques.[173] While the earlier studies of Weber and Morris[941,942] have shown a positive correlation between the degree of decrease in pyruvate carboxylase activity and the rate of growth of the hepatomas, such a correlation was not observed in the studies of Chang and Morris.[173] The decrease in pyruvate carboxylase activity in hepatoma cells is consistent with the observation that, where measured, the rate of gluconeogenesis is also decreased when compared to that in normal cells.[941]

Decreases in activity of hepatic pyruvate carboxylase have been reported in cases of Friedriech's ataxia,[238,239] Alpers infantile cerebral polio dystrophy,[842] and Reye's syndrome.[690] The decrease in pyruvate carboxylase activity in Reye's syndrome may result from damage to the mitochondria since the activities of a number of other mitochondrial enzymes are also decreased.[690]

In a small number of diseases there is evidence that abnormal concentrations of physiological modifiers or abnormal modifiers of pyruvate carboxylase activity are present. It is proposed that this results in an alteration in pyruvate carboxylase activity, which contributes to the clinical symptoms of the disease. Examples of some of these diseases are summarized in Table 13.

V. DRUGS, HEAVY METALS, AND ANTIMETABOLITES

There is some evidence that a number of drugs and antimetabolites alter flux from pyruvate to oxaloacetate (Table 14). In some cases this is due to the action of a metabolic product or products formed from the drug or antimetabolite (Table 14). These agents may either activate or inhibit pyruvate carboxylase, or cause a change in total amount of enzyme activity (Table 14).

Table 13
DISEASES IN MAN IN WHICH METABOLITES WHICH ARE NONPHYSIOLOGICAL INHIBITORS OF PYRUVATE CARBOXYLASE ARE FORMED

Disease	Proposed inhibitors of pyruvate carboxylase activity	Tissue	Ref.
Phenylketonuria	Phenylpyruvate	Brain	635
		Liver	842
Metabolic acidosis	H^+	Liver	425
3-Hydroxy-3-methyl-glutaryl-CoA lyase deficiency	3-Hydroxy-3-methyl-glutaric acid 3-Methylglutaconic acid	Liver	742

Table 14
SOME DRUGS AND ANTIMETABOLITES WHICH, WHEN ADMINISTERED IN VIVO, HAVE BEEN SHOWN TO MODIFY THE ACTIVITY OF PYRUVATE CARBOXYLASE

Agent	Proposed mechanism of action on pyruvate carboxylase	Tissue	Ref.
Drugs			
Dichloroacetate	Inhibition by oxalate	Liver	210, 221, 497
2-Chloropropionate			210
Ascorbate, isoascorbate, dehydroascorbate	Inhibition by oxalate	Liver	835
Acetazolamide	Decrease in total amount of enzyme	Liver	150
Clofibrate	Increase in total amount of enzyme	Kidney cortex	521
Oxalate	Inhibition by oxalate	Liver	222, 997
Antimetabolites			
1,2,3-Benzinetricaboxylate	Inhibition by 1,2,3-benzene-tricarboxylate	Liver	832
Hypoglycin	Inhibition by buturyl-CoA, isovalery-CoA, and methy-lenecyclopropyl-acetyl-CoA	Liver	6, 102, 826
Benzoic acid	Formation of benzoyl-CoA and decrease in acetyl-CoA	Liver	310
Avenaciolide	Decrease in total amount of enzyme	Liver	974
Aroclor 1254 (polychlorimated biphenyl mixture) and other chlorinated hydrocarbons	Decrease in total amount of enzyme	Liver of diabetic rat	555
		Liver of chicken	813—815
Aflatoxin B_1	Increase in total amount of enzyme	Liver of chicken	497

Since ethanol has effects on hepatic metabolism, including inhibition of gluconeogenesis,[473,884] a number of investigators have studied the effect of this agent on pyruvate carboxylase activity. Although some reports (e.g., Stifel et al.[824]) indicate that ethanol consumption by rats leads to a decrease in pyruvate carboxylase activity in the liver, others (e.g., Chan[165]) have observed no change in the total amount of pyruvate carboxylase in the liver and kidney.

Studies with fatty and cirrhotic livers from rats with chronic ethanol toxicity have shown that the infusion of oleate in a perfused liver preparation does not increase the rate of gluconeogenesis, whereas this and other fatty acids stimulate gluconeogenesis in control perfused livers (Chapter 5, Section I.C.3.a).[394] It was concluded that in fatty and cirrhotic livers pyruvate carboxylase is maximally activated by acetyl-CoA derived from endogenous sources.[394] This activation, which results from the increased concentration of free fatty acids in the liver, contrasts with the observed inhibition of the overall gluconeogenic pathway during hepatic ethanol metabolism. The latter effect is proposed to be due to inhbition of flux at other steps of the pathway, including a decrease in the availability of cytoplasmic oxaloacetate due to the reduction of this acid to malate in the presence of an increased ratio of NADH:NAD.[473,884] Thus, in the livers of chronic alcoholic patients or experimental animals with chronic ethanol toxicity, the inhibitory effects of ethanol oxidation on other steps of the gluconeogenic pathway appear to predominate.

Heavy metal salts have been shown to alter the activity of pyruvate carboxylase in vivo.[11,561-564,652,823] Both acute (1 hr) and chronic (e.g., 45 days) administration of cadmium chloride to rats (e.g., 0.25 to 60 mg/kg/day) leads to an increase in the total amount of pyruvate carboxylase activity in extracts of the liver and kidney.[175,561-564] These changes are associated with increases in the concentration of glucose in the plasma, the amount of cyclic AMP present in the liver, and the activity of adenyl cyclase.[175,561] A decrease in the total amount of pyruvate carboxylase activity in the kidney is observed in rats which have received mercuric chloride (e.g., 6 days after a single i.p. dose of 2 mg/kg).[652]

The i.v. administration of lead acetate (10 mg per rat) leads to a 95% inhibition of the rate of pyruvate carboxylation in isolated liver mitochondria.[11] A close correlation between the decrease in the inhibition of pyruvate carboxylation and a decrease in the mitochondrial ATP concentration was observed. Since there was no change in the total pyruvate carboxylase activity, it was concluded that lead inhibits flux from pyruvate to oxaloacetate by decreasing the concentration of ATP.[11] Subacute or chronic ingestion of lead causes an increase in the amount of pyruvate carboxylase activity in extracts of the kidney. In the chronic state, this increase was found to be associated with an increase in the activity of adenyl cyclase.[823]

VI. CONCLUSIONS

The two main diseases in which a decrease in pyruvate carboxylase activity is the primary defect are: the inherited deficiency of the enzyme caused by a mutation in the gene which specifies the amino acid sequence of the enzyme protein[693] and a decrease in the amount of holopyruvate carboxylase. The latter may be caused by mutations in the genes which specify synthesis of holocarboxylase synthetase[317,846] biotinidase,[981] or an enzyme involved in cellular biotin transport,[271,728,869] as in BRMCD in man, or by a dietary deficiency of biotin combined with one or more other factors, as in fatty liver and kidney syndrome in chickens[68] and possibly some other animals.[955] The sites of these and other defects are summarized in Figure 11.

In the vertebrates, pyruvate carboxylase and acetyl-CoA carboxylase are the two biotin-dependent enzymes with major roles in intermediary metabolism. Of these, pyruvate carboxylase appears to be more susceptible than acetyl-CoA carboxylase to a lack of available biotin caused by a decrease in (1) holocarboxylase synthetase activity, (2) the steady-state intracellular concentration of free biotin,[280,581,583,585,586] or (3) the amount of biotin available in the diet.[24] In many of the diseases discussed there appears to be only a partial deficiency in the amount of holopyruvate carboxylase. Moreover, in most circumstances, other reactions can partially compensate for the loss of gluconeogenesis from three-carbon precursors and decreased anaplerotic synthesis in the

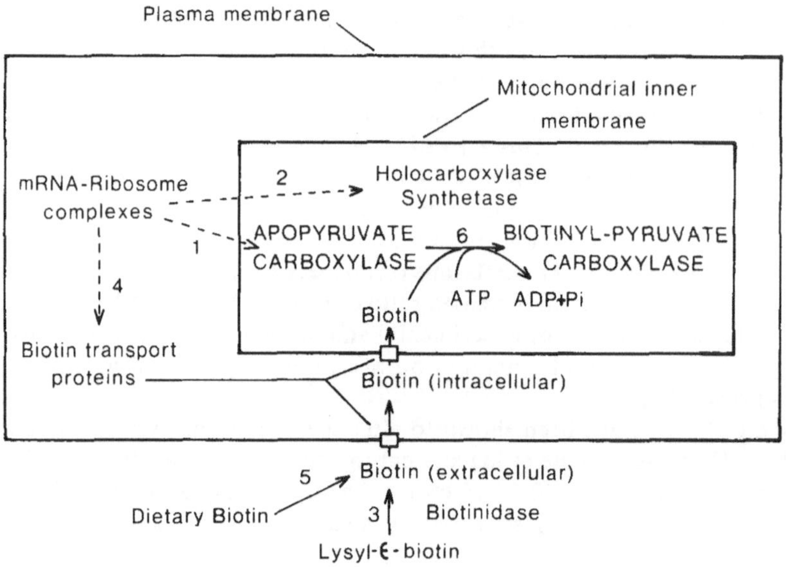

FIGURE 11. A schematic representation of the reactions involved in the formation of normal holopyruvate carboxylase (biotinyl-pyruvate carboxylase) and the sites, or proposed sites, at which genetic or metabolic factors (summarized in Table 1) act to alter the amount or activity of holopyruvate carboxylase. The sites are: (1) synthesis of the apopyruvate carboxylase protein;[693] (2) synthesis of the holocarboxylase synthetase protein;[317,846] (3) synthesis of the biotinidase protein;[981] (4) synthesis of biotin transport proteins;[271,728,869] (5) the supply of biotin from the diet (Table 1); (6) alteration of the activity of the holoenzyme by changes in the concentration of substrates or modifiers (Table 1).

absence of significant hepatic pyruvate carboxylase activity.[34,68,230] Thus, the affected individual can often survive for some time unless subjected to stress or other factors which inhibit alternative pathways of oxaloacetate formation and increase the demand for this metabolite.

The work conducted during the past 20 years on the role of pyruvate carboxylase in diseased states has chiefly been directed towards defining the nature of the major diseases caused by deficiency of the enzyme in man and other vertebrates. Further experiments are likely to involve investigation of the molecular details of the genetic defects and a search for better methods of diagnosis and treatment. These may include studies of the nature of abnormal protein(s) synthesized in children with an inherited deficiency of the enzyme, the precise nature of the enzyme defects in BRMCD, methods for the diagnosis of these diseases in carriers and in the fetus, the factors which influence the absorption and availability of biotin, the possible involvement of holopyruvate carboxylase deficiency in other diseases (e.g., SIDS), and methods to ensure adequate provision of biotin for poultry and other stock.

REFERENCES

1. Achuta-Murthy, P. N. and Mistry, S. P., Some aspects of biotin binding to protein catalysed by biotin-deficient chicken liver preparations, *Proc. Soc. Exp. Biol. Med.*, 145, 565, 1974.
2. Achuta-Murthy, P. N. and Mistry, S. P., Biotin, *Prog. Food Nutr. Sci.*, 2, 405, 1977.
3. Achuta-Murthy, P. N., Shen, S. C., and Mistry, S. P., Effect of fasting and refeeding high-carbohydrate, high-fat and high-protein diets on the activities of some enzymes of carbohydrate metabolism in rat kidney, *Fed. Proc. Fed. Am. Soc. Exp. Biol.*, 36, 1172, 1977.
4. Adam, P. A. J. and Haynes, R. C., Jr., Control of hepatic mitochondrial CO_2 fixation by glucagon, epinephrine, and cortisol, *J. Biol. Chem.*, 244, 6444, 1969.
5. Adiaenssens, K., Taghon, E., Karcher, D., Lavinha, F., and Bourgeois, N., Defective pyruvate carboxylase hyperpyruvate hyperalaninemia in an infant with severe neurological disease, in *New Research Data on Brain Metabolism and Demyelinating Encephalomyelitis*, Schmidt, R. M., Ed., Johann Ambrosius Barth, Leipzig, East Germany, 1974, 62.
6. Al-Bassam, S. S. and Sherratt, H. S. A., The antagonism of the toxicity of hypoglycin by glycine, *Biochem. Pharmacol.*, 30, 2817, 1981.
7. Alberts, A. W., Nervi, A. M., and Vagelos, P. R., Acetyl CoA carboxylase. II. Demonstration of biotin-protein and biotin-carboxylase subunits, *Proc. Natl. Acad. Sci. U.S.A.*, 63, 1319, 1969.
8. Allan, E. H., Chisholm, A. B., and Titheradge, M. A., Hormonal stimulation of mitochondrial pyruvate carboxylase in filipin-treated hepatocytes, *Biochem. J.*, 212, 417, 1983.
9. Allen, S. H. G. and Henrikson, K. P., Purification and subunit structure of propionyl coenzyme A carboxylase of *Mycobacterium smegmatis*, *J. Biol. Chem.*, 264, 5888, 1979.
10. Al-ssum, R. M. and White, P. J., Activities of anaplerotic enzymes and acetyl coenzyme A carboxylase in biotin-deficient *Bacillus megaterium*, *J. Gen. Microbiol.*, 100, 203, 1977.
11. Amatruda, J. M., Staton, A. J., and Kiesow, L. A., Inhibition of carbon dioxide fixation by lead acetate in rat liver mitochondria, *Biochem. J.*, 166, 75, 1977.
12. Anderson, J. W., Pyruvate carboxylase and phosphoenolpyruvate carboxykinase in rat intestinal mucosa, *Biochim. Biophys. Acta*, 208, 165, 1970.
13. Anderson, J. W. and Rosendall, A. F., Gluconeogenesis in jejunal mucosa of guinea pig, *Biochim. Biophys. Acta*, 304, 384, 1973.
14. Andia-Waltenbaugh, A. M., Kimura, S., Wood, J., Divakaran, P., and Friedmann, N., Effects of glucagon, insulin and cyclic-AMP on mitochondrial calcium uptake in the liver, *Life Sci.*, 23, 2437, 1978.
15. Andia-Waltenbaugh, A. M., Tate, C. A., and Friedmann, N. K., The effect of glucagon on the kinetics of hepatic mitochondrial calcium uptake, *Mol. Cell. Biochem.*, 36, 177, 1981.
16. Angus, C. W., Lane, M. D., Rosenfeld, P. J., and Kelly, T. J., Increase in translatable mRNA for mitochondrial pyruvate carboxylase during differentiation of 3T3 preadipocytes, *Biochem. Biophys. Res. Commun.*, 103, 1216, 1981.
17. Appleby, G., Coldbeck, J., Holdsworth, E. S., and Wadman, H., β-Carboxylase enzymes in marine phytoplankton and isolation and purification of pyruvate carboxylase from *Amphidinium carterae* (Dinophyceae), *J. Phycol.*, 16, 290, 1980.
18. Aprille, J. R., Net uptake of adenine nucleotides by newborn rat liver mitochondria, *Arch. Biochem. Biophys.*, 207, 157, 1981.
19. Aprille, J. R. and Asimakis, G. K., Postnatal development of rat liver mitochondria: state 3 respiration, adenine nucleotide translocase activity, and the net accumulation of adenine nucleotides, *Arch. Biochem. Biophys.*, 201, 564, 1980.
20. Aprille, J. R., Nosek, M. T., and Brennan, W. A., Jr., Adenine nucleotide content of liver mitochondria increases after glucagon treatment of rats or isolated hepatocytes, *Biochem. Biophys. Res. Commun.*, 108, 834, 1982.
21. Aprille, J. R., Yaswen, P., and Rulfs, J., Acute postnatal regulation of pyruvate carboxylase activity by compartmentation of mitochondrial adenine nucleotides, *Biochem. Biophys. Acta*, 675, 143, 1981.
22. Ardawi, M. A. M. and Newsholme, E. A., Maximum activities of some enzymes of glycolysis, the tricarboxylic acid cycle and ketone body and glutamine utilization pathways in lymphocytes of the rat, *Biochem. J.*, 208, 743, 1982.
23. Arinze, J. C. and Mistry, S. P., Hepatic acetyl CoA carboxylase, propionyl CoA carboxylase and pyruvate carboxylase activities during embryonic development and growth in chickens, *Proc. Soc. Exp. Biol. Med.*, 135, 553, 1970.
24. Arinze, J. C. and Mistry, S. P., Activities of some biotin enzymes and certain aspects of gluconeogenesis during biotin deficiency, *Comp. Biochem. Physiol.*, 38B, 285, 1971.
25. Armbruster, D. A. and Rudolph, F. B., Rat liver pyruvate carboxylase: inhibition by chromium nucleotide complexes, *J. Biol. Chem.*, 251, 320, 1976.

26. Armston, A. E., Halestrap, A. P., and Scott, R. D., The nature of the changes in liver mitochondrial function induced by glucagon treatment of rats. The effects of intramitochondrial volume, aging and benzyl alcohol, *Biochim. Biophys. Acta*, 681, 429, 1982.
27. Ash, D. E. and Schramm, V. L., Determination of free and bound manganese (II) in hepatocytes from fed and fasted rats, *J. Biol. Chem.*, 257, 9261, 1982.
28. Ashman, L. K. and Keech, D. B., Sheep kidney pyruvate carboxylase. Studies on the coupling of adenosine triphosphate hydrolysis and CO_2 fixation, *J. Biol. Chem.*, 250, 14, 1975.
29. Ashman, L. K., Keech, D. B., Wallace, J. C., and Nielsen, J., Sheep kidney pyruvate carboxylase. Studies on its activation by acetyl coenzyme A and characteristics of its acetyl coenzyme independent reaction, *J. Biol. Chem.*, 247, 5818, 1972.
30. Ashman, L. K., Wallace, J. C., and Keech, D. B., Desensitization of pyruvate carboxylase against acetyl CoA stimulation by chemical modification, *Biochem. Biophys. Res. Commun.*, 51, 924, 1973.
31. Ashworth, J. M., Kornberg, H. L., and Ward, R. L., The role of phosphopyruvate carboxylase in *Escherichia coli, Biochem. J.*, 94, 28P, 1965.
32. Atkin, B., Carrier detection of pyruvate carboxylase deficiency in fibroblasts and lymphocytes, *Pediatr. Res.*, 13, 1101, 1979.
33. Atkin, B., Buist, N., Leiter, A., and Utter, M., Cross reactive material (C.R.M.)-positive hepatic pyruvate carboxylase deficiency associated with renal tubular acidosis (RTA), *Am. J. Hum. Genet.*, 27, 14A, 1975.
34. Atkin, B. M., Buist, N. R. M., Utter, M. F., Leiter, A. B., and Banker, B. Q., Pyruvate carboxylase deficiency and lactic acidosis in a retarded child without Leigh's disease, *Pediatr. Res.*, 13, 109, 1979.
35. Atkin, B., Utter, M. F., and Weinberg, M. B., Pyruvate carboxylase and phosphoenolpyruvate carboxykinase activity in leukocytes and fibroblasts from a patient with pyruvate carboxylase deficiency, *Pediatr. Res.*, 13, 38, 1979.
36. Atkinson, D. E., Adenine nucleotides as stochiometric coupling agents in metabolism and as regulatory modifiers: the adenylate energy charge, in *Metabolic Pathways*, Vol. 5, Vogel, H. J., Ed., Academic Press, New York, 1981, 1.
37. Attwood, P. V., Johannssen, W., and Wallace, J. C., The effects of enzyme concentration and acetyl CoA on the quaternary structure of chicken liver pyruvate carboxylase, *Biochem. J.*, submitted.
38. Attwood, P. V. and Keech, D. B., Pyruvate carboxylase, *Curr. Top. Cell Regul.*, 23, 1, 1984.
39. Attwood, P. V. and Wallace, J. C., unpublished observations.
40. Atwal, A. S., Robblee, A. R., and Milligan, L. P., Effects of season, dietary protein supplement and biotin on performance and pyruvate carboxylase activity in livers of poults, *Poultry Sci.*, 51, 2074, 1972.
41. Avrameas, S. and Ternynck, T., Peroxidase-labelled antibody and Fab conjugates with enhanced intracellular penetration, *Immunochemistry*, 8, 1175, 1971.
42. Baal, E. G., Gabreels, F. J., Renter, W. O., Hommes, F. A., Gijsberts, T. H., Lamers, K. J., and Kok, J. C., A patient with pyruvate carboxylase deficiency in the liver. Treatment with aspartic-acid and thiamin, *Dev. Med. Child. Neurol.*, 23, 521, 1981.
43. Bacchi, C. J., Ciaccio, E. I., Kaback, D. B., and Hutner, S. H., Oxaloacetate production via carboxylation in *Crithidia fasciculata* preparations, *J. Protozool.*, 17, 305, 1970.
44. Baddams, H. M., Chang, L. B. F., and Barritt, G. J., Evidence that glucagon acts on the liver to decrease mitochondrial calcium stores, *Biochem. J.*, 210, 73, 1983.
45. Baird, G. D., Hibbitt, K. G., Hunter, G. D., Lund, P., Stubbs, M., and Krebs, H. A., Biochemical aspects of ketosis, *Biochem. J.*, 107, 683, 1968.
46. Bais, R., Pyruvate Carboxylase: Physical and Kinetic Studies, Ph.D. thesis, University of Adelaide, Australia, 1974, 47.
47. Bais, R., Greenwell, P., Wallace, J. C., and Keech, D. B., Influence of sodium dodecyl sulfate on the sedimentation velocity of proteins, *FEBS Lett.*, 41, 53, 1974.
48. Bais, R. and Keech, D. B., The magnesium ion (Mg^{2+}) activation of sheep kidney pyruvate carboxylase, *J. Biol. Chem.*, 247, 3255, 1972.
49. Bais, R., Wallace, J. C., and Keech, D. B., Evidence for identical single polypeptide chain monomers of sheep kidney pyruvate carboxylase, *Proc. Aust. Biochem. Soc.*, 7, 1, 1974.
50. Ballard, F. J., Filsell, O. H., and Jarrett, I. G., Effects of carbohydrate availability on lipogenesis in sheep, *Biochem. J.*, 226, 193, 1972.
51. Ballard, F. J. and Hanson, R. W., The citrate cleavage pathway and lipogenesis in rat adipose tissue: replenishment of oxaloacetate, *J. Lipid Res.*, 8, 73, 1967.
52. Ballard, F. J. and Hanson, R. W., Phosphoenolpyruvate carboxykinase and pyruvate carboxylase in developing rat liver, *Biochem. J.*, 104, 866, 1967.
53. Ballard, F. J. and Hanson, R. W., Purification of phosphoenolpyruvate carboxykinase from the cytosol fraction of rat liver and the immunochemical demonstration of differences between this enzyme and the mitochondrial phosphoenolpyruvate carboxykinase, *J. Biol. Chem.*, 244, 5625, 1969.

54. Ballard, F. J., Hanson, R. W., and Kronfeld, D. S., Factors controlling the concentration of mitochondrial oxaloacetate in liver during spontaneous bovine ketosis, *Biochem. Biophys. Res. Commun.*, 30, 100, 1968.
55. Ballard, F. J., Hanson, R. W., and Kronfeld, D. S., Gluconeogenesis and lipogenesis in tissue from ruminant and non-ruminant animals, *Fed. Proc. Fed. Am. Soc. Exp. Biol.*, 28, 218, 1969.
56. Ballard, F. J., Hanson, R. W., Kronfeld, D. S., and Raggi, F., Metabolic changes in liver associated with spontaneous ketosis and starvation in cows, *J. Nutr.*, 95, 160, 1968.
57. Ballard, F. J., Hanson, R. W., and Reshef, L., Immunochemical studies with soluble and mitochondrial pyruvate carboxylase activities from rat tissues, *Biochem. J.*, 119, 735, 1970.
58. Ballard, F. J. and Oliver, I. T., Glycogen metabolism in embryonic chicken and neonatal rat liver, *Biochim. Biophys. Acta*, 71, 578, 1963.
59. Balnave, D., Berry, M. N., and Cumming, R. B., Clinical signs of fatty liver and kidney syndrome in broilers and their alleviation by the short-term use of biotin or animal tallow, *Br. Poultry Sci.*, 18, 749, 1977.
60. Balnave, D. , Cumming, R. B., and Sutherland, T. M., The role of dietary fat in alleviating fatty liver and kidney syndrome in broilers, *Aust. Vet. J.*, 52, 433, 1976.
61. Balnave, D., Cumming, R. B., and Sutherland, T. M., A biochemical explanation for the fatty liver and kidney syndrome of broilers: its alleviation by the short-term use of dietary fat, *Br. J. Nutr.*, 38, 319, 1977.
62. Balnave, D. and Pearce, J., Aspects of lipid and carbohydrate metabolism, dietary biotin and fatty liver and kidney syndrome in broilers, *Br. Poultry Sci.*, 17, 627, 1976.
63. Balnave, D., Wolfenden, J., Ball, F. M., Cumming, R. B., and Leng, R. A., Studies of fatty liver and kidney syndrome in chickens: dynamics of glucose metabolism, *Br. J. Nutr.*, 38, 329, 1977.
64. Banaszak, L. J. and Bradshaw, R. A., Malate dehydrogenases, in *The Enzymes*, Vol. 11A, 3rd ed., Boyer, P. D., Ed., Academic Press, New York, 1975, 369.
65. Bandurski, R. S. and Greiner, C. M., The enzymatic synthesis of oxalacetate from phosphoryl-enolpyruvate and carbon dioxide, *J. Biol. Chem.*, 204, 781, 1953.
66. Bannister, D. W., The biochemistry of fatty liver and kidney syndrome. Biotin-mediated restoration of hepatic gluconeogenesis *in vitro* and its relationship to pyruvate carboxylase activity, *Biochem. J.*, 156, 167, 1976.
67. Bannister, D. W., Hepatic gluconeogenesis in chicks: effect of biotin on gluconeogenesis in biotin-deficiency and fatty liver and kidney syndrome, *Comp. Biochem. Physiol.*, 53B, 575, 1976.
68. Bannister, D. W., Recent advances in avian biochemistry: the fatty liver and kidney syndrome, *Int. J. Biochem.*, 10, 193, 1979.
69. Bannister, D. W. and Cleland, M. E., The biochemistry of fatty liver and kidney syndrome of the fowl (*Gallus domesticus*): reduced gluconeogenesis from precursors that are independent of pyruvate carboxylase, *Int. J. Biochem.*, 8, 869, 1977.
70. Bannister, D. W. and Evans, A. J., The fatty liver and kidney syndrome in poultry: biochemical aspects, in *Proc. Roche Symp. Cost/Effectiveness of Micronutrient Supplementation*, Hoffman-La Roche & Co., Basel, 1974, 63.
71. Bannister, D. W., Evans, A. J., and Whitehead, C. C., Evidence for a lesion in carbohydrate metabolism in fatty liver and kidney syndrome in chicks, *Res. Vet. Sci.*, 18, 149, 1975.
72. Bannister, D. W. and Whitehead, C. C., Presence of pyruvate carboxylase in the blood of the domestic fowl (*Gallus domesticus*) and an assay procedure, *Int. J. Biochem.*, 7, 619, 1976.
73. Barden, R. E., Fung, C-H., Utter, M. F., and Scrutton, M. C., Pyruvate carboxylase from chicken liver; steady state kinetic studies indicate a "two-site" ping-pong mechanism, *J. Biol. Chem.*, 247, 1323, 1972.
74. Barden, R. E. and Scrutton, M. C., Pyruvate carboxylase from chicken liver. Effects of univalent and divalent cations on catalytic activity, *J. Biol. Chem.*, 249, 4829, 1974.
75. Barden, R. E., Taylor, B. L., Isohashi, F., Frey, W. H., Zander, G., Lee, J. C., and Utter, M. F., Structural properties of pyruvate carboxylases from chicken liver and other sources, *Proc. Natl. Acad. Sci. U.S.A.*, 72, 4308, 1975.
76. Barra, D., Bossa, F., Doonan, S., Fahmy, H. M. A., Hughes, G. J., Kakoz, K. Y., Martini, F., and Petruzzelli, R., The structure of mitochondrial aspartate aminotransferase from pig heart and comparison with that of the cytoplasmic isozyme, *FEBS Lett.*, 83, 241, 1977.
77. Barritt, G. J., Keech, D. B., and Ling, A.-M., Apparent cooperative effect of acetyl CoA on sheep kidney pyruvate carboxylase, *Biochem. Biophys. Res. Commun.*, 24, 476, 1966.
78. Barritt, G. J., Thorne, R. F. W., and Hughes B. P., Effects of hormones and $N^6O^{2'}$-dibutyryl-adenosine 3′:5′-cyclic monophosphate, administered *in vivo*, on phosphate transport and metabolism in isolated rat liver mitochondria, *Biochem J.*, 172, 577, 1978.
79. Barritt, G. J., Zander, G. L., and Utter, M. F., The regulation of pyruvate carboxylase activity in gluconeogenic tissues, in *Gluconeogenesis Its Regulation in Mammalian Species*, Hanson, R. W. and Mehlman, M. A., Eds., John Wiley & Sons, New York, 1976, 3.

80. Bartels, P. D. and Sestoft, L., Thyroid hormone-induced changes in gluconeogenesis and ketogenesis in perfused rat liver, *Biochim. Biophys. Acta*, 633, 56, 1980.
81. Bartlett, K., Vitamin-responsive inborn errors of metabolism, in *Advances in Clinical Chemistry*, Vol. 23, Latner, A. L. and Schwartz, M. K., Eds., Academic Press, New York, 1983, 142.
82. Bartlett, K., Ng, H., and Leonard, J. V., A combined defect of three mitochondrial carboxylases presenting as biotin-responsive 3-methylcrotonyl glycinuria and 3-hydroxyisovaleric aciduria, *Clin. Chim. Acta*, 100, 183, 1980.
83. Bartnicki-Garcia, S. and Nickerson, W. J., Induction of yeast-like development in *Mucor* by carbon dioxide, *J. Bacteriol.*, 84, 829, 1962.
84. Baryshnikova, L. M. and Loginova, N. V., Carboxylases of coryneform bacteria, *Mikrobiologiya*, 48, 965, 1979.
85. Bas, S., Favarger, P., and Rous, S., Attempt to estimate the degree of saturation of mitochondrial pyruvate carrier and pyruvate carboxylase by their substrate in liver of living mice, *Experientia*, 34, 905, 1978.
86. Bas, S., Favarger, P., and Rous, S., Regulation of pyruvate carboxylase activity by pyruvate in mice liver, *Int. J. Biochem.*, 10, 117, 1979.
87. Bas, S., Gerlach, J., Favarger, P., and Rous, S., Studies on the degree of saturation of gluconeogenic and lipogenic enzymes *in vivo*, *Experientia*, 31, 716, 1975.
88. Baumgartner, R., Suormala, T., Wick, H., Bachmann, C., and Jaggi, K. H., Biotin dependency causing multiple carboxylase deficiency *in vivo*, *Pediatr. Res.*, 15, 1189, 1981.
89. Beaty, N. B. and Lane, M. D., Acetyl CoA carboxylase. Rapid purification of the chicken liver enzyme and steady state kinetic analysis of the carboxylase-catalysed reaction, *J. Biol. Chem.*, 257, 924, 1982.
90. Behrisch, H. W., Gluconeogenesis in skeletal muscle of the Alaskan King Crab, *Paralithodes camtschatica*, *Fed. Proc. Fed. Am. Soc. Exp. Biol.*, 35, 213, 1976.
91. Belo, P. S., Romsos, D. R., and Leveille, G. A., Influence of diet on glucose tolerance on the rate of glucose utilization and on gluconeogenic activities in the dog, *J. Nutr.*, 106, 1465, 1976.
92. Benedict, C. R. and Rinne, R. W., Glutamic acid synthesis from acetate units and bicarbonate in extracts of photosynthetic bacteria, *Biochem. Biophys. Res. Commun.*, 14, 474, 1964.
93. Benke, P. J. and Brooks, T., Pyruvate carboxylase induction in tissue culture in disorders associated with primary lactic-acidosis, *Clin. Res.*, 26, 812A, 1978.
94. Berens, R. L. and Marr, J. J., *Leishmania donovani* and *L. braziliensis:* function of pyruvate kinase, *Exp. Parasitol.*, 43, 407, 1977.
95. Berger, M. and Wood, H. G., Purification of the subunits of transcarboxylase by affinity chromatography on avidin-Sepharose, *J. Biol. Chem.*, 250, 927, 1975.
96. Berger, M. and Wood, H. G., Immunochemistry of the subunits of transcarboxylase, *J. Biol. Chem.*, 251, 7021, 1976.
97. Bergman, E. N., Glucose metabolism in ruminants as related to hypoglycaemia and ketosis, *Cornell Vet.*, 63, 341, 1973.
98. Berndt, J., Messner, B., Turkki, T., and Weissmann, E., Optical assay of pyruvate carboxylase in crude liver homogenates, *Anal. Biochem.*, 86, 154, 1978.
99. Bhagavan, H. N. and Coursin, D. B., Depletion of biotin from brain and liver in biotin deficiency, *J. Neurochem.*, 17, 289, 1970.
100. Beilicki, L. and Krieglstein, J., The effect of anesthesia on brain mitochondrial hexokinase, *Naunyn-Schmiedeberg's Arch. Pharmacol.*, 298, 229, 1977.
101. Billington, D., Kean, E. A., Osmundsen, H., and Sherratt, H. S. A., The effects of pent-4-enoate and methylene cyclopropylacetate on some enzymes of β-oxidation in rat liver mitochondria in vivo and in vitro, IRCS,. *Libr. Compend.*, 2, 1712, 1974.
102. Billington, D., Osmundsen, H., and Sherratt, H. S. A., Mechanisms of the metabolic disturbances caused by hypoglycin and by pent-4-enoic acid *in vitro* studies, *Biochem Pharmacol.*, 27, 2879, 1978.
103. Birnbaum, J., Coenzyme repression of acetyl-CoA carboxylase by (+)-biotin, *Arch. Biochem, Biophys.*, 132, 436, 1969.
104. Bitsch, R., Dersi, A., and Hötzel, D., Effect of different biotin supply in rats, *Nutr. Metab.*, 20, 163, 1976.
105. Bitsch, R., Dersi, A., and Hötzel, D., Effect of different biotin supply in rats, *Nutr. Metab.*, 21, (Suppl. 1), 27, 1977.
106. Blackmore, P. F., Dehaye, J.-P., and Exton, J. H., Studies on α-adrenergic activation of hepatic glucose output. The role of mitochondrial calcium release in α-adrenergic activation of phosphorylase in perfused rat liver, *J. Biol. Chem.*, 254, 6945, 1979.
107. Blair, J. McD., A reinterpretation of the kinetics of pyruvate carboxylase, *FEBS Lett.*, 2, 245, 1969.
108. Blaschkowski, H. P., Knappe, J., and Wieland, Th., S-Ethyl-coenzyme A and acetonyldethio-coenzyme A. Interactions with pyruvate carboxylase and phosphotransacetylase, *FEBS Lett.*, 98, 81, 1979.
109. Blass, J. P., Disorders of pyruvate metabolism, *Neurology*, 29, 280, 1979.

110. Blass, J. P., Cederbaum, S. D., and Dunn, H. G., Biochemical abnormalities in Leigh's disease, *Lancet*, 1, 1237, 1976.
111. Bloom, S. J. and Johnson, M. J., The pyruvate carboxylase of *Aspergillus niger*, *J. Biol. Chem.*, 237, 2718, 1962.
112. Blundell, T. L. and Johnson, L. N., *Protein Crystallography*, Academic Press, London, 1976, 1.
113. Bock, P. F. and Frieden, C., Another look at the cold lability of enzymes, *Trends Biochem. Sci.*, 3, 100, 1978.
114. Boeckx, R. L., and Dakshinamurti, K., Biotin-mediated protein biosynthesis, *Biochem J.*, 140, 549, 1974.
115. Bonjour, J. P., Biotin in man's nutrition and therapy — a review, *Int. J. Vit. Nutr. Res.*, 47, 107, 1977.
116. Bonnemere, C., Hamilton, J. A., Steinrauf, L. K., and Knappe, J., Structure of the bis-p-bromoanilide of carbon dioxide biotin, *Biochemistry*, 4, 240, 1965.
117. Böttger, I., Kriegel, H., and Wieland, O., Fluctuation of hepatic enzymes important in glucose metabolism in relation to thyroid function, *Eur. J. Biochem.*, 13, 253, 1970.
118. Böttger, I., Wieland, O., Brdiczka, D., and Pette, D., Intracellular localization of pyruvate carboxylase and phosphoenolpyruvate carboxykinase in rat liver, *Eur. J. Biochem.*, 8, 113, 1969.
119. Bowen, C. E., Rauschen, E., and Ingraham, L. L., The basicity of biotin, *Arch. Biochem. Biophys.*, 125, 865, 1968.
120. Bowman, I. B. R., Srivastava, H. K., and Flynn, I. W., Adaptations in oxidative metabolism during the transformation of *Trypanosoma rhodesiense* from bloodstream into culture form, in *Comparative Biochemistry of Parasites*, Van den Bossche, Ed., Academic Press, New York, 1972, 329.
121. Bradshaw, R. A., personal communications.
122. Brand, M. D. and De Selincourt, C., Effects of glucagon and Na^+ on the control of extramitochondrial free Ca^{2+} concentration by mitochondria from liver and heart, *Biochem. Biophys. Res. Commun.*, 92, 1377, 1980.
123. Brawand, F., Folly, G., and Walter, P., Relationship between extra- and intramitochondrial ATP/ADP ratios in rat liver mitochondria, *Biochim. Biophys. Acta*, 590, 285, 1980.
124. Brazier, J. and Wallace, J. C., unpublished observations.
125. Brdiczka, D., Pette, D., Brunner, G., and Miller, F., Kompartimentierte Verteilung von Enzymen in Rattenlebermitochondrien, *Eur. J. Biochem.*, 5, 294, 1968.
126. Brech, W., Shrago, E., and Wilken, D., Studies on pyruvate carboxylase in rat and human liver, *Biochim. Biophys. Acta.*, 201, 145, 1970.
127. Brennan, W. A., Jr., and Aprille, J. R., Regulation of hepatic gluconeogenesis by rapid compartmentation of mitochondrial adenine nucleotides in the newborn rabbit, *Comp. Biochem, Physiol.*, 77B, 35, 1984.
128. Bridgeland, E. S. and Jones, K. M., Formation of dicarboxylic acids and phosphoenol pyruvate in *Arthrobacter globiformis*, *Biochem. J.*, 104, 9P, 1967.
129. Brocks, D. G., Siess, E. A., and Wieland, O. H., Distinctive roles of oleate and glucagon in gluconeogenesis, *Eur. J. Biochem.*, 113, 39, 1980.
130. Brooks, D. E., Activity and androgenic control of enzymes associated with the tricarboxylic acid cycle, lipid oxidation and mitochondrial shuttles in the epididymis and epididymal spermatozoa of the rat, *Biochem. J.*, 174, 741, 1978.
131. Brooks, D. E., Metabolic activity in the epididymis and its regulation by androgens, *Physiol. Rev.*, 61, 515, 1981.
132. Brooks, P. H., Smith, D. A., and Irwin, V. C. R., Biotin-supplementation of diets; the incidence of foot lesions, and the reproductive performance of sows, *Vet. Rec.*, 101, 46, 1977.
133. Bruice, J. C. and Hegarty, A. F., Biotin-bound CO_2 and the mechanism of enzymatic carboxylation reactions, *Proc. Natl. Acad. Sci. U.S.A.*, 65, 805, 1970.
134. Brunette, M. G., Delvin, E., Hazel, B., and Scriver, C. R., Thiamine-responsive lactic acidosis in a patient with deficient low-Km pyruvate carboxylase activity in liver, *Pediatrics*, 50, 702, 1972.
135. Bryant, C., The utilization of carbon dioxide by *Moniezia expansa:* aspects of metabolic regulation, in *Comparative Biochemistry of Parasites*, Van den Bossche, Ed., Academic Press, New York, 1972, 49.
136. Bryant, C., Carbon dioxide utilisation, and the regulation of respiratory metabolic pathways in parasitic helminths, *Adv. Parasitol.*, 13, 35, 1975.
137. Bryla, J., Harris, E. J., and Plumb, J. A., The stimulatory effect of glucagon and dibutyryl cyclic AMP on ureogenesis and gluconeogenesis in relation to the mitochondrial ATP content, *FEBS Lett.*, 80, 443, 1977.
138. Bryla, J. and Niedzwiecka, A., Relationship between pyruvate carboxylation and citrulline synthesis in rat liver mitochondria: the effect of ammonia and energy, *Int. J. Biochem.*, 10, 235, 1979.
139. Bryla, J., Smith, C. M., and Williamson, J. R., Control of phosphoenolpyruvate synthesis by substrate level phosphorylation in guinea pig liver mitochondria, *J. Biol. Chem.*, 248, 4003, 1973.

140. Buckley, J. J., Libor, S., and Sundaram, T. K., Biotin subunits of acetyl CoA carboxylase and pyruvate carboxylase from a thermophilic *Bacillus, Arch. Biochem. Biophys.*, 192, 396, 1969.
141. Budd, K., Bicarbonate fixation by cell-free extracts and by mycelium of *Neocosmospora vasinfecta, J. Gen. Microbiol.*, 67, 99, 1971.
142. Bueding, E. and Saz, H. J., Pyruvate kinase and phosphoenolpyruvate carboxykinase activities of *Ascaris* muscle, *Hymenolepis diminuta* and *Schistosoma mansoni, Comp. Biochem. Physiol.*, 24, 511, 1968.
143. Buist, N. R. M., Is pyruvate carboxylase involved in the renal tubular reabsorption of bicarbonate?, *J. Inher. Metab. Dis.*, 3, 113, 1980.
144. Burri, B. J., Sweetman, L. and Nyhan, W. L., Mutant holocarboxylase synthetase: evidence for the enzyme defect in early infantile biotin-responsive multiple carboxylase deficiency, *J. Clin. Invest.*, 68, 1491, 1981.
145. Burton, D. R., Forsen, S., Karlstrom, G., and Dwek, R. A., Proton relaxation enhancement (PRE) in biochemistry: a critical survey, *Prog. NMR Spectrosc.*, 13, 1, 1979.
146. Buxton, R. S., A heat-sensitive lysis mutant of *Bacillus subtilis* 168 with a low activity of pyruvate carboxylase, *J. Gen. Microbiol.*, 105, 175, 1978.
147. Bykovskaya, S. V., Loginova, N. V., Gargarina, V. A., Malashenko, Yu. R., and Trotsenko, Yu. A., Influence of carbon dioxide on the methylotrophic metabolism of *Candida boidinii, Mikrobiologiya*, 47, 953, 1978.
148. Callikan, S. and Girard, J., Perinatal development of gluconeogenic enzymes in rabbit liver, *Biol. Neonate*, 36, 78, 1979.
149. Cannon, B. and Nedergaard, J., The physiological role of pyruvate carboxylation in hamster brown adipose tissue, *Eur. J. Biochem.*, 94, 419, 1979.
150. Cao, T. P. and Rous, S., Action of acetazolamide on liver pyruvate carboxylase and acetyl coenzyme A carboxylase activities in mice, *Experientia*, 34, 907, 1978.
151. Caplow, M., Studies of the mechanism of biotin catalysis, *J. Am. Chem. Soc.*, 87, 5774, 1965.
152. Caplow, M. and Yager, M., Studies on the mechanism of biotin catalysis, II, *J. Am. Chem. Soc.*, 89, 4513, 1967.
153. Cassady, A. I., Wells, J. R. E., and Wallace, J. C., Isolation and analysis of chicken genomal pyruvate carboxylase clones, *Proc. Aust. Biochem, Soc.*, 15, 65, 1983.
154. Caste, P. G. and Hartman, R. E., Carbon dioxide fixation by yeast cells of *Mucor rouxii, Mycologia*, 69, 423, 1977.
155. Cataldi de Flombaum, M. A., Cannata, J. J. B., Cazzulo, J. J., and Segura, E. L., CO_2-fixing enzymes in *Trypanosoma cruzi, Comp. Biochem. Physiol.*, 58B, 67, 1977.
156. Cazzulo, J. J., Claisse, L. M., and Stoppani, A. O. M., Carboxylase levels and carbon dioxide fixation in baker's yeast, *J. Bacteriol.*, 96, 623, 1968.
157. Cazzulo, J. J. and Stoppani, A. O. M., Purification and properties of pyruvate carboxylase from baker's yeast, *Arch. Biochem. Biophys.*, 121, 596, 1967.
158. Cazzulo, J. J. and Stoppani, A. O. M., The regulation of yeast pyruvate carboxylase by acetyl-coenzyme A and L-aspartate, *Arch. Biochem, Biophys.*, 127, 563, 1968.
159. Cazzulo, J. J. and Stoppani, A. O. M., Effects of adenosine phosphates and nicotinamide nucleotides on pyruvate carboxylase from baker's yeast, *Biochem. J.*, 112, 755, 1969.
160. Cazzulo, J. J., Sundaram, T. K., Dilks, S. N., and Kornberg, H. L., Synthesis of pyruvate carboxylase from its apoenzyme and (+)-biotin in *Bacillus stearothermophilus*. Purification and properties of the apoenzyme and the holoenzyme synthetase, *Biochem J.*, 122, 653, 1971.
161. Cazzulo, J. J., Sundaram, T. K., and Kornberg, H. L., Regulation of pyruvate carboxylase formation from the apo-enzyme and biotin in a thermophilic bacillus, *Nature (London)*, 223, 1137, 1969.
162. Cazzulo, J. J., Sundaram, T. K., and Kornberg, H. L., Mechanism of pyruvate carboxylase formation from the apo-enzyme and biotin in a thermophilic bacillus, *Nature (London)*, 227, 1103, 1970.
163. Cazzulo, J. J., Sundaram, T. K., and Kornberg, H. L., Properties and regulation of pyruvate carboxylase from *Bacillus stearothermophilus, Proc. R. Soc. London Ser. B* 176, 1, 1970.
164. Chacin, J., Park, O. H., Harris, J. B., and Alonso, D., Role of oxaloacetate in lipoate effect on frog gastric mucosa, *Am. J. Physiol.*, 231, 209, 1976.
165. Chan, A. W. K., Changes in renal gluconeogenic enzymes during chronic ethanol intake, *Fed. Proc. Fed. Am. Soc. Exp. Biol.*, 38, 377, 1979.
166. Chan, T. M., Bacon, C. B., and Hill, S. A., Glucagon stimulation of liver mitochondrial CO_2 fixation utilizing pyruvate generated inside the mitochondria, *J. Biol. Chem.*, 254, 8730, 1979.
167. Chan, T. M. and Freedland, R. A., The effect of propionate on the metabolism of pyruvate and lactate in the perfused rat liver, *Biochem. J.*, 127, 539, 1972.
168. Chance, B., Holmes, W., Higgins, J. C., and Connelly, C. M., Localisation of interaction sites in multi-component transfer systems: theorems derived from analogues, *Nature (London)*, 182, 1190, 1958.
169. Chandler, C. S. and Ballard, F. J., personal communication.

170. Chandler, C. S. and Ballard, F. J., Inhibition of pyruvate carboxylase degradation and total protein breakdown by lysosomotropic agents in 3T3-L1 cells, *Biochem. J.*, 210, 845, 1983.
171. Chang, A. Y. and Schneider, D. I., Hepatic enzyme activities in streptozotocin-diabetic rats before and after insulin treatment, *Diabetes*, 20, 71, 1971.
172. Chang, L. O., The development of pyruvate carboxylase in rat liver mitochondria, *Pediatr. Res.*, 11, 6, 1977.
173. Chang, L. O. and Morris, H. P., Enzymatic and immunological studies on pyruvate carboxylase in livers and liver tumours, *Cancer Res.*, 33, 2034, 1973.
174. Chang, S-C., Pyruvate metabolism in the lobster nerve as affected by the partial pressure of carbon dioxide: observations on the synthesis of acetylcholine and on metabolic compartmentation, *J. Neurochem.*, 19, 461, 1972.
175. Chapatwala, K. D., Rajanna, B., and Desaiah, D., Cadmium induced changes in gluconeogenic enzymes in rat kidney and liver, *Drug Chem. Toxicol.*, 3, 407, 1980.
176. Chapman-Smith, A., Studies on the Interaction of Sheep Liver Pyruvate Carboxylase with the Allosteric Effector, Acetyl-CoA, Ph.D. thesis, University of Adelaide, Australia, 1981, 42.
177. Chappell, J. B. and Crofts, A. R., *Regulation of Metabolic Processes in Mitochondria*, Gager, J. M., Papa, S., Quagliariello, E., and Slater, E. C., Eds., Elsevier, Amsterdam, 1966, 292.
178. Charles, A. M., Butler, B. J., and Willer, D., Pyruvate carboxylase from the facultative chemolithotrophs *Thiobacillus* A2 and *Thiobacillus novellus*, *FEMS Microbiol. Lett.*, 8, 9, 1980.
179. Charles, B. M., Hosking, G., Green, A., Pollitt, R., Bartlett, K., and Taitz, L. S., Biotin-responsive alopecia and developmental regression, *Lancet*, 2, 118, 1979.
180. Chen, C. H. and Awapara, J., Intracellular distribution of enzymes catalysing succinate production from glucose in *Rangia* mantle, *Comp. Biochem. Physiol.*, 30, 727, 1969.
181. Chern, C.-K., Kusaka, I., and Fukui, S., Significance of pyruvate carboxylase in sugar metabolism of *Agrobacterium tumefaciens*, *Agric. Biol. Chem.*, 40, 771, 1976.
182. Cheung, Y.-F. and Walsh, C., Studies on the intramolecular and intermolecular kinetic isotope effects in pyruvate carboxylase catalysis, *Biochemistry*, 15, 3749, 1976.
183. Chiang, G. S. and Mistry, S. P., Activities of pyruvate carboxylase and propionyl-CoA carboxylase in rat tissues during biotin administration, *Proc. Soc. Exp. Biol. Med.*, 146, 21, 1974.
184. Chlebowski, R. T. and Adam, P. A. J., Glucose production in the newborn dog. II. Evaluation of autonomic and enzymatic control in the isolated perfused canine liver, *Pediatr. Res.*, 9, 821, 1975.
185. Claret-Berthon, B., Claret, M., and Mazet, J. L., Fluxes and distribution of calcium in rat liver cells: kinetic analysis and identification of pools, *J. Physiol. (London)*, 272, 529, 1977.
186. Clark, M. G., Bloxham, D. P., Holland, P. C., and Lardy, H. A., Estimation of fructose diphosphatase-phosphofructokinase substrate cycle in the flight muscle of *Bombus affinis*, *Biochem. J.*, 134, 589, 1973.
187. Clarke, G. A. and Hartman, R. E., Carbon dioxide fixation in *Helminthosporium cynodontis*, *J. Gen. Microbiol.*, 92, 237, 1976.
188. Cleland, W. W., The kinetics of enzyme-catalysed reactions with two or more substrates, *Biochim. Biophys. Acta*, 67, 104, 1963.
189. Cleland, W. W., Steady state kinetics, in *The Enzymes*, Vol. 2, 3rd ed., Boyer, P. D., Ed., Academic Press, New York, 1970, 1.
190. Clements, P. R., Pyruvate Carboxylase: the Mode of Action of Acetyl CoA, PhD. thesis, University of Adelaide, Australia, 1977, 101.
191. Clements, P. R., Campbell, A. J., and Keech, D. B., The molecular basis for activation of pyruvate carboxylase by acetyl CoA, *Proc. Aust. Biochem. Soc.*, 11, 13, 1978.
192. Cohen, N. D., Beegen, H., Utter, M. F., and Wrigley, N. G., A re-examination of the electron microscopic appearance of pyruvate carboxylase from chicken liver, *J. Biol. Chem.*, 254, 1740, 1979.
193. Cohen, N. D., Duc, J. A., Beegen, H., and Utter, M. F., Quaternary structure of pyruvate carboxylase from *Pseudomonas citronellolis*, *J. Biol. Chem.*, 254, 9262, 1979.
194. Cohen, N. D., Utter, M. F., Wrigley, N. G., and Barrett, A. N., Quaternary structure of yeast pyruvate carboxylase: biochemical and electron microscope studies, *Biochemistry*, 18, 2197, 1979.
195. Cohen, P., The role of protein phosphorylation in neural and hormonal control of cellular activity, *Nature (London)*, 296, 613, 1982.
196. Coleman, D. L. and Hummel, K. P., Studies with the mutation, diabetes, in the mouse, *Diabetologia*, 3, 238, 1967.
197. Coleman, D. L. and Hummel, K. P., The mutation, diabetes, in the mouse, *Proc. Congr. Int. Diabetics*, 813, 1968.
198. Coleman, R. A., Reed, B. C., Mackall, J. C., Student, A. K., Lane, M. D., and Bell, R. M., Selective changes in microsomal enzymes of triacylglycerol phosphatidylcholine, and phosphatidylethanolamine biosynthesis during differentiation of 3T3-L1 preadipocytes, *J. Biol. Chem.*, 253, 7256, 1978.
199. Colen, A. H., Prough, R. A., and Fisher, H. F., The mechanism of glutamate dehydrogenase reaction. IV. Evidence for random and rapid binding of substrate and coenzyme in the burst phase, *J. Biol. Chem.*, 247, 7905, 1972.

200. Coll, K. E., Joseph, S. K., Corkey, B. E., and Williamson, J. R., Determination of matrix free Ca^{2+} concentration and kinetics of Ca^{2+} efflux in liver and heart mitochondria, *J. Biol. Chem.*, 257, 8696, 1982.
201. Colowick, S. P. and Womack, F. C., Binding of diffusible molecules by macromolecules: rapid measurement by rate of dialysis, *J. Biol. Chem.*, 244, 774, 1969.
202. Cooper, T. G. and Benedict, C. R., The participation of acetyl-CoA in pyruvate carboxylase, *Biochem. Biophys. Res. Commun.*, 22, 285, 1966.
203. Cooper, T. G. and Benedict, C. R., Regulation of pyruvate carboxylase by coenzyme A and acyl coenzyme A thioesters, *Biochemistry*, 7, 3032, 1968.
204. Cooper, T. G. and Benedict, C. R., The participation of acetyl-CoA in pyruvate carboxylase, *Biochem. Biophys. Res. Commun.*, 22, 285, 1966.
205. Cooper, R. A. and Kornberg, H. L., Net formation of phosphoenolpyruvate from pyruvate by *Escherichia coli*, *Biochim. Biophys. Acta*, 104, 618, 1965.
206. Cornish-Bowden, A., Assessment of protein sequence identity from amino acid composition data, *J. Theor. Biol.*, 65, 735, 1977.
207. Cornish-Bowden, A., How reliably do amino acid composition comparisons predict sequence similarities between proteins?, *J. Theor, Biol.*, 76, 369, 1979.
208. Cowan, M. J., Wara, D. W., Packman, S., and Ammann, A. J., Multiple biotin-dependent carboxylase deficiencies associated with defects in T-cell and B-cell immunity, *Lancet*, 2, 115, 1979.
209. Cowey, C. B., de la Higuera, M., and Adron, J. W., The effect of dietary composition and of insulin on gluconeogenesis in rainbow trout *(Salmo gairdneri)*, *Br. J. Nutr.*, 38, 385, 1977.
210. Crabb, D. W. and Harris, R. A., Mechanism responsible for the hypoglycemic actions of di-chloro acetate and 2-chloro propionate, *Arch. Biochem. Biophys.*, 198, 145, 1979.
211. Crabtree, B., Higgins, S. J., and Newsholme, E. A., The activities of pyruvate carboxylase, phosphoenolpyruvate carboxylase and fructose diphosphatase in muscles from vertebrates and invertebrates, *Biochem. J.*, 130, 391, 1972.
212. Crabtree, B., Taylor, D. J., Coombs, J. E., Smith, R. A., Templer, S. P., and Smith, G. H., The activities and intracellular distribution of enzymes of carbohydrate, lipid and ketone-body metabolism in lactating mammary glands from ruminants and non-ruminants, *Biochem, J.*, 196, 747, 1981.
213. Dahm, C. H., Jellinek, M., Minagawa, J., and Rao, P. S., Effects of estradiol-17β on some gluconeogenic enzymes in rat liver, *Biol. Reprod.*, 17, 734, 1978.
214. Davis, E. J., Spydevold, O., and Bremer, J., Pyruvate carboxylase and propionyl-CoA carboxylase as anaplerotic enzymes in skeletal muscle mitochondria, *Eur. J. Biochem.*, 110, 255, 1980.
215. Davis-van Thienen, W. and Davis, E. J., The effects of energetic steady state, pyruvate concentration, and octanoyl-(−)-carnitine on the relative rates of carboxylation and decarboxylation of pyruvate by rat liver mitochondria, *J. Biol. Chem.*, 256, 8371, 1981.
216. DeDuve, C. and Baudhuin, P., Peroxisomes, *Physiol. Rev.*, 46, 323, 1966.
217. DeDuve, C., Pressman, B. C., Gianetto, R., Wattiaux, R., and Appelmans, F., Tissue fractionation studies. VI. Intracellular distribution patterns of enzymes in rat-liver tissue, *Biochem. J.*, 60, 604, 1955.
218. De Groot, C. J. and Hommes, F. A., Further speculation on the pathogenesis of Leigh's encephalomyelopathy, *J. Pediatr.*, 82, 541, 1973.
219. De Groot, C. J., Jonxis, J. H. P., and Hommes, F. A., Further studies on Leigh's encephalomyelopathy, in *Organic Acidurias, Proc. Ninth Symp. Soc. Study of Inborn Errors of Metabolism*, Stern, J. and Toothill, C., Eds., Churchill Livingstone, London, 1972, 40.
220. Delisle, G. and Fritz, I. B., Interrelations between hepatic fatty acid oxidation and gluconeogenesis: a possible regulatory role of carnitine palmityltransferase, *Biochemistry*, 58, 790, 1967.
221. Demaugre, F., Cepanec, C., and Leroux, J. P., Characterization of oxalate as a catabolite of dichloroacetate responsible for the inhibition of gluconeogenesis and pyruvate carboxylation in rat liver cells, *Biochem. Biophys. Res. Commun.*, 85, 1180, 1978.
222. Dennis, S. C., DeBuysere, M., Scholz, R., and Olson, M. S., Studies on the relationship between ketogenesis and pyruvate oxidation in isolated rat liver mitochondria, *J. Biol. Chem.*, 253, 2229, 1978.
223. Denton, R. M. and Halestrap, A. P., Regulation of pyruvate metabolism in mammalian tissues, *Essays Biochem.*, 15, 37, 1979.
224. Denton, R, M., McCormack, J. G., and Edgell, N. J., Role of calcium ions in the regulation of intramitochondrial metabolism. Effects of Na^+, Mg^{2+} and ruthenium red on the Ca^{2+}-stimulated oxidation of oxoglutarate and on pyruvate dehydrogenase activity in intact rat heart mitochondria, *Biochem. J.*, 190, 107, 1980.
225. Denton, R. M., Randle, P. J., and Martin, B. R., Stimulation by calcium ions of pyruvate dehydrogenase phosphate phosphatase, *Biochem. J.*, 128, 161, 1972.
226. Denton, R. M., Richards, D. A., and Chin, J. G., Calcium ions and the regulation of NAD^+-linked isocitrate dehydrogenase from the mitochondria of rat heart and other tissues, *Biochem. J.*, 176, 899, 1978.

227. Deodhar, A. D. and Mistry, S. P., Gluconeogenesis in biotin deficiency: *in vivo* synthesis of pyruvate holocarboxylase in biotin deficient rat liver, *Biochem. Biophys. Res. Commun.*, 34, 755, 1969.
228. Desjardins, P. R. and Dakshinamurti, K., Acetyl-CoA holocarboxylase synthesis in biotin deficient rat adipose tissue, *Arch. Biochem. Biophys.*, 142, 292, 1971.
229. De Titta, G. J., Edmonds, J. W., Stallings, W., and Donohue, J., Molecular structure of biotin. Results of two independent crystal structure investigations., *J. Am. Chem. Soc.*, 98, 1920, 1976.
230. DeVivo, D. C., Haymond, M. W., Leckie, M. P., Bussmann, Y. L., McDougal, D. B., and Pagliara, A. S., The clinical and biochemical implications of pyruvate carboxylase deficiency, *J. Clin. Endocrinol. Metab.*, 45, 1281, 1977.
231. de Zwaan, A. and de Bont, A. M. Th., Phosphoenolpyruvate carboxykinase from adductor muscle of the sea mussel *Mytilus edulis* L., *J. Comp. Physiol.*, 96, 85, 1975.
232. de Zwaan, A., Kluytmans, J. H. F. M., and Zandee, D. I., Facultative anaerobiosis in molluscs, *Biochem. Soc. Symp.*, 41, 133, 1976.
233. de Zwaan, A. and van Marrewijk, W. J. A., Intracellular localization of pyruvate carboxylase, phosphoenolpyruvate carboxykinase and "malic enzyme" and the absence of glyoxylate cycle enzymes in the sea mussel (*Mytilus edulis* L.), *Comp. Biochem. Physiol.*, 44B, 1057, 1973.
234. de Zwaan, A. and Zandee, D. I., The utilization of glycogen and accumulation of some intermediates during anaerobiosis in *Mytilus edulis*, *Comp. Biochem. Physiol.*, 43B, 47, 1972.
235. de Zwaan, A. and Zandee, D. I., Body distribution and seasonal changes in the glycogen content of the common sea mussel *Mytilus edulis*, *Comp. Biochem. Physiol.*, 43A, 53, 1972.
236. Dhand, U. K., Jeacock, M. K., Shepherd, D. A. L., Smith, E. M., and Varnam, G. C. E., Activities of enzymes concerned with pyruvate, oxaloacetate, citrate, acetate and acetoacetate in placental cotyledons of sheep, *Biochim. Biophys. Acta*, 222, 216, 1970.
237. Diesterhaft, M. D. and Freese, E., Role of pyruvate carboxylase, phosphoenolpyruvate carboxykinase, and malic enzyme during growth and sporulation of *Bacillus subtilis*, *J. Biol. Chem.*, 248, 6062, 1973.
238. Dijkstra, U. J., Willems, J. L., Joosten, E. M. G., and Gabreëls, F. J., A deficiency of pyruvate carboxylase in Friedreichs ataxia, *Clin. Neurol. Neurosurg.*, 84, 137, 1982.
239. Dijkstra, U. J., Willems, J. L., Joosten, E. M. G., and Gabreëls, F. J. M., Friedreich ataxia and low pyruvate carboxylase activity in liver and fibroblasts, *Ann. Neurol.*, 13, 325, 1983.
240. Divjak, S. and Mor, J.-R., On the activity of carbon dioxide fixation in growing yeast, *Arch. Mikrobiol.*, 94, 191, 1973.
241. Dobson, C. M., The solution structure of lysozyme, in *NMR in Biology*, Dwek, R. A., Campbell, I. D., Richards, R. E., and Williams, R. J. P., Eds., Academic Press, London, 1977, 63.
242. Dodgson, J. B., Strommer, H., and Engel, J. D., Isolation of the chickem β-globin gene and a linked embryonic β-like globin gene from a chicken recombinant library, *Cell*, 17, 879, 1979.
243. Doolittle, R. F., Sequencing peptides and proteins lacking free α-amino groups, *Mol. Biol. Biochem. Biophys.*, 25, 38, 1977.
244. Dugal, B. S., Zur intrazellularen Lokalisation von Pyruvat-Carboxylase in der Huhnerleber, *Hoppe-Seyler's Z. Physiol. Chem.*, 353, 1373, 1972.
245. Dugal, B. S., Determination of the distribution of pyruvate carboxylase in rat liver, *FEBS Lett.*, 26, 97, 1972.
246. Dugal, B. S., Intracellular localization of pyruvate carboxylase in mammalian liver, *Exp. Cell Res.*, 89, 63, 1974.
247. Dugal, B. S. and Gopel, G., Effect of calcium ions on pyruvate carboxylase from pigeon liver, *Biochim. Biophys. Acta*, 410, 407, 1975.
248. Duggleby, R. G., Attwood, P. V., Wallace, J. C., and Keech, D. B., Avidin is a slow-binding inhibitor of pyruvate carboxylase, *Biochemistry*, 21, 3364, 1982.
249. Dwek, R. A., *Nuclear Magnetic Resonance (NMR) in Biochemistry: Applications to Enzyme Systems*, Clarendon Press, Oxford, 1974, 242.
250. Easterbrook-Smith, S. B., Campbell, A. J., Keech, D. B., and Wallace, J. C., The atypical velocity response by pyruvate carboxylase to increasing concentrations of acetyl coenzyme A, *Biochem. J.*, 179, 497, 1979.
251. Easterbrook-Smith, S. B., Hudson, P. J., Goss, N. H., Keech, D. B., and Wallace, J. C., Pyruvate carboxylase: mechanism of the second partial reaction, *Arch. Biochem. Biophys.*, 176, 709, 1976.
252. Easterbrook-Smith, S. B., Wallace, J. C., and Keech, D. B., Pyruvate carboxylase: affinity labelling of the MgATP binding site, *Eur. J. Biochem.*, 62, 125, 1976.
253. Easterbrook-Smith, S. B., Wallace, J. C., and Keech, D. B., A reappraisal of the reaction pathway of pyruvate carboxylase, *Biochem. J.*, 169, 225, 1978.
254. Ebels, E. J., Blokzijl, E. J., and Troelstra, J. A., A Wernicke-like encephalomyelopathy in children (Leigh), an inborn error of metabolism, *Helv. Paediatr. Acta*, 20, 310, 1965.
255. Edwards, E. M., Dhand, U. K., Jeacock, M. K., and Shepherd, D. A. L., Activities of enzymes concerned with pyruvate and oxaloacetate metabolism in the heart and liver of developing sheep, *Biochim. Biophys. Acta.*, 399, 217, 1975.

256. Edwards, E. M., Dhand, U. K., Jeacock, M. K., and Shepherd, D. A., Pyruvate and oxaloacetate metabolism in the kidney cortex of developing sheep, *Biol. Neonate*, 30, 49, 1976.
257. Edwards, E. M., Rottenbury, J. M., Varnam, G. C. E., Dhand, U. K., Jeacock, M. J., and Shepherd, D. A. L., Enzyme activities in the sheep placenta during the last three months of pregnancy, *Biochim. Biophys. Acta*, 497, 133, 1977.
258. Elder, J. A. and Lehninger, A. L., Respiration-dependent transport of carbon dioxide into rat liver mitochondria, *Biochemistry*, 12, 976, 1973.
259. Ephrussi, B., Hottinguer, H., and Chimenes, A. M., Action de l'acriflavine sur les levures. I. La mutation "petite colonie", *Ann. Inst. Pasteur*, 76, 351, 1949.
260. Ermolaeva, L. P. and Mil'man, L. S., Peculiarities of the activity of "satellite" enzymes of gluconeogenesis in loach oocytes and embryos, *Biokhimiya*, 41, 874, 1976.
261. Evans, A. J., Bannister, D. W., and Whitehead, C. C., Some aspects of lipid metabolism in fatty liver and kidney syndrome in chicks, *Res. Vet. Sci.*, 18, 26, 1975.
262. Evans, H. J. and Wood, H. G., The mechanism of the pyruvate, phosphate dikinase reaction, *Proc. Natl. Acad. Sci. U.S.A.*, 61, 1448, 1968.
263. Exton, J. H., Harper, S. C., Tucker, A. L., and Ho, R.-J., Effects of insulin on gluconeogenesis and cyclic AMP levels in perfused livers from diabetic rats, *Biochim. Biophys. Acta*, 329, 23, 1974.
264. Exton, J. H., Mallette, L. E., Jefferson, L. S., Wong, E. H. A., Friedmann, N., Miller, T. B., Jr., and Park, C. R., The hormonal control of hepatic gluconeogenesis, *Rec. Prog. Horm. Res.*, 26, 411, 1970.
265. Exton, J. H. and Park, C. R., Control of gluconeogenesis in liver. III. Effects of L-lactate, pyruvate, fructose, glucagon, epinephrine, and adenosine 3',5'-monophosphate on gluconeogenic intermediates in the perfused rat liver, *J. Biol. Chem.*, 244, 1424, 1969.
266. Fairlamb, A., Biochemistry of trypanosomiasis and rational approaches to chemotherapy, *Trends Biochem. Sci.*, 7, 249, 1982.
267. Fall, R. R., Alberts, A. W., and Vagelos, P. R., Analysis of bacterial biotin proteins, *Biochim. Biophys. Acta*, 379, 496, 1975.
268. Fearnley, I. M., Structural Aspects of Pyruvate Carboxylase, B.Sc. Hons. thesis, University of Adelaide, Australia, 1980, 17.
269. Feir, H. A. and Suzuki, I., Pyruvate carboxylase of *Aspergillus niger:* kinetic study of a biotin-containing carboxylase, *Can. J. Biochem.*, 47, 697, 1969.
270. Feldman, G. L., Hsia, Y. E., and Wolf, B., Biochemical characterization of biotin responsive multiple carboxylase deficiency. Heterogeneity within the *bio* genetic complementation group, *Am. J. Hum. Genet.*, 33, 692, 1981.
271. Feldman, G. L. and Wolf, B., Genetic complementation in heterokaryons of pyruvate carboxylase (PC) deficient human fibroblasts, *Am. J. Hum. Genet.*, 31, 47A, 1979.
272. Feldman, G. L. and Wolf, B., Measurement of pyruvate carboxylase activity in amniotic fluid cells, *Pediatr. Res.*, 14, 153, 1980.
273. Feldman, G. L. and Wolf, B., Evidence for two genetic complementation groups in pyruvate carboxylase-deficient human fibroblast cell lines, *Biochem. Genet.*, 18, 617, 1980.
274. Feldman, G. L. and Wolf, B., Deficient acetyl-CoA carboxylase activity in multiple carboxylase deficiency, *Clin. Chim. Acta*, 111, 147, 1981.
275. Felicioli, R. A., Gabriella, F., and Rossi, C. A., Pyruvate carboxylase activity of mammalian brain cortex, *Life Sci.*, 6, 133, 1967.
276. Felicioli, R. A., Gabrielli, F., and Rossi, C. A., The synthesis of phosphoenolpyruvate in the gluconeogenesis. Enzyme levels of chick embryo livers, *Eur. J. Biochem.*, 3, 19, 1967.
277. Ferry, P., Pegorier, J.-P., Williamson, D. H., and Girard, J., Interactions *in vivo* between oxidation of nonesterified fatty-acids and gluconeogenesis in the new born rat, *Biochem J.*, 182, 593, 1979.
278. Filsell, O. H., Jarrett, I. G., Taylor, P. H., and Keech, D. B., Effects of fasting, diabetes and glucocorticoids on gluconeogenic enzymes in the sheep, *Biochem. Biophys. Acta*, 184, 54, 1969.
279. Finol, J. and Chacin, J., Gluconeogenesis in toad gastric mucosa, *Comp. Biochem. Physiol.*, 65B, 651, 1980.
280. Fischer, A., Munnich, A., Saudubray, J. M., Mamas, S., Coude, F-X., Charpentier, C., Dray, F., Frezal, J., and Griscelli, C., Biotin-responsive immunoregulatory dysfunction in multiple carboxylase deficiency, *J. Clin. Immunol.*, 2, 35, 1982.
281. Flint, A. P. F. and Denton, R. M., The role of nicotinamide-adenine dinucleotide phosphate-dependent malate dehydrogenase and isocitrate dehydrogenase in the supply of reduced nicotinamide adenine dinucleotide phosphate for steroidogenesis in the superovulated rat ovary, *Biochem. J.*, 117, 73, 1970.
282. Foldes, M. and Barritt, G. J., Regulation by calcium ions of pyruvate carboxylation, pyruvate transport, and adenine nucleotide transport in isolated rat liver mitochondria, *J. Biol. Chem.*, 252, 5372, 1977.
283. Foote, J. L., Christner, J. E., and Coon, M. J., Biotin- and adenosine triphosphate-dependent activation of propionyl apocarboxylase, *Biochim. Biophys. Acta*, 67, 676, 1963.

284. Fowler, A. V. and Zabin, I., The amino acid sequence of β-galactosidase of *Escherichia coli, Proc. Natl. Acad. Sci. U.S.A.*, 74, 1507, 1977.
285. Freedland, R. A. and Krebs, H. A., The effect of thyroxine treatment on the rate of gluconeogenesis in the perfused rat liver, *Biochem. J.*, 104, 45p, 1976.
286. Freedman, A. D. and Kohn, L., Pyruvate metabolism and control: factors affecting pyruvate carboxylase activity, *Science*, 145, 58, 1964.
287. Frey, W. H. and Utter, M. F., Binding of acetyl CoA to chicken liver pyruvate carboxylase, *J. Biol. Chem.*, 252, 51, 1977.
288. Freytag, S. O., Regulation of the turnover of pyruvate carboxylase in 3T3-L-1 cells, *Fed. Proc. Fed. Am. Soc. Exp. Biol.*, 39, 1683, 1980.
289. Freytag, S. O. and Utter, M. F., Induction of pyruvate carboxylase apoenzyme and holoenzyme in 3T3-L1 cells during differentiation, *Proc. Natl. Acad. Sci. U.S.A.*, 77, 1321, 1980.
290. Freytag, S. O. and Utter, M. F., Regulation of the synthesis and degradation of pyruvate carboxylase in 3T3-L1 cells, *J. Biol. Chem.*, 258, 6307, 1983.
291. Fridovich, I., Superoxide dismutases, *Adv. Enzymol.*, 41, 35, 1974.
292. Friedman, B., Goodman, E. H., Saunders, H. L., Kostos, V., and Weinhouse, S., Regulation of metabolism in the liver. Estimation of pyruvate recycling during gluconeogenesis in perfused rat liver, *Metabolism*, 20, 2, 1971.
293. Friedmann, N. and Rasmussen, H., Calcium, manganese and hepatic gluconeogenesis, *Biochim. Biophys. Acta*, 222, 41, 1970.
294. Frigg, M. and Torhorst, J., Autoradiographic localisation of ^{3}H-biotin in chick tissues, *Int. J. Vitam. Nutr. Res.*, 52, 417, 1982.
295. Frigg, M. and Wick, H., Effect of graded biotin levels in the diet on liver pyruvate carboxylase (EC-6.4.1.1) of chicks fed *ad-libitum* and after starvation, *Int. J. Vitam. Nutr. Res.*, 47, 57, 1977.
296. Fritz, I. B. and Lee, L., Role of the acetyl-CoA to CoA-SH ratio in influencing ketogenesis from various substrates by isolated liver mitochondria, in *Regulation of Hepatic Metabolism*, Lundquist, F. and Tystrup, N., Eds., Munksgaard, Copenhagen, 1974, 224.
297. Fuller, R. C. and Kornberg, H. L., A possible route for malate oxidation by *Chromatium, Biochem. J.*, 79, 8P, 1961.
298. Fuller, R. C., Smillie, R. M., Sisler, E. C., and Kornberg, H. L., Carbon metabolism in *Chromatium, J. Biol. Chem.*, 236, 2140, 1961.
299. Fung, C. H., The Regulatory Properties and Reaction Mechanism of Chicken Liver Pyruvate Carboxylase, Ph.D. thesis, Case Western Reserve University, Cleveland, 1971, 97.
300. Fung, C. H., Gupta, R. K., and Mildvan, A. S., Magnetic resonance studies of the proximity and spatial arrangement of propionyl coenzyme A and pyruvate on a biotin metallo-enzyme: transcarboxylase, *Biochemistry*, 15, 85, 1975.
301. Fung, C.-H., Gupta, R. K., and Mildvan, A. S., Magnetic resonance studies of the proximity and spatial arrangement of propionyl CoA and pyruvate on a biotin-metalloenzyme, transcarboxylase, *Biochemistry*, 15, 85, 1976.
302. Fung, C-H., Mildvan, A. S., Allerhand, A., Komoroski, R., and Scrutton, M. C., Interaction of pyruvate with pyruvate carboxylase and pyruvate kinase as studied by paramagnetic effects on ^{13}C relaxation rates, *Biochemistry*, 12, 620, 1973.
303. Fung, C.-H., Mildvan, A. S., and Leigh, J. S. Jr., Electron and nuclear magnetic studies of the interaction of pyruvate with transcarboxylase, *Biochemistry*, 13, 1160, 1974.
304. Gailiusis, J., Rinne, R. W., and Benedict, C. R., Pyruvate oxaloacetate exchange reaction in baker's yeast, *Biochim. Biophys. Acta*, 92, 595, 1964.
305. Garber, A. J. and Hanson, R. W., The control of phosphoenolpyruvate formation by rabbit liver mitochondria. The relationship between phosphoenolpyruvate, malate, and aspartate formation and the regulation of intramitochondrial oxaloacetate, *J. Biol. Chem.*, 246, 5555, 1971.
306. Garrison, J. C. and Borland, M. K., Regulation of mitochondrial pyruvate carboxylation and gluconeogenesis in rat hepatocytes via an α-adrenergic, adenosine 3′:5′-monophosphate-independent mechanism, *J. Biol. Chem.*, 254, 1129, 1979.
307. Garrison, J. C., Borland, M. K., Florio, V. A., and Twible, D. A., The role of calcium ion as a mediator of the effects of angiotensin II, catecholamines and vasopressin on the phosphorylation and activity of enzymes in isolated hepatocytes, *J. Biol. Chem.*, 254, 7147, 1979.
308. Garrison, J. C. and Haynes, R. C., Jr., The hormonal control of gluconeogenesis by regulation of mitochondrial pyruvate carboxylation in isolated rat liver cells, *J. Biol. Chem.*, 250, 2769, 1975.
309. Garthoff, L., Wolf, G., and Mehlman, M. A., Subcellular distribution of pyruvate carboxylase and phosphoenolpyruvate carboxykinase in dog liver and kidney, *Proc. Soc. Exp. Biol. Med.*, 141, 532, 1972.
310. Gatley, S. J., Inhibition by benzoic-acid of ureogenesis and gluconeogenesis in isolated rat hepatocytes, *Fed. Proc. Fed. Am. Soc. Exp. Biol.*, 36, 672, 1977.

311. Gaudry, M., Munnich, A., Ogier, H., Marsac, C., Marquet, A., Saudubray, J. M., Mitchell, G., Causse, M., and Frezal, J., Deficient liver biotinidase activity in multiple carboxylase deficiency, *Lancet*, 2, 397, 1983.
312. Gerwin, B. I., Jacobsen, B. E., and Wood, H. G., Transcarboxylase. VIII. Isolation and properties of a biotin-carboxyl carrier protein, *Proc. Natl. Acad. Sci. U.S.A.*, 64, 1315, 1969.
313. Gevers, W. and Krebs, H. A., The effects of adenine nucleotides on carbohydrate metabolism in pigeon liver homogenates, *Biochem. J.*, 98, 720, 1966.
314. Gibbons, E. and Perham, R. N., Kinetic and molecular properties of citraconyl-aldolase. The reversible denaturation and hybridisation of the native and modified enzymes, *Biochem. J.*, 139, 331, 1974.
315. Giles, I. G., Poat, P. C., and Munday, K. A., The regulation of pyruvate kinase in the hepatopancreas of *Carcinus maenas*, *Comp. Biochem. Physiol.*, 55B, 423, 1976.
316. Girard, J. R., Ferre, P., Kervran, A., Pegorier, J. P., and Assan, R., Role of the insulin/glucagon ratio in the changes of hepatic metabolism during development of the rat, in *Glucagon: Its Role in Physiology and Clinical Medicine*, Foa, P. P., Bajaj, J. S., and Foa, N. L., Eds., Springer-Verlag, New York, 1977, 563.
317. Gnheim, H. K. and Bartlett, K., Mechanism of biotin-responsive combined carboxylase deficiency, *Lancet*, 1, 1187, 1982.
318. Gnheim, H. K., Noy. G., and Bartlett, K., Biotin-dependent carboxylases and cultured human fibroblasts, *Biochem. Soc. Trans.*, 9, 405, 1981.
319. Godinot, C. and Lardy, H. A., Biosynthesis of glutamate dehydrogenase in rat liver. Demonstration of its microsomal localization and hypothetical mechanism of transfer to mitochondria, *Biochemistry*, 12, 2051, 1973.
320. Gomez-Puyou, A., Tuena de Gomez-Payou, M., Klapp, M., and Carafoli, E., The effect of calcium on the translocation of adenine nucleotides in rat liver mitochondria, *Arch. Biochem. Biophys.*, 194, 399, 1979.
321. Gompertz, D., Draffan, G. H., Watts, J. L., and Hull, D., Biotin-responsive β-methylcrotonylglycinuria, *Lancet*, 2, 22, 1971.
322. Goodall, G. J., Ph.D. thesis, University of Adelaide, Adelaide, South Australia, 1981.
323. Goodall, G. J., Baldwin, G. S., Wallace, J. C., and Keech, D. B., Factors that influence the translocation of the N-carboxybiotin moiety between the two sub-sites of pyruvate carboxylase, *Biochem, J.*, 199, 603, 1981.
324. Goodall, G. J., Prager, R., Wallace, J. C., and Keech, D. B., A mechanism for the transfer of the carboxyl group from 1′-N-carboxybiotin to acceptor substrates by biotin-containing enzymes, *FEBS Lett.*, 163, 6, 1983.
325. Gorin, E., Tal-Or, Z., and Shafrir, Eds., Glyceroneogenesis in adipose tissue of fasted, diabetic and triamcinolone treated rats, *Eur. J. Biochem.*, 8, 370, 1969.
326. Goss, N. H., Pyruvate Carboxylase: Physico-Chemical Aspects, Ph.D. thesis, University ofAdelaide, South Australia, Australia, 1978.
327. Goss, J. A., Cohen, N. D., and Utter, M. F., Characterization of the subunit structure of pyruvate carboxylase from *Pseudomonas citronellolis*, *J. Biol. Chem.*, 256, 11819, 1981.
328. Goss, N. H., Dyer, P. Y., Keech, D. B., and Wallace, J. C., An electron microscopic study of pyruvate carboxylase, *J. Biol. Chem.*, 254, 1734, 1979.
329. Gottschalk, E-M., Mayer, F., Klostermann, A., and Seubert, W., Determination of molecular weight and molecular structure of rat-liver pyruvate carboxylase, *Eur. J. Biochem.*, 64, 411, 1976.
330. Graafmans, W. D. J., Effect of blue light on metabolism in *Penicillium isariiforme*, *J. Gen. Microbiol.*, 101, 157, 1977.
331. Grant, P. T., Sargent, J. R., and Ryley, J. F., Respiratory systems in the *Trypanosomidae*, *Biochem. J.*, 81, 200, 1961.
332. Gratzner, H. G., Ahmad, P. M., Zegadlo, J., and Ahmad, F., Immunofluorescent localization of acetyl-CoA carboxylase, fatty acid synthetase and pyruvate carboxylase during adipocyte conversion of 3T3 fibroblasts, *Cell Biol. Int. Rep.*, 4, 497, 1980.
333. Green, A., Some problems in diagnosis and pathogenesis of Leigh's encephalomyelopathy, *Arch. Dis. Child.*, 52, 514, 1977.
334. Green, H. K. and Kehinde, O., Sublines of mouse 3T3 cells that accumulate lipid, *Cell*, 1, 113, 1974.
335. Green, M. J. and Toms, E. J., The properties of subunits of avidin coupled to Sepharose, *Biochem. J.*, 133, 687, 1973.
336. Green, N. M., Avidin. The use of [^{14}C] biotin for kinetic studies and for assay, *Biochem. J.*, 89, 585, 1963.
337. Green, N. M., Avidin, in *Advances in Protein Chemistry*, Vol. 29, Anfinsen, C. B., Edsall, J. T., and Richards, F. M., Eds., Academic Press, New York, 1975, 85.
337a. Green, N. M., Konieczny, L., Toms, E. J., and Valentine, R. C., The use of bifunctional biotinyl compounds to determine the arrangement of subunits in avidin, *Biochem. J.*, 125, 781, 1971.
338. Green, N. M., Valentine, R. C., Wrigley, N. G., Ahmad, F., Jacobson, B., and Wood, H. G., Transcarboxylase. XI. Electron microscopy and subunits structure, *J. Biol. Chem.*, 247, 6284, 1972.

339. Greenbaum, A. L., Gumaa, K. A., and McLean, P., The distribution of hepatic metabolites and the control of the pathways of carbohydrate metabolism in animals of different dietary and hormonal status, *Arch. Biochem. Biophys.*, 143, 617, 1971.

340. Gregolin, C., Ryder, E., Kleinschmidt, A. K., Wann, R. C., and Lane, M. D., Molecular characteristics of liver acetyl CoA carboxylase, *Proc. Natl. Acad. Sci. U.S.A.*, 56, 148, 1966.

341. Greville, G. D., Intracellular compartmentation and the citric acid cycle, in *Citric Acid Cycle — Control and Compartmentation*, Lowenstein, J. M., Ed., Marcel Dekker, New York, 1969, 1.

342. Grimm, F. C. and Doherty, D. G., Properties of the two forms of malic dehydrogenase from beef heart, *J. Biol. Chem.*, 236, 1980, 1961.

343. Gröbe, H., Von Bassewitz, D. B., Dominick, H.-C., and Pfeiffer, R. A., Subacute necrotizing encephalomyelopathy: clinical, ultrastructural, and therapeutic studies in an infant, *Acta Paediatr. Scand.*, 64, 755, 1975.

344. Grover, W. D., Auerbach, V. H., and Patel, M. S., Biochemical studies and therapy in subacute necrotizing encephalomepelopathy (Leigh's syndrome), *J. Pediatr.*, 81, 39, 1972.

345. Gruskin, A. B., Patel, M. S., Linshaw, M., Ettenger, R., Huff, D., and Grouer, W., Renal function studies and kidney pyruvate carboxylase in subacute necrotising encephalomyelopathy (Leigh's Syndrome), *Pediatr. Res.*, 7, 832, 1973.

346. Guchhait, R. B., Polakis, S. E., Dimroth, P., Stoff, E., Moss, J., and Lane, M. D., Acetyl CoA carboxylase system in *Escherichia coli*. Purification and properties of biotin carboxylase, carboxyl transferase and carboxyl carrier protein components by affinity chromatography on avidin-Sepharose, *J. Biol. Chem.*, 249, 6633, 1974.

347. Guchhait, R. B., Polakis, S. E., Hollis, D., Fenselau, C., and Lane, M. D., Acetyl coenzyme A carboxylase systems of *Escherichia coli*. Site of carboxylation of biotin and enzymatic reactivity of 1′N-(ureido)carboxybiotin derivatives, *J. Biol. Chem.*, 249, 6646, 1974.

348. Guise, K. S., Quan, F., Tsui, F., and Gravel, R. A., Isolation of human biotin-dependent carboxylase cDNA clones, *J. Cell. Biochem.*, Suppl. 7A, 162, 1983.

349. Gul, B. and Dils, R., Pyruvate carboxylase in lactating rat and rabbit mammary gland, *Biochem. J.*, 111, 263, 1969.

350. Gurr, J. A. and Jones, K. M., Purification and characterization of pyruvate carboxylase from *Arthrobacter globiformis*, *Arch. Biochem. Biophys.*, 179, 444, 1977.

351. Guynn, R. W., Veloso, D., Harris, R. L., Lawson, J. W. R., and Veech, R. L., Ethanol administration and the relationship of malonyl-coenzyme A concentrations to the rate of fatty acid synthesis in rat liver, *Biochem. J.*, 136, 639, 1973.

352. Guynn, R. W., Veloso, D., and Veech, R. L., The concentration of malonyl-coenzyme A and the control of fatty acid synthesis *in vivo*, *J. Biol. Chem.*, 247, 7325, 1972.

353. Haarasilta, S., Ezymatic determination of biotin, *Anal. Biochem.*, 87, 306, 1978.

354. Haarasilta, S., On the Regulation of Anaplerotic and Gluconeogenic Enzymes in Baker's Yeast *(Saccharomyces cerevisiae)*, Ph.D. thesis, University of Helsinki, 1981.

355. Haarasilta, S. and Oura, E., On the activity and regulation of anaplerotic and gluconeogenetic enzymes during the growth process of baker's yeast, *Eur. J. Biochem.*, 52, 1, 1975.

356. Haarasilta, S. and Oura, E., Effect of aeration on the activity of gluconeogenetic enzymes in *Saccharomyces cerevisiae* growing under glucose limitation, *Arch. Microbiol.*, 106, 271, 1975.

357. Haarasilta, S., Oura, E., and Suomalainen, H., Pyruvate holo- and apocarboxylase content of biotin-deficient baker's yeast and the characteristics of the holoenzyme formation in permeabilized cells, *Arch. Microbiol.*, 122, 121, 1979.

358. Haarasilta, S. and Taskinen, L., Location of three key enzymes of gluconeogenesis in baker's yeast, *Arch. Microbiol.*, 113, 159, 1977.

359. Halestrap, A. P., The mitochondrial pyruvate carrier. Kinetics and specificity for substrates and inhibitors, *Biochem. J.*, 148, 85, 1975.

360. Halestrap. A. P., Stimulation of pyruvate transport in metabolizing mitochondria through changes in the transmembrane pH gradient induced by glucagon treatment of rats, *Biochem. J.*, 172, 389, 1978.

361. Halestrap, A. P., Hormonal control of mitochondrial respiratory chain activity, *Short-Term Regulation of Liver Metabolism*, Hue, L. and Van de Werve, G., Eds., Elsevier/North-Holland, Amsterdam, 1981, 389.

362. Halestrap, A. P., The nature of the stimulation of the respiratory chain of rat liver mitochondria by glucagon pretreatment of animals, *Biochem. J.*, 204, 37, 1982.

363. Halestrap, A. P. and Denton, R. M., The specificity and metabolic implications of the inhibition of pyruvate transport in isolated mitochondria and intact tissue preparations by α-cyano-4-hydroxycinnamate and related compounds, *Biochem. J.*, 148, 97, 1975.

364. Hall, H. C., Sordahl, L. A., and Stefko, P. L., The effect of insulin on oxidative phosphorylation in normal and diabetic mitochondria, *J. Biol. Chem.*, 235, 1536, 1960.

365. Halsall, G. B., Atassi-Gandhi sedimentation coefficient and molecular weight relationships, *Nature (London)*, 215, 880, 1967.
366. Hammond, K. D. and Balinsky, D., Activities of key gluconeogenic enzymes and glycogen synthase in rat and human livers, hepatomas, and hepatoma cell cultures, *Cancer Res.*, 38, 1317, 1978.
367. Hansen, T. L. and Christensen, E., Studies on pyruvate carboxylase from human diploid fibroblasts and amniotic fluid cells, The Society for the Study of Inborn Errors of Metabolism, 16 Annual Meeting, Bristol, England, 1978.
368. Hansen, T. L. and Christensen, E., Studies on pyruvate carboxylase from cultured human fibroblast and amniotic fluid cells, *J. Inher. Metab. Dis.*, 2, 23, 1979.
369. Hansen, T. L. and Christensen, E., Pyruvate carboxylase activity in amniotic fluid cells, *Pediatr. Res.*, 14, 1412, 1980.
370. Hansen, T. L., Christensen, E., and Brandt, N. J., Pyruvate carboxylase activity in cultured human fibroblasts and amniotic fluid cells, Abstract, A 1-8 052, 11 FEBS Meeting, Copenhagen, August, 1977.
371. Hansen, T. L., Christensen, E., and Brandt, N. J., Studies on pyruvate carboxylase, pyruvate decarboxylase and lipoamide dehydrogenase in subacute necrotizing encephalomyelopathy, *Acta Pediatr. Scand.*, 71, 263, 1982.
372. Hansen, T. L., Christensen, E., Willems, J. L., and Trijbels, J. M. F., A mutation of pyruvate carboxylase in fibroblasts from a patient with severe, chronic lactic acidemia, *Clin. Chim. Acta*, 131, 39, 1983.
373. Hansford, R. G. and Castro, F., Effect of micromolar concentrations on free calcium ions on the reduction of heart mitochondrial NAD(P) by 2-oxoglutarate, *Biochem. J.*, 198, 525, 1981.
374. Harano, Y., de Palma, R. G., and Miller, M., Fatty acid oxidation, citric acid cycle activity, and morphology of mitochondria in diabetic rat liver, *Proc. Soc. Exp. Biol. Med.*, 131, 913, 1969.
375. Harris, E. J., Berent, C., and Wimhurst, J., Stimulation of mitochondrial pyruvate carboxylation by Mn^{2+} and the distribution of the products between mitochondrion and medium, *FEBS Lett.*, 6, 93, 1970.
376. Harris, E. J. and van Dam, K., Changes of total water and sucrose space accompanying induced ion uptake or phosphate swelling of rat liver mitochondria, *Biochem. J.*, 106, 759, 1968.
377. Hartman, R. E., Carbon dioxide fixation by extracts of *Streptococcus faecalis* var. *liquefaciens*, *J. Bacteriol.*, 102, 341, 1970.
378. Hartman, R. E. and Keen, N. T., Enzymes catalyzing anaplerotic carbon dioxide fixation in *Verticillium albo-atrum*, *Phytopathology*, 63, 947, 1973.
379. Hartman, R. E. and Keen, N. T., The pyruvate carboxylase of *Verticillium albo-atrum*, *J. Gen. Microbiol.*, 81, 15, 1974.
380. Harvey, R. P., Krieg, P. A., Robins, A. J., Coles, L. S., and Wells, J. R. E., Non-tandem arrangement and divergent transcription of chicken histone genes, *Nature (London)*, 294, 49, 1981.
381. Haschemeyer, R. H., Electron microscopy of enzymes, *Adv. Enzymol.*, 33, 71, 1970.
382. Hatch, M. D., Mechanism and function of the C_4 pathway of photosynthesis, in *Photosynthesis and Photorespiration*, Hatch, M. D., Osmond, C. B., and Slater, R. O., Eds., Wiley-Interscience, New York, 1971, 139.
383. Hatch, M. D. and Slack, C. R., A new enzyme for the interconversion of pyruvate and phosphoenolpyruvate and its role in the C_4 dicarboxylic acid pathway of photosynthesis, *Biochem. J.*, 106, 141, 1968.
384. Hayashi, S. and Ooshiro, Z., Gluconeogenesis in isolated liver cells of the eel, *Anguilla japonica*, *J. Comp. Physiol.*, 132, 343, 1979.
385. Haynes, R. C., The fixation of carbon dioxide by rat liver mitochondria and its relation to gluconeogenesis, *J. Biol. Chem.*, 240, 4103, 1965.
386. Haynes, R. C., Hormonal regulation of pyruvate metabolism in rat liver mitochondria, in *Energy Metabolism and the Regulation of Metabolic Processes in Mitochondria*, Mehlman, M. A. and Hanson, R. W., Eds., Academic Press, New York, 1972, 239.
387. Haynes, R. C., Jr., Mechanisms of hormonal control of gluconeogenesis, *Metabolism*, 25(1), 1361, 1976.
388. Haynes, R. C., Jr., Garrison, J. C., and Yamazaki, R. K., Comparison of effects of glucagon and valinomycin on rat liver mitochondria and cells, *Mol. Pharmacol.*, 10, 381, 1974.
389. Heard, G. S., Hood, R. L., and Johnson, A. R., Hepatic biotin and the sudden infant death syndrome, *Med. J. Aust.*, 2, 305, 1983.
390. Hegarty, A. F., Bruice, T. C., and Benkovic, S. J., Biotin and the nucleophilicity of 2-methoxy-2-imidazoline toward the sp^2 carbonyl carbon, *Chem. Commun.*, 1173, 1969.
391. Helmrath, T. A. and Bieber, L. L., Development of gluconeogenesis in neonatal pig liver, *Am. J. Physiol.*, 227, 1306, 1974.
392. Henderson, J. R. and Lamonds, M. R., Effects of D_2O on citric acid and pyruvate carboxylase formation by *Aspergillus niger*, *Arch. Biochem. Biophys.*, 115, 187, 1966.

393. Hendrick, D. and Moller, F., Pyruvate carboxylase in the developing chick embryo, *Comp. Biochem. Physiol.*, 45B, 197, 1973.
394. Henley, K. S., Laughrey, E. G., and Clancy, P. E., Gluconeogenesis in the fatty and cirrhotic liver of the rat. The effect of oleate and ethanol, in *Regulation of Hepatic Metabolism*, Lundquist, F. and Tygstrup N., Eds., Academic Press, New York, 1974, 401.
395. Henning, H.-V., Seiffert, I., and Seubert, W., Cortisol induzierter Anstieg der Pyruvatcarboxylaseaktivitat in der Rattenleber, *Biochim. Biophys. Acta*, 77, 345, 1963.
396. Henning, H. V. and Seubert, W., Zum Mechanismus der Gluconeogenese und ihrer Steuerung. I. Quantitative Bestimmung der Pyruvatcarboxylase in Rohextracten der Rattenleber, *Biochem. Z.*, 340, 160, 1964.
397. Henning, H. V., Stumpf, B., Ohly, B., and Seubert, W., On the mechanism of gluconeogenesis and its regulation. III. The glucogenic capacity and the activities of pyruvate carboxylase and PEP-carboxylase of rat kidney and rat liver after cortisol treatment and starvation, *Biochem. Z.*, 344, 274, 1966.
398. Henrikson, K. P., Allen, S. H. G., and Maloy, W. L., An avidin monomer affinity column for the purification of biotin-containing enzymes, *Anal. Biochem.*, 94, 366, 1979.
399. Hensgens, H. E. S. J., Verhoeven, A. J., and Meijer, A. J., The relationship between intramitochondrial N-acetylglutamate and activity of carbamyl-phosphate synthetase (ammonia). The effect of glucagon, *Eur. J. Biochem.*, 107, 197, 1980.
400. Henslee, J. G. and Srere, P. A., Resolution of rat mitochondrial matrix proteins by two-dimensional polyacrylamide gel electrophoresis, *J. Biol. Chem.*, 254, 5488, 1979.
401. Herberg, L., Major, E., Hennigs, U., Gruneklee, D., Freytag, G., and Gries, F. A., Differences in the development of the obese-hyperglycemic syndrome in obob and NZO mice, *Diabetologia*, 6, 292, 1970.
402. Herrera, E., Knopp, R. H., and Freinkel, N., Carbohydrate metabolism in pregnancy. VI. Plasma fuels, insulin, liver composition, gluconeogenesis and nitrogen metabolism during late gestation in the fed and fasted rat, *J. Clin. Invest.*, 48, 2260, 1969.
403. Hewick, R. M., Hunkapiller, M. W., Hood, L. E., and Dreyer, W. J., A gas-liquid solid phase peptide and protein sequenator, *J. Biol. Chem.*, 256, 7990, 1981.
404. Higa, A. I., Milrad de Forchetti, S. R., and Cazzulo, J. J., CO_2-fixing enzymes in *Pseudomonas fluorescens*, *J. Gen. Microbiol.*, 93, 69, 1976.
405. Hillier, A. J. and Jago, G. R., The metabolism of [^{14}C]bicarbonate by *Streptococcus lactis:* the fixation of [^{14}C]bicarbonate by pyruvate carboxylase, *J. Dairy Res.*, 45, 445, 1978.
406. Hochachka, P. W. and Guderley, H. E., Forms and functions of gluconeogenic enzymes in arthropods, in *Isozymes*, Vol. 2, Markert, C. L., Ed., Academic Press, New York, 1975, 519.
407. Hochachka, P. W., Moon, T. W., Mustafa, T., and Storey, K. B., Metabolic sources of power for mantle muscle of a fast swimming squid, *Comp. Biochem. Physiol.*, 52B, 151, 1975.
408. Hochachka, P. W. and Mustafa, T., Invertebrate facultative anaerobiosis, *Science*, 178, 1065, 1972.
409. Hoek, J. B., Nicholls, D. G., and Williamson, J. R., Determination of the mitochondrial protonmotive force in isolated hepatocytes, *J. Biol. Chem.*, 255, 1458, 1980.
410. Hogeboom, G. H., Fractionation of cell components of animal tissues, in *Methods in Enzymology*, Vol. 1, Colowick, S. P. and Kaplan, N. O., Eds., Academic Press, New York, 1955, 16.
411. Holden, J. T., The composition of microbial amino acid pools, in *Amino Acid Pools*, Holden, J. T., Ed., Elsevier, Amsterdam, 1962, 73.
412. Holdsworth, E. S. and Bruck, K., Enzymes concerned with β-carboxylation in marine phytoplankter. Purification and properties of phosphoenolpyruvate carboxykinase, *Arch. Biochem. Biophys.*, 182, 87, 1977.
413. Hollenberg, P. F., Flashner, M., and Coon, M. J., Role of lysyl-ε-amino groups in adenosine diphosphate binding and catalytic activity of pyruvate kinase, *J. Biol. Chem.*, 246, 946, 1971.
414. Holt, J. G., Ed,. *The Shorter Bergey's Manual of Determinative Bacteriology*, 8th ed., Williams & Wilkins, Baltimore, 1979.
415. Hommes, F. A., Discussion of paper by Cooper, J. R. and Pincus, J. H., Thiamine triphosphate deficiency in Leigh's disease (subacute necrotizing encephalomyelopathy), in *Inborn Errors of Metabolism*, Hommes, F. A. and Van den Berg, C. J., Eds., Academic Press, New York, 1973, 127.
416. Hommes, F. A., Polman, H. A., and Reenink, J. D., Leighs encephalomyelopathy: an inborn error of gluconeogenesis, *Arch. Dis. Child.*, 43, 423, 1968.
417. Hommes, F. A., Schrijver, J., and Dias, T. R., Pyruvate carboxylase deficiency, studies on patients and on an animal model system, *Symposia of the Society for the Study of Inborn Errors of Metabolism*, Burman, D., Holton, J. B., and Pennock, C. A., Eds., MTP Press, London, 1978, 16, 269.
418. Hood, R. L., Johnson, A. R., Fogerty, A. C., and Pearson, J. A., Fatty liver and kidney syndrome in chicks. II. Biochemical role of biotin, *Aust. J. Biol. Sci.*, 29, 429, 1976.
419. Hopkirk, T. J. and Bloxam, D. P., Studies on the biosynthesis of hepatic pyruvate kinase and its correlation with enhanced lipogenesis in meal-trained rats, *Biochem. J.*, 182, 383, 1979.

420. Hopner, T. and Knappe, J., Einbau von Biotin in β-Methylcrotonyl-CoA-carboxylase durch Holocarboxylase-synthetase, *Biochem. Z.*, 342, 190, 1965.
421. Horiike, K. and McCormick, D. B., Correlations between biological activity and number of functional groups chemically modified, *J. Theor. Biol.*, 79, 403, 1979.
422. Houghton, M., Stewart, A. G., Dael, S. M., Emtage, J. S., Eaton, M. R. W., Smith, J. C., Patel, T. P., Lewes, H. M., Ponter, A. G., Birch, J. R., Cartwright, T., and Carey, N. H., The amino-terminal sequence of human fibroblast interferon as deduced from reverse transcripts obtained using synthetic oligonucleotide primers, *Nucl. Acids Res.*, 8, 1913, 1980.
423. Hudson, P. J., Haley, J., Cronk, M., Shine, J., and Niall, H., Molecular cloning and characterisation of cDNA sequences coding for rat relaxin, *Nature (London)*, 291, 127, 1981.
424. Hudson, P. J., Keech, D. B., and Wallace, J. C., Pyruvate carboxylase: affinity labelling of the pyruvate binding site, *Biochem. Biophys. Res. Commun.*, 65, 231, 1975.
425. Iles, R. A., Cohen, R. D., Rist, A. H., and Baron, P. G., The mechanism of inhibition by acidosis of gluconeogenesis from lactate in rat liver, *Biochem. J.*, 164, 185, 1977.
426. Irias, J. J., Olmsted, M. R., and Utter, M. F., Pyruvate carboxylase. Reversible inactivation by cold, *Biochemistry*, 8, 5136, 1969.
427. Ishibashi, H. and Cottam, G. L., Glucagon-stimulated phosphorylation of pyruvate kinase in hepatocytes, *J. Biol. Chem.*, 253, 8767, 1978.
428. Jensen, C. B., Sistare, F. D., Hamman, H. C., and Haynes, R. C., Jr., Stimulation of mitochondrial functions by glucagon treatment. Evidence that effects are not artifacts of mitochondrial isolation, *Biochem. J.*, 210, 819, 1983.
429. Johannssen, W., Attwood, P. V., Wallace, J. C., and Keech, D. B., Localisation of the active site of pyruvate carboxylase by electron microscopic examination of avidin-enzyme complexes, *Eur. J. Biochem.*, 133, 201, 1983.
430. Johnson, A. R., Hood, R. L., and Emery, J. L., Biotin and the sudden infant death syndrome, *Nature (London)*, 285, 159, 1980.
431. Johnson, A. R., Hood, R. L., Pearson, J. A., and Fogerty, A. C., The role of biotin in the stress induced death of chickens exhibiting fatty liver and kidney syndrome, *Proc. Nutr. Soc.*, 35, 130A, 1976.
432. Johnson, M. E. M., Das, N. M., Butcher, F. R., and Fain, J. N., The regulation of gluconeogenesis in isolated rat liver cells by glucagon, insulin, dibutyrylcyclic adenosine monophosphate, and fatty acids, *J. Biol. Chem.*, 247, 3229, 1972.
433. Jones, C. T. and Ashton, I. K., The appearance, properties, and functions of gluconeogenic enzymes in the liver and kidney of the guinea pig during fetal and early neonatal development, *Arch. Biochem. Biophys.*, 174, 506, 1976.
434. Kawachi, T. and Rudney, H., Solubilisation and purification of β-hydroxy-β-methylglutaryl coenzyme A reductase from rat liver, *Biochemistry*, 9, 1700, 1970.
435. Kaziro, Y., Dehoa, S., Warner, R. C., and Chen, J.-Y., Metabolism of propionic acid in animal tissues. VIII. Crystalline propionyl carboxylase, *J. Biol. Chem.*, 236, 1917, 1961.
436. Kaziro, Y., Hass, L. F., Boyer, P. D., and Ochoa, S., Mechanism of the propionyl carboxylase reaction. II. Isotopic exchange and tracer experiments, *J. Biol. Chem.*, 237, 1460, 1962.
437. Keech, D. B. and Barritt, G. J., Allosteric activation of sheep kidney pyruvate carboxylase by magnesium ion (Mg^{2+}) and the magnesium adenosine triphosphate ion ($MgATP^{2-}$), *J. Biol. Chem.*, 242, 1983, 1967.
438. Keech, D. B. and Farrant, R. K., The reactive lysine residue at the allosteric site of sheep kidney pyruvate carboxylase, *Biochim. Biophys. Acta*, 151, 493, 1968.
439. Keech, D. B. and Utter, M. F., Pyruvate carboxylase. II. Properties, *J. Biol. Chem.*, 238, 2609, 1963.
440. Keech, D. B. and Wallace, J. C., How significant is homotropic cooperativity in terms of metabolic regulation?, *Trends Biochem. Sci.*, 6, 182, 1981.
441. Khew, Y-S., Baldwin, G. S., Snoswell, M. A., and Wallace, J. C., unpublished observations.
442. Khew, Y-S., Johannssen, W., Attwood, P. V., Wallace, J. C., and Keech, D. B., Studies on dilution inactivation of sheep liver pyruvate carboxylase, *Proc. Aust. Biochem. Soc.*, 15, 2, 1983.
443. Kimmich, G. A. and Rasmussen, H., Regulation of pyruvate carboxylase activity by calcium in intact rat liver mitochondria, *J. Biol. Chem.*, 244, 190, 1969.
444. Klapper, M. H., Rapid binding of PMB to non-SH groups on hexmerythrin, *Biochem. Biophys. Res. Commun.*, 38, 172, 1970.
445. Klein, R. A., Linstead, D. J., and Wheeler, M. V., Carbon dioxide fixation in trypanosomatids, *Parasitology*, 71, 93, 1975.
446. Kleinsek, D. A. and Porter, J. W., An alternate method of purification and properties of rat liver β-hydroxy-β-methylglutaryl coenzyme A reductase, *J. Biol. Chem.*, 254, 7591, 1979.
447. Klotz, I. M. and Darnall, D. W., Protein subunits: a table (2nd ed.), *Science*, 166, 126, 1969.

448. Kluger, R. and Adawadkar, P. D., A reaction proceeding through intramolecular phosphorylation of a urea. A chemical mechanism for enzymic carboxylation of biotin involving cleavage of adenosine 5′-triphosphate, *J. Am. Chem. Soc.*, 98, 3741, 1976.
449. Kluger, R., Davis, P. P., and Adawadkar, P. D., Mechanism of urea participation in phosphate ester hydrolysis. Mechanistic and stereochemical criteria for enzymic formation and reaction of phosphorylated biotin, *J. Am. Chem. Soc.*, 101, 5995, 1979.
450. Knappe, J., Mechanism of biotin action, *Ann. Rev. Biochem.*, 39, 757, 1970.
451. Knappe, J., Biederbick, K., and Brummer, W., On the structure of CO_2-biotin enzymes, *Angew. Chem.*, 1, 401, 1962.
452. Knappe, J., Ringelmann, E., and Lynen, F., Zur biochemischen Function des Biotins. III. Die chemische Konstitution des enzymatisch gebildeten Carboxybiotins, *Biochem. Z.*, 335, 168, 1961.
453. Knappe, J., Wenger, B., and Wiegand, U., Zur Konstitution der carboxylierten β-Methyl-crotonyl-carboxylase (CO_2-Biotinenzym), *Biochem. Z.*, 337, 232, 1963.
454. Knull, H. R., Taylor, W. F., and Wells, W. W., Effects of energy metabolism on *in vivo* distribution of hexokinase in brain, *J. Biol. Chem.*, 248, 5414, 1973.
455. Knull, H. R., Taylor, W. F., and Wells, W. W., Insulin effects on brain energy metabolism and the related hexokinase distribution, *J. Biol. Chem.*, 249, 6930, 1974.
456. Kohler, G-H., Rindt, K-P., and Ohmann, E., Die Bildung aktiver Pyruvat-Carboxylase aus Apoenzym und Biotin in *Rhodopseudomonas spheroides, Biochem. Physiol. Pflanzen,* 169, 99, 1976.
457. Kohn, H., Model studies on the mechanism of biotin dependent carboxylations, *J. Am. Chem. Soc.*, 98, 3690, 1976.
458. Kondrup, J. and Grunnet, N., The effect of acute and prolonged ethanol treatment on the contents of coenzyme A, carnitine and their derivatives in rat liver, *Biochem. J.*, 132, 373, 1973.
459. Kornacker, M. S. and Ball, E. G., Citrate cleavage in adipose tissue, *Proc. Natl. Acad. Sci. U.S.A.*, 54, 899, 1965.
460. Kornberg, H. L., Anaplerotic sequences and their role in metabolism, in *Essays in Biochemistry,* Vol. 2, Campbell, P. N. and Greville, G. D., Eds., Academic Press, London, 1966, 1.
461. Kornberg, H. L. and Krebs, H. A., Synthesis of cell constituents from C_2-units by a modified tricarboxylic acid cycle, *Nature (London),* 179, 988, 1957.
462. Kornberg, H. L. and Madsen, N. B., Synthesis of C_4-dicarboxylic acids from acetate by a "glyoxylate bypass" of the tricarboxylic acid cycle, *Biochim. Biophys. Acta,* 24, 651, 1957.
463. Kornberg, H. L. and Madsen, N. B., The metabolism of C_2 compounds in micro-organisms. III. Synthesis of malate from acetate via the glyoxylate cycle, *Biochem. J.*, 68, 549, 1958.
464. Koshland, D. E., Jr., Nemethy, G., and Filmer, D., Comparison of experimental binding data and theoretical models in proteins containing sub-units, *Biochemistry,* 5, 365, 1966.
465. Kosow, D. P., Huang, S. G., and Lane, M. D., Propionyl holocarboxylase synthesis. I. Preparation and properties of the enzyme system, *J. Biol. Chem.*, 237, 3633, 1962.
466. Kosow, D. P. and Lane, M. D., Restoration of biotin-deficiency-induced depression of propionyl carboxylase activity *in vivo* and *in vitro, Biochem. Biophys. Res. Commun.*, 4, 92, 1961.
467. Kosow, D. P. and Lane, M. D., Propionyl apocarboxylase activation catalysed by cell-free enzyme extracts, *Biochem. Biophys. Res. Commun.*, 5, 191, 1961.
468. Kosow, D. P. and Lane, M. D., Propionyl holocarboxylase formation: covalent bonding of biotin to apo-carboxylase lysyl ε-amino groups, *Biochem. Biophys. Res. Commun.*, 7, 439, 1962.
469. Krampitz, L. O. and Werkman, C. H., The enzymic decarboxylation of oxaloacetate, *Biochem. J.*, 35, 595, 1941.
470. Krassner, J. M., Cytochromes, lactic dehydrogenase and transformation in *Leishmania, J. Protozool.*, 13, 286, 1966.
471. Krebs, H. A., Considerations concerning the pathways of synthesis in living matter, *Bull. Johns Hopkins Hosp.*, 95, 19, 1954.
472. Krebs, H. A., The regulation of release of ketone-bodies by liver, *Adv. Enz. Regul.*, 4, 339, 1966.
473. Krebs, H. A., The effects of ethanol on the metabolic activities of the liver, *Adv. Enz. Reg.*, 6, 467, 1967.
474. Krebs, E. G. and Beavo, J. A., Phosphorylation-dephosphorylation of enzymes, *Annu. Rev. Biochem.*, 48, 923, 1979.
475. Krebs, H. A., Cornell, N. W., Lund, P., and Hems, R., Isolated liver cells as experimental material, in *Regulation of Hepatic Metabolism,* Lundquist, F. and Tystrup, N., Eds., Munksgaard, Copenhagen, 1974, 726.
476. Krebs, H. A. and Kornberg, H. L., A survey of the energy transformations in living matter, *Ergeb. Physiol. Biol. Chem. Exp. Pharmakol.*, 49, 212, 1957.
477. Krebs, H. A., Speake, R. N., and Hems, R., Acceleration of renal gluconeogenesis by ketone bodies and fatty acids, *Biochem. J.*, 94, 712, 1965.
478. Kremer, B. P. and Berks, R., Photosynthesis and carbon metabolism in marine and freshwater distoms, *Z. Pflanzenphysiol.*, 87, 149, 1978.

479. Krieg, P. A., Robins, A. J., Gait, M. J., Titmas, R. C., and Wells, J. R. E., Chicken histone H5: selection of a cDNA recombinant using an extended synthetic primer, *Nucl. Acids Res.*, 10, 1495, 1982.
480. Krulwich, T. A., Sharon, B. I., and Perrin, L. S., Natural paucity of anaplerotic enzymes: basis for dependence of *Arthrobacter pyridinolis* on L-malate for growth, *J. Bacteriol.*, 127, 179, 1976.
481. Kuri-Harcuch, W. and Green, H., Increasing activity of enzymes on pathway of triacylglycerol synthesis during adipose conversion of 3T3 cells, *J. Biol. Chem.*, 252, 2158, 1977.
482. Kurokawa, K. and Rasmussen, H., Ionic control of renal gluconeogenesis. I. The interrelated effect of calcium and hydrogen ions, *Biochim. Biophys. Acta*, 313, 17, 1973.
483. Lachica, V. F. and Hartman, P. A., Inhibition of CO_2 fixation in group D *streptococci*, *Can. J. Microbiol.*, 15, 57, 1969.
484. Landriscina, C., Papa, S., Coratelli, P., Mazzarella, L., and Quagliariello, E., Enzymatic activities of the matrix and inner membrane of pigeon-liver mitochondria, *Biochim. Biophys. Acta*, 205, 136, 1970.
485. Lane, M. D. and Lynen, F., The biochemical function of biotin. VI. Chemical structure of the carboxylated active site of propionyl carboxylase, *Proc. Natl. Acad. Sci. U.S.A.*, 49, 379, 1963.
486. Lane, M. D., Rominger, K. L., Young, D. L., and Lynen, F., The enzymatic synthesis of holotranscarboxylase from apotranscarboxylase and (+)biotin. II. Investigation of the reaction mechanism, *J. Biol. Chem.*, 239, 2865, 1964.
487. Lane, M. D., Young, D. L., and Lynen, F., The enzymatic synthesis of holotranscarboxylase from apotranscarboxylase and (+)biotin. I. Purification of the apoenzyme and synthetase; characteristics of the reaction, *J. Biol. Chem.*, 239, 2858, 1964.
488. Lardy, H. A., Gluconeogenesis: pathways and hormonal regulation, *Harvey Lect.*, 60, 261, 1966.
489. Lardy, H. A., Paetkau, V., and Walter, P., Paths of carbon in gluconeogenesis and lipogenesis: the role of mitochondria in supplying precursors of phosphoenolpyruvate, *Proc. Natl. Acad. Sci. U.S.A.*, 53, 1410, 1965.
490. Larmour. R. and Marchant, R., Carbon dioxide fixation and conidiation in *Fusarium culmorum* grown in continuous culture, *J. Gen. Microbiol.*, 99, 59, 1977.
491. Losada, M., Canovas, J. L., and Ruiz-Amil, M., Oxaloaceate, citramalate, and glutamate formation from pyruvate in baker's yeast, *Biochem. Z.*, 340, 60, 1964.
492. Laursen-Jones, A. P., Fatty liver and kidney syndrome in chickens: a review, *Proc. XIX World Vet. Congr.*, 3, 1165, 1971.
493. Lehnert, W., Niederhoff, H., and Junder, A., A case of biotin-responsive 3-methylcrotonylglycine and 3-hydroxyisovaleric aciduria, *Eur. J. Pediatr.*, 132, 107, 1979.
494. Leigh, D. J., Subacute necrotizing encephalomyelopathy in an infant, *J. Neurol. Neurosurg. Psychiatr.*, 14, 216, 1951.
495. Leighton, F., Poole, B., Beaufay, H., Baudhuin, P., Coffey, J. W., Fowler, S., and DeDuve, C., The large-scale separation of peroxisomes, mitochondria and lysosomes from the livers of rats injected with Triton WR-1339, *J. Cell Biol.*, 37, 482, 1968.
496. Leiter, A. B., Weinberg, M., Isohashi, F., Utter, M. F., and Linn, T., Relationship between phosphorylation and activity of pyruvate dehydrogenase in rat liver mitochondria and the absence of such a relationship for pyruvate carboxylase, *J. Biol. Chem.*, 253, 2716, 1978.
497. Leroux, J. P., Demaugre, F., Buc, H. A., and Saudubray, J. M., Effect of dichloro acetate and 2-chloro propionate in the carbohydrate of isolated hepatocytes therapeutic applications, *C.R. Slances Soc. Biol. Fil.*, 174, 506, 1980.
498. Leroux, J. P., Marsac, C., and Saudubray, J. M., Congenital enzymopathic hyper lactic-acidemia, *Ann. Biol. Clin.*, 34, 151, 1976.
499. Lerud, R. and Finnerty, W. R., CO_2 fixation in a hydrocarbon utilizing microorganism, *Bacteriol. Proc.*, 129, 1967.
500. Levy, M., Fishman, L., and Shenkein, I., Mouse submaxillary gland proteases, in *Methods in Enzymology*, Vol. 19, Colowick, S. P. and Kaplan, N. O., Eds., Academic Press, New York, 1970, 672.
501. Lewis, A. J. and Miller, J. D. A., The tricarboxylic acid pathway in *Desulfovibrio*, *Can. J. Microbiol.*, 23, 916, 1977.
502. Liao, U. and Anderson, D. E., Regulation at the phosphoenolpyruvate branchpoint in *Azotobacter vinelandii:* phosphoenolpyruvate carboxylase, *J. Bacteriol.*, 106, 31, 1971.
503. Libor, S. M., Sundaram, T. K., and Scrutton, M. C., Pyruvate carboxylase from a thermophilic *Bacillus.* Studies on the specificity of activation by acyl derivatives of coenzyme A and on the properties of catalysis in the absence of activator, *Biochem. J.*, 169, 543, 1978.
504. Libor, S., Sundaram, T. K., Warwick, R., Chapman, J. A., and Grundy, S. M. W., Pyruvate carboxylase from a thermophilic *Bacillus:* some molecular characteristics, *Biochemistry*, 18, 3647, 1979.
505. Libor, S., Warwick, R., and Sundaram, T. K., Pyruvate carboxylase from *Bacillus stearothermophilus:* molecular size, biotin content and subunit composition, *Biochem. Soc. Trans.*, 3, 309, 1975.
506. Libor, S., Warwick, R., and Sundaram, T. K., Binding behavior of pyruvate carboxylase from *Bacillus stearothermophilus* on Sepharose-avidin, *FEBS Lett.*, 57, 34, 1975.

507. Ling, A-M. and Keech, D. B., Pyruvate carboxylase from sheep kidney: purification and some properties of the enzyme, *Enzymologia*, 30, 367, 1966.

508. Loginova, N. V., Sokolov, A. P., and Trotsenko, Yu. A., Properties of pyruvate carboxylase from the facultative methylotroph *Pseudomonas oleovorans*, *Biochemistry* (Transl.), 47, 1375, 1982.

509. Loginova, N. V. and Trotsenko, Yu. A., Pyruvate and phosphoenolpyruvate carboxylases in methylotrophs, *Mikrobiologiya*, 48, 202, 1979.

510. Lohr, J. E., Fatty liver and kidney syndrome in New Zealand chickens, *N. Z. Vet. J.*, 23, 167, 1975.

511. Lopes-Cardozo, M., Klazinga, W., and van den Bergh, S. G., Evidence for a homogeneous pool of acetyl-CoA in rat-liver mitochondria, *Eur. J. Biochem.*, 83, 635, 1978.

512. Lorber, V., Lifson, N., Sakami, W., and Wood, H. G., Conversion of propionate to liver glycogen in the intact rat, studied with isotopic propionate, *J. Biol. Chem.*, 183, 531, 1950.

513. Lorber, V., Lifson, N., Wood, H. G., Sakami, W., and Shreeve, W. W., Conversion of lactate to liver glycogen in the intact rat, studied with isotopic lactate, *J. Biol. Chem.*, 183, 517, 1950.

514. Losada, M., Canovas, J. L., and Ruiz-Amil, M., Oxaloacetate, citramalate and glutamate formation from pyruvate in baker's yeast, *Biochem. Z.*, 340, 60, 1964.

515. Lynen, F., Structures of biotin enzymes, in *Lipmann Symposium: Energy, Regulation and Biosynthesis in Molecular Biology*, Richter, D., Ed., de Gruyter, Berlin, 1974, 671.

516. Lynen, F., Knappe, J., Lorch, E., Jutting, G., Ringelmann, E., and Lachance, J.-P., Zur biochemischen Function des Biotins. II. Reinigung and Wirkung weise der β-Methyl-crotonyl-carboxylase, *Biochem. Z.*, 335, 123, 1961.

517. Lynen, F. and Rominger, K. L., Formation of acetyl CoA carboxylase from its apocarboxylase, *Fed. Proc. Fed. Am. Soc. Exp. Biol.*, 22, 537, 1963.

518. Mackall, J. C. and Lane, M. D., Role of pyruvate carboxylase in fatty acid synthesis: alterations during preadipocyte differentiation, *Biochem. Biophys. Res. Commun.*, 79, 720, 1977.

519. Mackall, J. C. and Lane, M. D., Changes in mammary gland acetyl-coenzyme A carboxylase associated with lactogenic differentiation, *Biochem. J.*, 162, 635, 1977.

520. Mackall, J. C., Student, A. K., Polakis, S. E., and Lane, M. D., Induction of lipogenesis during differentiation in a "preadipocyte" cell line, *J. Biol. Chem.*, 251, 6462, 1976.

521. Mackerer, C. R. and Haettinger, J. R., Renal gluconeogenesis in clofibrate-treated rats, *J. Pharmacol. Exp. Ther.*, 204, 683, 1978.

522. Madappally, M. M. and Mistry, S. P., Synthesis of chicken liver pyruvate holocarboxylase *in vivo* and *in vitro*, *Biochim. Biophys. Acta*, 215, 316, 1970.

523. Maesaka, H., Komiya, K., Misugi, K., and Tada, K., Hyperalaninemia hyperpyruvicemia and lactic-acidosis due to pyruvate carboxylase deficiency of the liver. Treatment with thiamine and lipoic acid, *Eur. J. Pediatr.*, 122, 159, 1976.

524. Maghunathan, R. and Foster, H. W., Jr., Gestational changes in pyruvate carboxylase of human amniotic fluid, *IRCS Med. Sci. (Biochem.)*, 11, 75, 1973.

525. Mahan, D. E., Mushahwar, I. K., and Koeppe, R.E., Purification and properties of rat brain pyruvate carboxylase, *Biochem. J.*, 145, 25, 1975.

526. Maloy, W. L., Bowien, B. U., Zwolinski, G. K., Kumar, K. G., Wood, H. G., Ericsson, L. H., and Walsh, K. A., Amino acid sequence of the biotinyl subunit from transcarboxylase, *J. Biol. Chem.*, 254, 11615, 1979.

527. Marco, R., Pestana, A., Sebastian, J., and Sols, A., Oxaloacetate metabolic crossroads in liver. Enzyme compartmentation and regulation of gluconeogenesis, *Mol. Cell. Biochem.*, 3, 53, 1974.

528. Marco, R., Sebastian, J., and Sols, A., Location of the enzymes of the oxalacetate metabolic crossroads in rat liver mitochondria, *Biochem. Biophys. Res. Commun.*, 34, 725, 1969.

529. Marsac, C., Augereau, Ch., Feldman, G., Wolf, B., Hansen, T. L., and Berger, R., Prenatal diagnosis of pyruvate carboxylase deficiency, *Clin. Chim. Acta*, 119, 121, 1982.

530. Marsac, C., Gaudry, M., Augereau, C., Moncion, A., Saudubray, J-M., Coudé, F-X., and Munnich, A., Biotin-dependant carboxylase activities in normal human and multicarboxylase deficient patient fibroblasts: relationship to the biotin content of the culture medium, *Clin. Chim. Acta*, 129, 119, 1983.

531. Marsac, C., Saudubray, J. M., Moncion, A., and Leroux, J. P., Development of gluconeogenic enzymes in the liver of human newborns, *Biol. Neonate*, 28, 317, 1976.

532. Martin, B. R. and Denton, R. M., The intracellular localization of enzymes in white-adipose-tissue-fat-cells and permeability properties of fat-cell mitochondria. Transfer of acetyl units and reducing power between mitochondria and cytoplasm, *Biochem. J.*, 117, 861, 1970.

533. Martin, E., Simon, M. W., Schaeffer, F. W., and Mukkada, A. J., Enzymes of carbohydrate metabolism in four human species of *Leishmania:* a comparative survey, *J. Protozool.*, 23, 600, 1976.

534. Mason, J. V. and Donaldson, W. E., Fatty acid synthesising systems in chick liver: influences of biotin deficiency and dietary fat, *J. Nutr.*, 102, 667, 1972.

535. Masters, C. J., Interactions between soluble enzymes and subcellular structure, *CRC Crit. Rev. Biochem.*, 11, 105, 1981.

536. Matsubara, T. and Tochino, Y., Depression of respiratory activities in the liver mitochondria of diabetic rats and the restorative action of insulin, *J. Biochem.*, 66, 397, 1969.
537. Maxon, W. D. and Johnson, M. J., Aeration studies on propagation of baker's yeast, *Ind. Eng. Chem.*, 45, 2554, 1953.
538. Mayer, F., Wallace, J. C., and Keech, D. B., Further electron microscopic studies on pyruvate carboxylase, *Eur. J. Biochem.*, 112, 265, 1980.
539. Mayobre, M. L., Favarger, P., and Rous, S., Dissimilar effects of biotin deficiency on pyruvate carboxylase and acetyl coenzyme A carboxylase, *Experientia (Basel)*, 31, 722, 1975.
540. Mazumder, R. and Ochoa, S., Methylmalonyl-CoA mutase from sheep liver, *Methods Enzymol.*, 13, 198, 1969.
541. McAllister, H. C. and Coon, M. J., Further studies on the properties of liver propionyl coenzyme A holocarboxylase synthetase and the specificity of holocarboxylase formation, *J. Biol. Chem.*, 241, 2855, 1966.
542. McCarthy, J. T. and Charles, A. M., CO_2 fixation by the facultative autotroph *Thiobacillus novellus* during autotrophy-heterotrophy interconversions, *Can. J. Microbiol.*, 20, 1577, 1974.
543. McClure, W. R. and Lardy, H. A., Rat liver pyruvate carboxylase. IV. Factors affecting the regulation *in vivo*, *J. Biol. Chem.*, 246, 3591, 1971.
544. McClure, W. R., Lardy, H. A., and Cleland, W. W., Rat liver pyruvate carboxylase: isotopic exchange studies of the first partial reaction, *J. Biol. Chem.*, 246, 3584, 1971.
545. McClure, W. R., Lardy, H. A., and Kneifel, H. P., Rat liver pyruvate carboxylase. I. Preparation, properties and cation specificity, *J. Biol. Chem.*, 246, 3569, 1971.
546. McClure, W. R., Lardy, H. A., Wagner, M., and Cleland, W. W., Rat liver pyruvate carboxylase: kinetic studies of the forward reaction, *J. Biol. Chem.*, 246, 3579, 1971.
547. McCormack, J. G. and Denton, R. M., The effects of calcium ions and adenine nucleotides on the activity of pig heart 2-oxoglutarate dehydrogenase complex, *Biochem. J.*, 180, 533, 1979.
548. McCormack, J. G. and Denton, R. M., Role of calcium ions in the regulation of intramitochondrial metabolism. Properties of the Ca^{2+}-sensitive dehydrogenases within intact uncoupled mitochondria from the white and brown adipose tissue of the rat, *Biochem. J.*, 190, 95, 1980.
549. McCullough, W. and Roberts, C. F., Genetic regulation of isocitrate lyase in *Aspergillus nidulans*, *J. Gen. Microbiol.*, 120, 67, 1980.
550. McDaniel, H. G., A comparison of effect of insulin on hepatic metabolites, gluconeogenesis, and ketogenesis, *Am. J. Physiol.*, 233, E13, 1977.
551. McGarry, J. D. and Foster, D. W., In support of the roles of malonyl-CoA and carnitine acyltransferase I in the regulation of hepatic fatty acid oxidation and ketogenesis, *J. Biol. Chem.*, 254, 8163, 1979.
552. McGarry, J. D., Stark, M. J., and Foster, D. W., Hepatic malonyl-CoA levels of fed, fasted, and diabetic rats as measured using a simple radioisotopic assay, *J. Biol. Chem.*, 253, 8291, 1978.
553. McGarry, J. D., Takabayashi, Y., and Foster, D. W., The role of malonyl-CoA in the coordination of fatty acid synthesis and oxidation in isolated rat hepatocytes, *J. Biol. Chem.*, 253, 8294, 1978.
554. McGurk, K. S. and Spivey, H. O., Ligand-induced conformational transitions and secondary structure composition of chicken liver pyruvate carboxylase, *Biochem. J.*, 177, 697, 1979.
555. Mehlman, M. A., Tobin, R. B., Friend, B., and Mackerer, C. R., The effects of a polychlorinated BI phenyl mixture aroclor 1254 on liver gluconeogenic enzymes of normal and alloxan diabetic rats, *Toxicology*, 5, 89, 1975.
556. Mehlman, M. A., Tobin, R. B., and Johnston, J. B., Metabolic control of enzymes in normal, diabetic, and diabetic insulin-treated rats utilizing 1,3 butanediol, *Metabolism*, 20, 149, 1971.
557. Meijer, A. J., Gimpel, J. A., Deleeuw, G., Tischler, M. E., Tager, J. M., and Williamson, J. R., Interrelationships between gluconeogenesis and ureogenesis in isolated hepatocytes, *J. Biol. Chem.*, 253, 2308, 1978.
558. Meijer, A. J., van Woerkom, G. M., Steinman, R., and Williamson, J. R., Inhibition by Ca^{2+} of carbamoylphosphate synthetase (ammonia), *J. Biol. Chem.*, 256, 3443, 1981.
559. Menahan, L. A. and Wieland, O., The role of endogenous lipid in gluconeogenesis and ketogenesis of perfused rat liver, *Eur. J. Biochem.*, 9, 182, 1969.
560. Menahan, L. A. and Wieland, O., The role of thyroid function in the metabolism of perfused rat liver with particular reference to gluconeogenesis, *Eur. J. Biochem.*, 10, 188, 1969.
561. Merali, Z., Kacew, S., and Singhal, R. L., Response of hepatic carbohydrate and cyclic AMP metabolism to cadmium treatment in rats, *Can. J. Physiol. Pharmacol.*, 53, 174, 1974.
562. Merali, Z., Kacew, S., and Singhal, R. L., Alterations in rat liver and kidney cortex metabolism after acute and chronic exposure to cadmium, *Toxicol. Appl. Pharmacol.*, 29, 140, 1974.
563. Merali, Z. and Singhal, R. L., Effect of selenium on cadmium induced toxicity in rats, *Toxicol. Appl. Pharmacol.*, 33, 127, 1975.
564. Merali, Z. and Singhal, R. L., Protective effect of selenium on certain hepatotoxic and pancreotoxic manifestations of subacute cadmium administration, *J. Pharmacol. Exp. Ther.*, 195, 58, 1975.

565. Merali, Z. and Singhal, R. L., Long-term effects of orally administered cadmium on neonatal rats, *International Congress on Toxicology*, Abstracts, Toronto, 11, 1977.
566. Mersmann, H. J., Glycolytic and gluconeogenic enzyme levels in pre- and post-natal pigs, *Am. J. Physiol.*, 220, 1297, 1971.
567. Mildvan, A. S. and Scrutton, M. C., Pyruvate carboxylase. X. The demonstration of direct coordination of pyruvate and α-ketobutyrate by the bound manganese and formation of enzyme-metal-substrate bridge complexes, *Biochemistry*, 6, 2978, 1967.
568. Mildvan, A. S., Scrutton, M. C., and Utter, M. F., Pyruvate carboxylase. VII. A possible role for tightly bound manganese, *J. Biol. Chem.*, 241, 3488, 1966.
569. Miller, A. L. and Atkinson, D. E., Response of yeast pyruvate carboxylase to the adenylate energy charge and other regulatory parameters, *Arch. Biochem. Biophys.*, 152, 531, 1972.
570. Milrad de Forchetti, S. R. and Cazzulo, J. J., Some properties of the pyruvate carboxylase from *Pseudomonas fluorescens, J. Gen. Microbiol.*, 93, 75, 1976.
571. Mirocha, C. J. and Rick, P. D., Carbon dioxide fixation in the dark as a nutritional factor in parasitism, in *The Dynamic Role of Molecular Constituents in Plant-Parasite Interaction,* Mirocha, C. J. and Uritani, I., Eds., American Phytopathological Society, St. Paul, Minn., 1967, 121.
572. Mizunaga, T., Kuraishi, H., and Aida, K., Levels of biotin enzymes during unbalanced growth and death of biotin-deficient yeast cells, *J. Gen. Appl. Microbiol.*, 19, 1, 1973.
573. Mizunaga, T., Kuraishi, H., Aida, K., and Uemura, T., The presence of apopyruvate carboxylase and synthesis of holopyruvate carboxylase in biotin requiring baker's yeast, *Agric. Biol. Chem.*, 34, 146, 1970.
574. Moon, T. W. and Johnston, I. A., Starvation and the activities of glycolytic and gluconeogenic enzymes in skeletal muscles and liver of the plaice, *Pleuronectes platessa, J. Comp. Physiol.*, 136, 31, 1980.
575. Mörikofer-Zwez, S., Kunin, A. S., and Walter, P., Effects of calcium and sucrose on pyruvate carboxylase activity in intact rat liver mitochondria, *J. Biol. Chem.*, 248, 7588, 1973.
576. Moss, J. and Lane, M. D., The biotin-dependent enzymes, *Adv. Enzymol.*, 35, 321, 1971.
577. Mosse, P. R. L., An investigation of gluconeogenesis in marine teleosts, and the effect of long-term exercise on hepatic gluconeogenesis, *Comp. Biochem. Physiol.*, 67B, 583, 1980.
578. Mottershead, D. G., Oliver, K. J., Brazier, J. A., and Wallace, J. C., Immunological cross-reaction of some biotin carboxylases, *Proc. Aust. Biochem. Soc.*, 16, 69, 1984.
579. Mueller, M. J. and Seitz, H. J., Starvation induced changes of hepatic glucose metabolism in hypothyroid and hyperthyroid rats *in vivo, J. Nutr.*, 111, 1370, 1981.
580. Mullhofer, G. and Kuntzen, O., Intracellular compartmentation in the perfused rat liver. Examination of the existence of a uniform mitochondrial oxaloacetate pool as an intermediate of both gluconeogenesis and the citric acid cycle, *Hoppe-Seyler's Z. Physiol. Chem.*, 353, 1461, 1972.
581. Munnich, A., Fischer, A., Saudubray, J. M., Grischelli, C., Coude, F. X., Ogier, H., Charpentier, C., and Frezal, J., Biotin-responsive immunoregulatory dysfunction in multiple carboxylase deficiency, *J. Inher. Metab. Dis.*, 4, 113, 1981.
582. Munnich, A., Saudubray, J. M., Carre, G., Coude, F. X., Ogier, H., Charpentier, C., and Frezal, J., Defective biotin absorption in multiple carboxylase deficiency, *Lancet,* II, 263, 1981.
583. Munnich, A., Saudubray, J. M., Cotisson, A., Coude, F. X., Ogier, H., Charpentier, C., Marsac, C., Carre, G., Bourgeay-Causse, M., and Frezal, J., Biotin dependent multiple carboxylase deficiency presenting as a congenital lactic acidosis, *Eur. J. Pediatr.*, 137, 203, 1981.
584. Munnich, A., Saudubray, J. M., Coude, F. X., Charpentier, C., Saurat, J. H., and Frezal, J., Fatty-acid-responsive alopecia in multiple carboxylase deficiency, *Lancet*, 1, 1080, 1980.
585. Munnich, A., Saudubray, J. M., Ogier, H., Coude, F. X., Marsac, C., Roccichioli, F., Labarthe, J. C., Cazenave, C., Laugier, J., Charpentier, C., and Frezal, J., Deficit multiple des carboxylases: une maladie metabolique vitamino-dependante curable par la biotine, *Arch. Fr. Pediatr.*, 38, 83, 1981.
586. Munnich, A., Saudubray, J. M., Ogier, H., Marsac, C., Charpentier, C., and Frezal, J., Alopecie, candidose chronique, retard mental et comas acidocetosiques a repitition curables par la biotine, *Ann. Med. Int.*, 131, 435, 1980.
587. Murad, S. and Freedland, R. A., Effect of thyroxine administration on enzymes associated with glucose metabolism in the liver, *Proc. Soc. Exp. Biol. Med.*, 124, 1176, 1967.
588. Murphy, J. V., Neurochemical changes in Leigh's disease, *J. Nutr. Sci. Vitaminol.*, 22, 69, 1976.
589. Murphy, J. V., Isohashi, F., Weinberg, M. B., and Utter, M. F., Pyruvate carboxylase deficiency: an alleged biochemical cause of Leigh's disease, *Pediatrics*, 68, 401, 1981.
590. Myers, D. E. and Utter, M. F., The enzyme synthesis of some potential photoaffinity analogs of benzoyl coenzyme A, *Anal. Biochem.*, 112, 23, 1981.
591. Nabeshima, S., Mishina, M., Tanaka, A., and Fukui, S., Effect of various compounds on isocitrate lyase formation in *Candida tropicalis, Agric. Biol. Chem.*, 41, 503, 1977.
592. Nakashima, K., Rudolph, F. B., Wakabayashi, J., and Lardy, H. A., Rat liver pyruvate carboxylase. V. Reversible dissociation by chloride salts of monovalent cations, *J. Biol. Chem.*, 250, 331, 1975.

593. Nakazawa, T., Asami, K., Suzuki, H., and Yukawa, O., Appearance of energy conservation system in rat liver mitochondria during development. The role of adenine nucleotide translocation, *J. Biochem. (Tokyo)*, 73, 397, 1973.
594. Negelein, E., unpublished data, 1943; Bucher, T. and Pfleiderer, G., Pyruvate kinase from muscle, in *Methods in Enzymology*, Vol. 1, Colowick, S. P. and Kaplan, N. O., Eds., Academic, New York, 1955, 435.
595. Nelson, P., Yarnell, G., and Wagle, S. R., Biochemical studies of the developing embryo. II. Studies on CO_2 fixation enzymes, *Arch. Biochem. Biophys.*, 114, 543, 1966.
596. Newaz, S. S. and Hersh, L. B., Reduced nicotinamide adenine dinucleotide-activated phosphoenolpyruvate carboxylase in *Pseudomonas* MA: potential regulation between carbon assimilation and energy production, *J. Bacteriol.*, 124, 825, 1975.
597. Newsholme, E. A., Brand, K., Lang, J., Stanley, J. C., and Williams, T., The maximum activities of enzymes that are involved in substrate cycles in liver and muscle of obese mice, *Biochem. J.*, 182, 621, 1979.
598. Newsholme, E. A., Crabtree, B., Higgins, S. J., Thornton, S. D., and Start, C., The activities of fructose diphosphase in flight muscles from the bumblebee and the role of this enzyme in heat generation, *Biochem. J.*, 128, 89, 1972.
599. Newsholme, E. A. and Gevers, W., Control of glycolysis and gluconeogenesis in liver and kidney cortex, *Vitam. Horm.*, 25, 1, 1967.
600. Newsholme, E. A. and Start, C., Regulation of carbohydrate metabolism in liver, in *Regulation in Metabolism*, Newsholme, E. A. and Start, C., Eds., John Wiley & Sons, London, 1973, chap. 6.
601. Nicholls, D. G., Shepherd, D., and Garland, P. B., A continuous recording technique for the measurement of carbon dioxide, and its application to mitochondrial oxidation and decarboxylation reactions, *Biochem. J.*, 103, 677, 1967.
602. Nisselbaum, J. S. and Bodansky, O., Immunological and kinetic properties of anionic and cationic glutamic - oxaloacetic transaminases separated from human heart and human liver, *J. Biol. Chem.*, 239, 4232, 1964.
603. Moosa, A. and Hughes, E. A., L-Glutamine therapy in Leigh's encephalomyelopathy, *Arch. Dis. Child.*, 49, 246, 1974.
604. Nordlie, R. C. and Lardy, H. A., Mammalian liver phosphoenolpyruvate carboxykinase activities, *J. Biol. Chem.*, 238, 2259, 1963.
605. Nordlie, R. C., Varricchio, F. E., and Holten, D. D., Effects of altered hormonal states and fasting on rat liver mitochondrial phosphoenolpyruvate carboxykinase levels, *Biochem. Biophys. Acta*, 97, 214, 1965.
606. Northrop, D. B., Transcarboxylase: kinetic analysis of the reaction mechanism, *J. Biol. Chem.*, 244, 5808, 1969.
607. Northrop, D. B. and Wood, H. G., Transcarboxylase: exchange reactions and kinetics of oxalate inhibition, *J. Biol. Chem.*, 244, 5820, 1969.
608. Numa, S., Bortz, W. M., and Lynen, F., Regulation of fatty acid synthesis at the acetyl CoA carboxylation step, *Adv. Enz. Reg.*, 3, 407, 1965.
609. Numa, S., Ringelmann, E., and Lynen, F., Zur biochemischen Funktion des Biotins. VIII. Die Bindung der Kohlensaure in der carboxylierten Acetyl-CoA-carboxylase, *Biochem. Z.*, 340, 228, 1964.
610. Oberholzer, V. G., Levin, B., Burgess, E. A., and Young, W. F., Methylmalonic aciduria. An inborn error of metabolism leading to chronic metabolic acidosis, *Arch. Dis. Child.*, 42, 492, 1967.
611. Obermayer, M. and Lynen, F., Structure of biotin enzymes, *Trends Biochem. Sci.*, 1, 169, 1976.
612. O'Brien, R. W., Chuang, D. T., Taylor, B. L., and Utter, M. F., Novel enzymic machinery for the metabolism of oxaloacetate, phosphoenolpyruvate, and pyruvate in *Pseudomonas citronellolis*, *J. Biol. Chem.*, 252, 1257, 1977.
613. O'Doherty, P. J. A., and Feltham, L. A. W., Glycolysis and gluconeogenesis in the giant scallop, *Placopecten magellanicus* (Gmelin), *Comp. Biochem. Physiol.*, 38B, 543, 1971.
614. Oliver, K. J., Rylatt, D. B., Bundeson, P., and Wallace, J. C., Study of the common structural determinants of the biotin carboxylases using monoclonal antibodies, *Proc. Aust. Biochem. Soc.*, 16, 29, 1984.
615. Osmani, S. A., Marston, F. A. O., Selmes, I. P., Chapman, A. G., and Scrutton, M. C., Pyruvate carboxylase from *Aspergillus nidulans*. Regulatory properties, *Eur. J. Biochem.*, 118, 271, 1981.
616. Osmani, S. A., Mayer, F., Marston, F. A. O., Selmes, I. P., and Scrutton, M. C., Pyruvate carboxylase from *Aspergillus nidulans:* effects of regulatory modifiers on the structure of the enzyme, *Eur. J. Biochem.*, 139, 509, 1984.
617. Osmani, S. A. and Scrutton, M. C., Activation of pyruvate carboxylase from *Aspergillus nidulans* by acetyl coenzyme A, *FEBS Lett.*, 135, 253, 1981.
618. Osmani, S. A. and Scrutton, M. C., The sub-cellular localisation of pyruvate carboxylase and of some other enzymes in *Aspergillus nidulans*, *Eur. J. Biochem.*, 133, 551, 1983.
619. Oura, E., Haarasilta, S., and Londesborough, J., Carbon dioxide fixation by baker's yeast in a variety of growth conditions, *J. Gen. Microbiol.*, 118, 51, 1980.

620. Overman, S. A. and Romano, A. H., Pyruvate carboxylase of *Rhizopus nigricans* and its role in fumaric acid production, *Biochem. Biophys. Res. Commun.*, 37, 457, 1969.
621. Owen, O. E., Patel, M. S., Block, B. S. B., Kreulen, T. H., Reichle, F. A., and Mozzoli, M. A., Gluconeogenesis in normal, cirrhotic, and diabetic humans, in *Gluconeogenesis*, Hanson, R. W. and Mehlman, M. A., Eds., John Wiley & Sons, New York, 1976, 533.
622. Packman, S. and Caswell, N., Biochemical differences between clinical classes of biotin responsiveness multiple carboxylase deficiency, *Am. J. Hum. Genet.*, 31, 49A, 1980.
623. Packman, S., Cowan, M. J., Golbus, M. S., Caswell, N. M., Sweetman, L., Burri, B. J., Nyhan, W. L., and Baker, H., Prenatal treatment of biotin-responsive multiple carboxylase deficiency, *Lancet*, 1, 1435, 1982.
624. Packman, S., Sweetman, L., Baker, H., and Wall, S., The neonatal form of biotin-responsive multiple carboxylase deficiency, *J. Pediatr.*, 99, 418, 1981.
625. Packman, S., Sweetman, L., and Wall, S., Biotin responsive multiple carboxylase deficiency in a child with congenital lactic acidosis, *Am. J. Hum. Genet.*, 31, 58A, 1979.
626. Packman, S., Sweetman, L., Yoshino, M., Baker, H., and Cowan, M., Biotin-responsive multiple carboxylase deficiency of infantile onset, *J. Pediatr.*, 99, 421, 1981.
627. Palacian, E., de Torrontegui, G., and Losada, M., Inhibition of yeast pyruvate carboxylase by L-aspartate and oxaloacetate, *Biochem. Biophys. Res. Commun.*, 24, 644, 1966.
628. Palacian, E. and Neet, K. E., Pyruvate carboxylase: inactivation by sulphydryl group reagents in the presence of certain inorganic ions, substrates, and modifiers and the effects of allosteric properties of the enzyme, *Biochim. Biophys. Acta*, 278, 297, 1972.
629. Palade, G. E., Electron microscopy of mitochondria and other cytoplasmic structures, in *Enzymes: Units of Biological Structure and Function*, Gaebler, O. H., Ed., Academic Press, New York, 1956, 185.
630. Parrilla, R., Ayuso-Parrilla, M. S., and Williamson, J. R., Use of endogenous triglycerides to support gluconeogenesis in the perfused rat liver, *Pflugers Arch.*, 366, 211, 1976.
631. Parrilla, R., Jimenez, I., and Ayuso-Parrilla, M. S., Glucagon and insulin control of gluconeogenesis in the perfused isolated rat liver. Effects on cellular metabolite distribution, *Eur. J. Biochem.*, 56, 375, 1975.
632. Parrilla, R., Jimenez, M.-I., and Ayuso-Parrilla, M. S., Cellular redistribution of metabolites during glucagon and insulin control of gluconeogenesis in the isolated perfused rat liver, *Arch. Biochem. Biophys.*, 174, 1, 1976.
633. Patel, M. S., The relative significance of CO_2-fixing enzymes in the metabolism of rat brain, *J. Neurochem.*, 22, 717, 1974.
634. Patel, M. S., The effect of ketone bodies on pyruvate carboxylation by rat brain mitochondria, *J. Neurochem.*, 23, 865, 1974.
635. Patel, M. S., Grover, W. D., and Auerbach, V. H., Pyruvate metabolism by homogenates of human brain: effects of phenylpyruvate and implications for the etiology of the mental retardation in phenylketonuria, *J. Neurochem.*, 20, 289, 1973.
636. Patel, M. S. and Tilghman, S. M., Regulation of pyruvate metabolism via pyruvate carboxylase in rat brain mitochondria, *Biochem. J.*, 132, 185, 1973.
637. Patty, L. and Smith, E. L., Reversible modification of arginine residues. Application to sequence studies by restriction of tryptic hydrolysis to lysine residues, *J. Biol. Chem.*, 250, 557, 1975.
638. Patzelt-Wenczler, R., Pauls, H., Erolman, E., and Schoner, W., Evidence for a sulfhydryl group in the ATP binding site of (sodium-potassium ion)-activated ATPase, *Eur. J. Biochem.*, 53, 301, 1975.
639. Payne, J. and Morris, J. G., Pyruvate carboxylase in *Rhodopseudomonas spheroides*, *J. Gen. Microbiol.*, 59, 97, 1969.
640. Pearce, J. and Balnave, D., A review of biotin deficiency and the fatty liver and kidney syndrome in poultry, *Br. Vet. J.*, 134, 598, 1978.
641. Pearce, J. and Balnave, D., Biotin deficiency and liver metabolism in relation to fatty liver and kidney syndrome, *Br. Poultry Sci.*, 19, 431, 1978.
642. Pearson, J. A., Johnson, A. R., Hood, R. L., and Fogerty, A. C., Fatty liver and kidney syndrome in chicks. I. Effect of biotin in diet, *Aust. J. Biol. Sci.*, 29, 419, 1976.
643. Peret, J. and Chanez, M., Influence of diet, cortisol and insulin on the activity of pyruvate carboxylase and phosphoenolpyruvate carboxykinase in the rat liver, *J. Nutr.*, 106, 103, 1976.
644. Pette, D., Mitochondrial enzyme activities, in *Regulation of Metabolic Processes in Mitochondria*, Tager, J. M., Papa, S., Quagliariello, E., and Slater, E. C., Eds., Elsevier, New York, 1966, 28.
645. Peters, R. F. and White, A. M., The existence of a glyconeogenic pathway in rat skin, *Biochem. J.*, 156, 465, 1976.
646. Pettit, F. H., Roche, T. E., and Reed, L. J., Function of calcium ions in pyruvate dehydrogenase phosphatase activity, *Biochem. Biophys. Res. Commun.*, 49, 563, 1972.
647. Peuhkurinen, K. J. and Hassinen, I. E., Pyruvate carboxylation as an anaplerotic mechanism in the isolated perfused rat heart, *Biochem. J.*, 202, 67, 1982.

648. Phibbs, P. V., Feary, T. W., and Blevins, W. T., Pyruvate carboxylase deficiency in pleiotropic carbohydrate-negative mutant strains of *Pseudomonas aeruginosa*, *J. Bacteriol.*, 118, 999, 1974.
649. Philippidis, H. and Ballard, F. J., The development of gluconeogenesis in rat liver. Experiments *in vivo*, *Biochem J.*, 113, 651, 1969.
650. Phillips, J. W. and Hird, F. J. R., Gluconeogenesis in vertebrate livers, *Comp. Biochem. Physiol.*, 57B, 127, 1977.
651. Phillips, N. F. B., Keech, D. B., and Wallace, J. C., Pyruvate carboxylase: mechanism of the first partial reaction, *Proc. Aust. Biochem. Soc.*, 14, 21, 1981.
652. Phillips, R., Yamauchi, M., Cote, M. G., and Plaa, G. L., Assessment of mercuric chloride induced nephro toxicity by *p*-amino hippuric-acid uptake and the activity of four gluconeogenic enzymes in rat renal cortex, *Toxicol. Appl. Pharmacol.*, 41, 407, 1977.
653. Pincus, J. H., Subacute necrotizing encephalomyelopathy (Leigh's disease): a consideration of clinical features and etiology, *Dev. Med. Child. Neurol.*, 14, 87, 1972.
654. Podos, S. M., Hyperpyruvicemia with hyper-alpha-alaninemia, *Arch. Ophthal. Mol.*, 83, 504, 1970.
655. Polakis, S. E., Guchhait, R. B., and Lane, M. D., On the possible involvement of a carbonyl phosphate intermediate in the adenosine triphosphate-dependent carboxylation of biotin, *J. Biol. Chem.*, 247, 1335, 1972.
656. Polakis, S. E., Guchhait, R. B., Zwergel, E. E., and Lane, M. D., Acetyl coenzyme A carboxylase system of *Escherichia coli:* studies on the mechanisms of the biotin carboxylase- and carboxytransferase-catalysed reactions, *J. Biol. Chem.*, 249, 6657, 1974.
657. Poston, H. A. and McCartney, T. H., Effect of dietary biotin and lipid on growth, stamina, lipid metabolism and biotin-containing enzymes in brook trout *(Salvelinus fontinalis)*, *J. Nutr.*, 104, 315, 1974.
658. Prescott, D. J. and Rabinowitz, J. L., The enzymatic carboxylation of propionyl CoA. Studies involving deuterated and tritiated substrates, *J. Biol. Chem.*, 243, 1551, 1968.
659. Prescott, L. M. and Campbell, J. W., Phosphoenolpyruvate carboxylase activity and glycogenesis in the flatworm *Hymenolepis diminuta*, *Comp. Biochem. Physiol.*, 14, 491, 1965.
660. Prichard, R. K. and Schofield, P. J., The metabolism of phosphoenolpyruvate in the adult liver fluke *Fasciola hepatica*, *Biochim. Biophys. Acta*, 170, 63, 1968.
661. Prinz, W. and Seubert, W., Effect of insulin on pyruvate carboxylase in alloxan diabetic animals, *Biochem. Biophys. Res. Commun.*, 16, 582, 1964.
662. Prior, R. L. and Scott, R.A., Ontogeny of gluconeogenesis in the bovine fetus: influence of maternal dietary energy, *Fed. Proc. Fed. Am. Soc. Exp. Biol.*, 36, 1140, 1977.
663. Prior, R. L. and Scott, R. A., Ontogeny of gluconeogenesis in the bovine fetus: influence of maternal dietary energy, *Dev. Biol.*, 58, 384, 1977.
664. Prpic, V., Spencer, T. L., and Bygrave, F. L., Stable enhancement of calcium retention in mitochondria isolated from rat liver after the administration of glucagon to the intact animal, *Biochem. J.*, 176, 705, 1978.
665. Rabier, D., Briand, P., Petit, F., Parvy, P., Kamoun, P., and Cathelineau, L., Acute effects of glucagon on citrulline biosynthesis, *Biochem. J.*, 206, 627, 1982.
666. Raghunathan, R. and Arinze, I. J., Perinatal development of gluconeogenesis in guinea-pig liver, *Int. J. Biochem.*, 8, 737, 1977.
667. Raghunathan, R. and Arinze, I. J., Fetal to neonatal development of gluconeogenic enzymes in the guinea-pig kidney, *Int. J. Biochem.*, 9, 279, 1978.
668. Raghunathan, R., Russell, J. D., and Arinze, I. J., Pyruvate carboxylase and phosphoenolpyruvate carboxykinase in cultured human fibroblasts, *J. Cell. Physiol.*, 92, 285, 1977.
669. Raijman, L. and Bartulis, T., Effect of ATP translocation on citrulline and oxaloacetate synthesis by isolated rat liver mitochondria, *Arch. Biochem. Biophys.*, 195, 188, 1979.
670. Raj, H. G. and Venkitasubramanian, T. A., Carbohydrate metabolism in aflatoxin B_1 toxicity, *Environ. Physiol. Biochem.*, 4, 181, 1974.
671. Ramirez, A. and Eyzaguirre, J., Pyruvate carboxylase of *Pseudomonas aeruginosa*. Purification and properties, *Arch. Biol. Med. Exp.*, 9, 20, 1973.
672. Rast, D., Stussi, H., and Zobrist, P., Self-inhibition of the *Agaricus bisporus* spore by CO_2 and/or γ-glutaminyl-4-hydroxybenzene and γ-glutaminyl-3,4-benzoquinone: a biochemical analysis, *Physiol. Plant.*, 46, 227, 1979.
673. Ray, P. D., Hepatic gluconeogenesis in the rabbit, in *Gluconeogenesis: Its Regulation in Mammalian Species*, Hanson, R. W. and Mehlman, M. A., Eds., John Wiley & Sons, New York, 1976, 293.
674. Reeck, G. R., Amino acid compositions of selected proteins, in *Handbook of Biochemistry and Molecular Biology*, Vol. 2, 2nd ed., Fasman, G. D., Ed., CRC Press, 1970, C-281.
675. Reed, L. J., Pyruvate dehydrogenase complex, in *Current Topics in Cellular Regulation*, Vol. 1, Horecker, B. L. and Stadtman, E. R., Eds., Academic Press, New York, 1969, 233.
676. Reed, G. H. and Scrutton, M. C., Pyruvate carboxylase from chicken liver. Magnetic resonance studies of the effect of substrates and inhibitors on the environment of the bound manganese, *J. Biol. Chem.*, 249, 6156, 1974.

677. Reeves, R. E., A new enzyme with the glycolytic function of pyruvate kinase, *J. Biol. Chem.*, 243, 3202, 1968.
678. Reichlin, M., Use of antibody in the study of protein structure, *Mol. Biol. Biochem. Biophys.*, 25, 55, 1977.
679. Reisner, A. H. and Westwood, N. H., Determination of the relatedness of the soluble surface proteins of *Trypanosoma brucei* and *Paramecium aurelia, Aust. J. Exp. Biol. Med. Sci.*, 55, 19, 1977.
680. Renner, E. D. and Bernlohr, R. W., Characterisation and regulation of pyruvate carboxylase of *Bacillus licheniformis, J. Bacteriol.*, 109, 764, 1972.
681. Retey, J. and Lynen, F., Zur biochemishen function des biotins. IX. Der sterische verhauf der carboxylierung von propionyl CoA, *Biochem. Z.*, 342, 256, 1965.
682. Richards, R. I., Shine, J., Ullrich, A., Wells, J. R. E., and Goodman, H. M., Molecular cloning and sequence analysis of adult chicken β-globin cDNA, *Nucl. Acids Res.*, 7, 1137, 1979.
683. Richards, R. I. and Wells, J. R. E., Chicken globin genes. Nucleotide sequence of cDNA clones coding for the α-globin expressed during hemolytic anemia, *J. Biol. Chem.*, 255, 9306, 1980.
684. Riou, J. P., Claus, T. H., and Pilkis, S. J., Stimulation by glucagon of *in vivo* phosphorylation of rat hepatic pyruvate kinase, *J. Biol. Chem.*, 253, 656, 1978.
685. Robinson, B. H., Development of gluconeogenic enzymes in the new born guinea-pig, *Biol. Neonate*, 29, 48, 1976.
686. Robinson, B. H., The postnatal development of gluconeogenic enzymes in guinea-pig kidney cortex, *FEBS Lett.*, 65, 319, 1976.
687. Robinson, B. H., Inborn errors of metabolism leading to lacticacidemia, *Trends Biochem. Sci.*, 7, 151, 1982.
688. Robinson, B. H., Inborn errors of pyruvate metabolism, *Biochem. Soc. Trans.*, 11, 623, 1983.
689. Robinson, G. A., Butcher, R. W., and Sutherland, E. W., *Cyclic AMP*, Academic Press, New York, 1971, 17.
690. Robinson, B. H., Gall, D. G., and Cutz, E., Deficient activity of hepatic pyruvate dehydrogenase and pyruvate carboxylase in Reye's syndrome, *Pediatr. Res.*, 11, 279, 1977.
691. Robinson, E. H. and Lovell, R. T., Essentiality of biotin for Channel catfish (*Ictalurus punctatus*) fed lipid and lipid-free diets, *J. Nutr.*, 108, 1600, 1978.
692. Robinson, B. H. and Oei, J., The effect of premature delivery on the development of gluconeogenic enzymes in the guinea-pig, *Pediatr. Res.*, 10, 819, 1976.
693. Robinson, B. H., Oei, J., Saunders, M., and Gravel, R., [^{3}H]Biotin-labeled proteins in cultured human skin fibroblasts from patients with pyruvate carboxylase deficiency, *J. Biol. Chem.*, 258, 6660, 1983.
694. Robinson, B. H., Taylor, J., and Sherwood, W. G., The genetic heterogeneity of lactic acidosis: occurrence of recognizable inborn errors of metabolism in a pediatric population with lactic acidosis, *Pediatr. Res.*, 14, 956, 1980.
695. Rognstad, R., Control of pyruvate kinase flux during gluconeogenesis in isolated liver cells, *Int. J. Biochem.*, 7, 403, 1976.
696. Rognstad, R. and Katz, J., The balance of pyridine nucleotides and ATP in adipose tissue, *Proc. Natl. Acad. Sci. U.S.A.*, 55, 1148, 1966.
697. Rognstad, R. and Katz, J., Gluconeogenesis in the kidney cortex. Quantitative estimation of carbon flow, *J. Biol. Chem.*, 247, 6047, 1972.
698. Rolleston, F. S., A theoretical background to the use of measured concentrations of intermediates in study of the control of intermediary metabolism, in *Current Topics in Cellular Regulation*, Vol. 5, Horecker, B. L. and Stadtman, E. R., Eds., Academic Press, New York, 1972, 47.
699. Rolph, T. P. and Jones, C. T., Pyruvate carboxylase activity and hypoglycaemia in the neonatal guinea pig, *Biochem. Soc. Trans.*, 9, 399, 1981.
700. Rolph, T. P. and Jones, C. T., Delayed development of gluconeogenic capacity and the appearance of hypoglycaemia in the newborn guinea-pig after intra-uterine growth restriction, *J. Dev. Physiol.*, 4, 1, 1982.
701. Ronson, C. W. and Primrose, S. B., Carbohydrate metabolism in *Rhizobium trifolii:* identification and symbiotic properties of mutants, *J. Gen. Microbiol.*, 112, 77, 1979.
702. Rose, I. A., O'Connell, E. L., and Solomon, F., Intermolecular tritium transfer in the transcarboxylation reaction, *J. Biol. Chem.*, 251, 902, 1976.
703. Ross, B. D., Hems, R., Freedland, R. A., and Krebs, H. A., Carbohydrate metabolism in rat liver, *Biochem. J.*, 105, 869, 1967.
704. Roth, K. S., Allan, L., Yang, W., Foreman, J. W., and Dakshinamurti, K., Serum and urinary biotin levels during treatment of holocarboxylase synthetase deficiency, *Clin. Chim. Acta*, 109, 337, 1981.
705. Roth, K. S., Cohn, R., Yandrasitz, J., Preti, G. T., Dodd, P., and Segal, S., Beta-methylcrotonic aciduria associated with lactic acidosis, *J. Pediatr.*, 88, 229, 1976.
706. Roth, K. S., Yang, W., Foreman, J. W., and Segal, S., Holocarboxylase synthetase deficiency: a biotin-responsive organic acidemia, *J. Pediatr.*, 96, 845, 1980.

707. Roth, K. S., Yang, W., Allan, L., Saunders, M., Gravel, R. A., and Dakshinamurti, K., Prenatal administration of biotin in biotin responsive multiple carboxylase deficiency, *Pediatr. Res.*, 16, 126, 1982.
708. Rowan, A. N., Newsholme, E. A., and Scrutton, M. C., Partial purification and some properties of pyruvate carboxylase from the flight muscle of the locust, *(Schistocerca gregaria), Biochim. Biophys. Acta*, 522, 270, 1978.
709. Ruderman, N. B., Aoki, T. T., and Cahill, G. F., Gluconeogenesis and its disorders in man, in *Gluconeogenesis: Its Regulation in Mammalian Species*, Hanson, R. W. and Mehlman, M. A., Eds., John Wiley & Sons, New York, 1976, 515.
710. Ruiz-Amil, M., de Torrontegui, E., Palacian, E., Catalina, L., and Losada, M., Properties and function of yeast pyruvate carboxylase, *J. Biol. Chem.*, 240, 3485, 1965.
711. Rulfs, J. and Aprille, J. R., Adenine nucleotide pool size, adenine nucleotide translocase activity, and respiratory activity in newborn rabbit liver mitochondria, *Biochim. Biophys. Acta*, 681, 300, 1982.
712. Rylatt, D. B., Keech, D. B., and Wallace, J. C., Pyruvate carboxylase: isolation of the biotin-containing tryptic peptide and the determination of its primary sequence, *Arch. Biochem. Biophys.*, 183, 113, 1977.
713. Ryley, J. F., Studies on the metabolism of the protozoa. VII. Comparative carbohydrate metabolism of eleven species of trypanosome, *Biochem. J.*, 62, 215, 1956.
714. Ryley, J. F., Studies on the metabolism of the protozoa. IX. Comparative metabolism of bloodstream and culture forms of *Trypanosoma rhodesiense, Biochem. J.*, 85, 211, 1962.
715. Sacktor, B., Biological oxidations and energetics in insect mitochondria, in *The Physiology of Insecta*, Vol. 4, 2nd ed., Rockstein, M., Ed., Academic Press, New York, 1974, 271.
716. Safer, D., Hainfeld, J., Wall, J. S., and Reardon, J. F., Biospecific labeling with undecagold: visualization of the biotin-binding site on avidin, *Science*, 218, 290, 1982.
717. Saggerson, E. D. and Evans, C. J., The activities and intracellular distribution of nicotinamide-adenine dinucleotide phosphate-malate dehydrogenase, phosphoenolpyruvate carboxykinase and pyruvate carboxylase in rat, guinea-pig and rabbit tissues, *Biochem. J.*, 146, 329, 1975.
718. Sagy, Z., Barzilay, Z., Barasa, V., Oren, M., Vardi, P., Cohen, B. E., and Gutman, A., Congenital lactic-acidosis associated with pyruvate carboxylase deficiency, *Isr. J. Med. Sci.*, 17, 1159, 1981.
719. Sahl, H. G. and Trüper, H. G., Enzymes of CO_2 fixation in *Chromatiaceae, FEMS Microbiol. Lett.*, 2, 129, 1977.
720. Salganicoff, L. and Koeppe, R. E., Subcellular distribution of pyruvate carboxylase, diphosphopyridine nucleotide isocitrate dehydrogenases, and malate enzyme in rat brain, *J. Biol. Chem.*, 243, 3416, 1968.
721. Sander, J. E., Malamud, N., Cowan, M. J., Packman, S., Ammann, A. J., and Wara, D. W., Intermittent ataxia and immunodeficiency with multiple carboxylase deficiencies: a biotin-responsive disorder, *Ann. Neurol.*, 8, 544, 1980.
722. Sander, J. E., Packman, S., and Townsend, J. J., Brain pyruvate carboxylase and the pathophysiology of biotin-dependent diseases, *Neurology*, 32, 878, 1982.
723. Sandmann, G. and Hilgenberg, W., Pyruvate carboxylase in *Phycomyces blakesleeanus*, Bgff., *Z. Pflanzenphysiol.*, 90, 415, 1978.
724. Sandmann, G. and Hilgenberg, W., The light-dependent intermediary metabolism of *Phycomyces blakesleeanus*, Bgff. I. The CO_2 fixation reaction, *Biochem. Physiol. Pflanzen*, 173, 390, 1978.
725. Sanger, F., Nicklen, S., and Coulson, A. R., DNA sequencing with chain-terminating inhibitors, *Proc. Natl. Acad. Sci. U.S.A.*, 74, 5463, 1977.
726. Saudubray, J. M., Marsac, C., Charpentier, C., Cathelineau, L., Besson Leaud, M., and Leroux, J. P., Neonatal congenital lactic-acidosis with pyruvate carboxylase deficiency in two siblings, *Acta Paediatr. Scand.*, 65, 717, 1976.
727. Saunders, M., Sherwood, W. G., Duthie, M., Surh, L., and Gravel, R., Evidence for a defect of holocarboxylase synthetase activity in cultured lymphoblasts from a patient with biotin-responsive multiple carboxylase deficiency, *Am. J. Hum. Genet.*, 34, 590, 1982.
728. Saunders, M., Sweetman, L., Robinson, B., Roth, K., Cohn, R., and Gravel, R. A., Biotin-response organicaciduria. Multiple carboxylase defects and complementation studies with propionicacidemia in cultured fibroblasts, *J. Clin. Invest.*, 64, 1695, 1979.
729. Saunders, M., Sweetman, L., Robinson, B., Roth, K., Cohn, S., Sherwood, G., and Gravel, R., Multiple carboxylase defects and complementation studies in biotin responsive organic aciduria, *Am. J. Hum. Genet.*, 31, 61A, 1979.
730. Savina, M. V. and Wojtczak, A. B., Enzymes of gluconeogenesis and the synthesis of glycogen from glycerol in various organs of the lamprey *(Lampetra fluviatilis), Comp. Biochem. Physiol.*, 57B, 185, 1977.
731. Saz, H. J., Comparative biochemistry of carbohydrates in nematodes and cestodes, in *Comparative Biochemistry of Parasites*, Van den Bossche, H., Ed., Academic Press, New York, 1972, 33.

732. Schachman, H. K., Ultracentrifugation, diffusion and viscometry, in *Methods in Enzymology*, Vol. 4, Colowick, S. P. and Kaplan, N. O., Eds., Academic Press, New York, 1957, 32.
733. Schaefer, P. C. and Veneziale, C. M., Regulation of pyruvate metabolism in rat-liver mitochondria by K^+ and P_i, *Eur. J. Biochem.*, 35, 18, 1973.
734. Schmer, G. and Kreil, G., Micro method for detection of formyl and acetyl groups in proteins, *Anal. Biochem.*, 29, 186, 1969.
735. Schnaitman, C. and Greenawalt, J. W., Enzymatic properties of the inner and outer membranes of rat liver mitochondria, *J. Cell. Biol.*, 38, 158, 1968.
736. Schrijver, J., Biotin Deficiency in the Rat as a Model for Reduced Pyruvate Carboxylase Activity. The Effect of Thiamin (a Contribution to the Study of Leigh's Disease), Ph.D. thesis, University of Groningen, The Netherlands, 1978, 21.
737. Schrijver, J., Dias, Th., and Hommes, F. A., Some biochemical observations on biotin deficiency in the rat as a model for human pyruvate carboxylase deficiency, *Nutr. Metab.*, 23, 179, 1979.
738. Schrijver, J. and Hommes, F. A., Biotin deficiency in the rat as a model for Leigh's syndrome, *Pediatr. Res.*, 10, 895, 1976.
739. Schrijver, J., Koster, J. F., and Hülsmann, W. C., Insignificance of gluconeogenesis in human blood platelets, *Eur. J. Clin. Invest.*, 5, 7, 1975.
740. Schulz, G. E. and Schirmer, R. H., *Principles of Protein Structure*, Cantor, C. R., Ed., Springer-Verlag, New York, 1979, 95.
741. Schulz, G. E. and Schirmer, R. H., *Principles of Protein Structure*, Cantor, C. R., Ed., Springer-Verlag, New York, 1979, 205.
742. Schutgens, R. B. H., and Wognum, P. J., Inhibition of gluconeogenesis in 3-hydroxy-3-methylglutaryl coenzyme-A-lyase deficiency, *J. Clin. Chem. Clin. Biochem.*, 19, 831, 1981.
743. Scrutton, M. C., Assay of enzymes of CO_2 metabolism, in *Methods in Microbiology*, Vol. 6A, Norris, J. R. and Ribbons, D. W., Eds., Academic Press, New York, 1971, 479.
744. Scrutton, M. C., Pyruvate carboxylase. Studies of activator-independent catalysis and of the specificity of activation by acyl derivatives of coenzyme A for the enzyme from rat liver., *J. Biol. Chem.*, 249, 7057, 1974.
745. Scrutton, M. C., Properties of the activation of phosphoenolpyruvate carboxylase from *Escherichia coli* by acyl derivatives of coenzyme A, *FEBS Lett.*, 48, 145, 1974.
746. Scrutton, M. C., A plea for a comparative approach to evaluation of the physiological significance of regulatory mechanisms observed in *in vitro* studies, *Med. Hypotheses*, 1, 207, 1975.
747. Scrutton, M. C., Fine control of the conversion of pyruvate (phosphoenolpyruvate) to oxaloacetate in various species, *FEBS Lett.*, 89, 1, 1978.
748. Scrutton, M. C. and Fatebene, F., Inhibition of rat liver pyruvate carboxylase by acetoacetyl-CoA, *FEBS Lett.*, 62, 220, 1976.
749. Scrutton, M. C. and Fung, C-H., Pyruvate carboxylase from chicken liver: effects of sulfate and other anions on catalytic activity and structural parameters, *Arch. Biochem. Biophys.*, 150, 636, 1972.
750. Scrutton, M. C. and Griffiths, J. R., Regulation of pyruvate carboxylase in vertebrates and microorganisms, in *Short-Term Regulation of Liver Metabolism*, Hue, L. and Van de Werve, G., Eds., Elsevier/North-Holland, New York, 1981, chap. 11.
751. Scrutton, M. C., Griminger, P., and Wallace, J. C., Pyruvate carboxylase. Bound metal content of the vertebrate liver enzyme as a function of diet and species, *J. Biol. Chem.*, 247, 3305, 1973.
752. Scrutton, M. C., Keech, D. B., and Utter, M. F., Pyruvate carboxylase: partial reactions and the locus of activation by acetyl CoA, *J. Biol. Chem.*, 240, 574, 1965.
753. Scrutton, M. C. and Mildvan, A. S., Pyruvate carboxylase. XI. Nuclear magnetic resonance studies of the bound manganese after interaction of the biotin residues with avidin, *Biochemistry*, 7, 1490, 1968.
754. Scrutton, M. C. and Mildvan, A. S., Pyruvate carboxylase: nuclear magnetic resonance studies of the enzyme-manganese-oxaloacetate and enzyme-manganese-pyruvate bridge complexes, *Arch. Biochem. Biophys.*, 140, 131, 1970.
755. Scrutton, M. C., Olmsted, M. R., and Utter, M. F., Pyruvate carboxylase from chicken liver, *Methods Enzymol.*, 13, 235, 1969.
756. Scrutton, M. C., Pearce, P. H., and Fatabene, F., Pyruvate carboxylase: effect of reaction components and analogues of acetyl coenzyme A on the rate of inactivation in the presence and absence of trinitrobenzene sulphonate, *Eur. J. Biochem.*, 76, 219, 1977.
757. Scrutton, M. C., Reed, G. H., and Mildvan, A. S., Application of physical methods to the study of enzymes containing bound manganese: problems and prospects, *Adv. Exp. Med. Biol.*, 40, 79, 1973.
758. Scrutton, M. C. and Taylor, B. L., Isolation and characterization of pyruvate carboxylase from *Azotobacter vinelandii* OP, *Arch. Biochem. Biophys.*, 164, 641, 1974.
759. Scrutton, M. C. and Utter, M. F., Pyruvate carboxylase. III. Some physical and chemical properties of the highly purified enzymes, *J. Biol. Chem.*, 240, 1, 1965.

760. Scrutton, M. C. and Utter, M. F., Pyruvate carboxylase. V. Interaction of the enzyme with adenosine triphosphate, *J. Biol. Chem.*, 240, 3714, 1965.
761. Scrutton, M. C. and Utter, M. F., Pyruvate carboxylase. IX. Some properties of the activation by certain acyl derivatives of Coenzyme A, *J. Biol. Chem.*, 242, 1723, 1967.
762. Scrutton, M. C. and Utter, M. F., The regulation of glycolysis and gluconeogenesis in animal tissues, *Annu. Rev. Biochem.*, 37, 249, 1968.
763. Scrutton, M. C., Utter, M. F., and Mildvan, A. S., Pyruvate carboxylase. VI. The presence of tightly bound manganese, *J. Biol. Chem.*, 241, 3480, 1966.
764. Scrutton, M. C. and White, M. D., Pyruvate carboxylase from rat liver: catalytic properties in the absence, and at low concentrations, of acetyl-CoA, *Biochem. Biophys. Res. Commun.*, 48, 85, 1972.
765. Scrutton, M. C. and White, M. D., Pyruvate carboxylase. Specific inactivation of acetyl Coenzyme A-dependent oxaloacetate synthesis during modification of the enzyme by trinitrobenzene sulphonic acid, *J. Biol. Chem.*, 248, 5541, 1973.
766. Scrutton, M. C. and White, M. D., Pyruvate carboxylase. Inhibition of the mammalian and avian liver enxymes by α-ketoglutarate and L-glutamate, *J. Biol. Chem.*, 249, 5405, 1974.
767. Scrutton, M. C. and White, M. D., Purification and properties of human liver pyruvate carboxylase, *Biochem. Med.*, 9, 271, 1974.
768. Scrutton, M. C. and Young, M. R., Pyruvate carboxylase, in *The Enzymes*, Vol. 6, 3rd ed., Boyer, P. D., Ed., Academic Press, New York, 1972, 1.
769. Scrutton, M. C., Young, M. R., and Utter, M. F., Pyruvate carboxylase from bakers yeast: the presence of bound zinc, *J. Biol. Chem.*, 245, 6220, 1970.
770. Seitz, H. J., Kaiser, M., Krone, W., and Tarnowski, W., Physiologic significance of glucocorticoids and insulin in the regulation of hepatic gluconeogenesis during starvation in rats, *Metabolism*, 25, 1545, 1976.
771. Seubert, W., Henning, H. V., Schonen, W., and L'Age, M., Effects of cortisol on the levels of metabolites and enzymes controlling glucose production from pyruvate, *Adv. Enzyme Regul.*, 6, 153, 1968.
772. Seubert, W. and Remberger, U., Reinigung und Wirkungsweise der Pyruvatcarboxylase aus *Pseudomonas citronellolis*, *Biochem. Z.*, 334, 401, 1961.
773. Seufert, D., Herleman, E.-M., Allbrecht, E., and Seubert, W., On the mechanism of gluconeogenesis and its regulation. VII. Purification and properties of pyruvate carboxylase from rat liver, *Hoppe-Seyler's Z. Physiol. Chem.*, 352, 459, 1971.
774. Shalteil, S. and Soria, M., Dinitrophenylation and thiolysis in the reversible labelling of a cysteine residue associated with the nicotinamide-adenine dinucleotide site of rabbit muscle glyceraldehyde-3-phosphate dehydrogenase, *Biochemistry*, 8, 4411, 1969.
775. Shen, C. S. and Mistry, S. P., Development of gluconeogenic, glycolytic and pentose-shunt enzymes in the chicken kidney, *Poultry Sci.*, 58, 663, 1979.
776. Shen, C. S. and Mistry, S. P., Effect of fasting and refeeding on hepatic and renal gluconeogenic enzymes in the chicken, *Poultry Sci.*, 58, 890, 1979.
777. Shen, C. S. and Mistry, S. P., Intracellular distribution of hepatic and renal gluconeogenic enzymes in embryonic and growing chickens, *Poultry Sci.*, 58, 1239, 1979.
778. Sherwood, W. G., Robinson, B. H., Brewster, T., Saunders, M., and Gravel, R. A., Lactic acidosis in biotin-responsive multiple carboxylase deficiency caused by holocarboxylase synthetase deficiency in early and late onset, *J. Pediatr.*, 101, 546, 1982.
779. Shows, T. B., Chapman, V. M., and Ruddle, F. A., Mitochondrial malate dehydrogenase and malic enzyme: Mendelian inherited variants in the mouse, *Biochem. Genet.*, 4, 704, 1970.
780. Shrago, E. and Lardy, H. A., Paths of carbon in gluconeogenesis and lipogenesis. II. Conversion of precursors to phosphoenolpyruvate in liver cytosol, *J. Biol. Chem.*, 241, 663, 1966.
781. Shrago, E., Lardy, H. A., Nordlie, R. C., and Foster, D. O., Metabolic and hormonal control of phosphoenolpyruvate carboxykinase and malic enzyme in rat liver, *J. Biol. Chem.*, 238, 3188, 1963.
782. Shulman, R. G., *Biological Applications of Magnetic Resonance*, Academic Press, New York, 1979, 1.
783. Siegel, L., Foote, J. L., Christner, J. E., and Coon, M. J., Propionyl-CoA holocarboxylase synthesis from biotinyl adenylate and the apocarboxylase in the presence of an activating enzyme, *Biochem. Biophys. Res. Commun.*, 13, 307, 1963.
784. Siegel, L., Foote, J. L., and Coon, M. J., The enzymatic synthesis of propionyl coenzyme A holocarboxylase from *d*-biotinyl 5′-adenylate and the apocarboxylase, *J. Biol. Chem.*, 240, 1025, 1965.
785. Siess, E. A., Brocks, D. G., Lattke, H. K., and Wieland, O. H., Effect of glucagon on metabolite compartmentation in isolated rat liver cells during gluconeogenesis from lactate, *Biochem. J.*, 166, 225, 1977.
786. Siess, E. A., Fahimi, F. M., and Wieland, O. H., Decrease by glucagon in hepatic succinyl-CoA, *Biochem. Biophys. Res. Commun.*, 95, 205, 1980.
787. Siess, E. A., Fahimi, F. M., and Wieland, O. H., Evidence that glucagon stabilizes rather than activates mitochondrial functions in rat liver, *Hoppe-Seyler's Z. Physiol. Chem.*, 362, 1643, 1981.

788. Siess, E. A. and Wieland, O. H., Improved response of isolated liver cells to glucagon in the presence of rat serum, *Biochem. Biophys. Res. Commun.*, 64, 323, 1975.
789. Sigel, H., McCormick, D. B., Griesser, R., Bernhard, P., and Wright, L. D., Metal ion complexes with biotin and biotin derivatives. Participation of sulphur in the orientation of divalent cations, *Biochemistry*, 8, 2687, 1969.
790. Sillen, L. G. and Martell, A. E., *Stability Constants*, Special Publ. No. 17, The Chemical Society, London, 1964, 362.
791. Simpson, J. W. and Awapara, J., Phosphoenolpyruvate carboxykinase activity in invertebrates, *Comp. Biochem. Physiol.*, 12, 457, 1964.
792. Simpson, J. W. and Awapara, J., The pathway of glucose degradation in some invertebrates, *Comp. Biochem. Physiol.*, 18, 537, 1966.
793. Sinsheimer, R. L., Recombinant DNA, *Ann. Rev. Biochem.*, 46, 415, 1977.
794. Siu, P. M. L. and Wood, H. G., Phosphoenolpyruvic carboxytransphosphorylase, a CO_2 fixation enzyme from propionic acid bacteria, *J. Biol. Chem.*, 237, 3044, 1962.
795. Skinner, V. M. and Armitt, S., Mutants of *Aspergillus nidulans* lacking pyruvate carboxylase, *FEBS Lett.*, 20, 16, 1972.
796. Smith, C. M., Bryla, J., Damon, S., LaNoue, K. F., and Williamson, J. R., Fluorometric assays for succinyl-CoA and propionyl-CoA in mitochondrial extracts, *Anal. Biochem.*, 51, 408, 1973.
797. Smith, R. W., The effects of pregnancy and lactation on the activities in rat liver of some enzymes associated with glucose metabolism, *Biochim. Biophys. Acta*, 411, 22, 1975.
798. Smith, S. B. and Prior, R. L., Evidence for a functional ATP citrate lyase NADP malate dehydrogenase pathway in bovine adipose tissue enzyme and metabolite levels, *Arch. Biochem. Biophys.*, 211, 192, 1981.
799. Snell, K., Pyruvate carboxylation in liver mitochondria of the developing neonatal rat, *Int. J. Biochem.*, 5, 463, 1974.
800. Snoswell, A. M. and Koundakjian, P. P., Relationships between carnitine and coenzyme A esters in tissues of normal and alloxan-diabetic sheep, *Biochem. J.*, 127, 133, 1972.
801. Snoswell, M. A., Chicken Liver Pyruvate Carboxylase: Mechanism of the First Partial Reaction, B.Sc. thesis, University of Adelaide, Australia, 1982.
802. Snoswell, M. A. and Wallace, J. C., unpublished observations.
803. Soboll, S., Scholz, R., Freisl, M., Elbers, R., and Heldt, H. W., Distribution of metabolites between mitochondria and cytosol of perfused liver, in *Use of Isolated Liver Cells and Kidney Tubules in Metabolic Studies*, Tager, J. M., Söling, H. D., and Williamson, J. R., Eds., North-Holland/American Elsevier, Amsterdam, 1976, 29.
804. Söling, H-D., and Kleineke, J., Species dependent regulation of hepatic gluconeogenesis in higher animals, in *Gluconeogenesis: its Regulation in Mammalian Species*, Hanson, R. W. and Mehlman, M. A., Eds., Wiley, New York, 1976, 369.
805. Söling, H. D., Kleineke, J., Willms, B., Janson, G., and Kuhn, A., Relationship between intracellular distribution of phospho*enol*pyruvate carboxykinase, regulation of gluconeogenesis, and energy cost of glucose formation, *Eur. J. Biochem.*, 37, 233, 1973.
806. Söling, H. D. and Volkmann, B., A sensitive assay for the determination of propionyl-CoA in biological material, *Anal. Biochem.*, 52, 305, 1973.
807. Söling, H. D., Willms, B., Friedrichs, D., and Kleineke, J.. Regulation of gluconeogenesis by fatty acid oxidation in isolated perfused livers of non-starved rats, *Eur. J. Biochem.*, 4, 364, 1968.
808. Söling, H. D., Willms, B., Kleineke, J., and Gehlhoff, M., Regulation of gluconeogenesis in the guinea pig liver, *Eur. J. Biochem.*, 16, 289, 1970.
809. Sols, A. and Marco, R., Concentrations of metabolites and binding sites. Implications in metabolic regulation, in *Current Topics in Cellular Regulation*, Vol. 2, Horecker, B. L. and Stadtman, E. R., Eds., Academic Press, New York, 1970, 227.
810. Somero, G. N. and Low, P. S., Temperature: a "shaping force" in protein evolution, *Biochem. Soc. Symp.*, 41, 33, 1976.
811. Spydevold, Ø., Davis, E. J., and Bremer, J., Replenishment and depletion of citric acid cycle intermediates in skeletal muscle. Indication of pyruvate carboxylation, *Eur. J. Biochem.*, 71, 155, 1976.
812. Spydevold, Ø., Greenbaum, A. L., Baquer, N. Z., and McLean, P., Adaptive responses of enzymes of carbohydrate and lipid metabolism to dietary alteration in genetically obese Zucker rats (fa/fa), *Eur. J. Biochem.*, 89, 329, 1978.
813. Srebocan, V., Plazonic, M., Pompe-Gotal, J., and Brmalj, V., On the biochemical mechanism of triazine herbicides toxicity: effect on the gluconeogenic activity in poultry liver, *Vet. Arch.*, 45, 273, 1975.
814. Srebocan, V., Pompe-Gotal, J., Brmalj, V., and Srebocan, E., Effect of pesticidal chlorinated hydocarbons on the enzymatic regulation of carbohydrate metabolism in poultry. II. Gluconeogenic response of insecticide-treated chicks to stress, *Vet. Arch.*, 50, 151, 1980.

815. Srebocan, V., Pompe-Gotal, J., Brhaaj, V., and Srebocan, E., Effect of pesticidal chlorinated hydocarbons on the enzymatic regulation of carbohydrate metabolism in poultry. III. Influence on the enzymatic regulation of gluconeogenesis in the liver of embryonic chickens, *Vet. Arch.*, 50, 221, 1980.
816. Stadtman, E. R., Allosteric regulation of enzyme activity, *Adv. Enzymol.*, 28, 41, 1966.
817. Stallings, W. C., Monti, C. T., Lane, M. D., and De Titta, G. J., Molecular structure and intermolecular interactions of N1′-methoxycarbonylbiotin methyl ester: a model for carboxybiotin, *Proc. Natl. Acad. Sci. U.S.A.*, 77, 1260, 1980.
818. Stan, H.-J., Pyruvate carboxylase from *Penicillium camemberti.* V. Regulation of the enzyme activity by the ratio of ATP and ADP, *Int. J. Biochem.*, 3, 573, 1972.
819. Stan, H.-J. and Schormuller, J., Regulation of pyruvate carboxylase of *Penicillium camemberti* by the adenylate system, *Biochem. Biophys. Res. Commun.*, 32, 289, 1968.
820. Steele, P., Edgar, J., and Duncan, G., Effects of biotin supplementation on incidence of acute death syndrome in broiler chickens, *Poultry Sci.*, 61, 909, 1982.
821. Steffens, G. J., Gunzler, W. A., Otting, F., Frankies, E., and Flohe, L., The complete amino acid sequence of low molecular mass urokinase from human urine, *Hoppe-Seyler's Z. Physiol. Chem.*, 363, 1043, 1982.
822. Sterling, K., Thyroid hormone action at the cell level, *N. Engl. J. Med.*, 300, 117, 1979.
823. Stevenson, A., Merali, Z., Kacew, S., and Singhal, R. L., Effects of subacute and chronic lead treatment on glucose homeostasis and renal cyclic AMP metabolism in rats, *Toxicology*, 6, 265, 1976.
824. Stifel, F. B., Greene, H. L., Lufkin, E. G., Wrensch, M. R., Hagler, L., and Herman, R. H., Acute effects of oral and intravenous ethanol on rat hepatic enzyme activities, *Biochim. Biophys. Acta*, 428, 633, 1976.
825. Stirk, J- H., Alberti, K. G. M. M., and Bartlett, K., The effects of large doses of biotin on leucocyte carboxylases in humans, *Biochem. Soc. Trans.*, 185, 11, 1983.
826. Stirk, J-H., Bartlett, K., and Sherratt, H. S. A., The effects of some short-chain fatty acids on pyruvate carboxylase activity in intact isolated rat liver mitochondria, *Biochem. Soc. Trans.*, 11, 286, 1983.
827. Stormer, B., Janssen, W., Reinauer, H., Staib, W., and Hollman, S., Enzymativitäten und Substratkonzentrationem in der Ratten leber unter den Bedingungen der Cortisol-Gluconeogenesis, *Z. Physiol. Chem.*, 351, 296, 1970.
828. Struck, E., Ashmore, J., and Wieland, O., Stimulierung der gluconeogenese durch langkettige fettsauren und glucagon, *Biochem. Z.*, 343, 107, 1965.
829. Struck, E., Ashmore, J., and Wieland, O., Pyruvate carboxylase activity and glucose production in isolated perfused rat liver, *Enzymol. Biol. Clin.*, 7, 38, 1966.
830. Stubbe, J. and Abeles, R. H., Biotin carboxylations — concerted or not concerted? This is the question!, *J. Biol. Chem.*, 252, 8338, 1977.
831. Stubbe, J., Fish, S., and Abeles, R. H., Are carboxylations involving biotin concerted or nonconcerted?, *J. Biol. Chem.*, 255, 236, 1980.
832. Stucki, J. W., Influence of 1,2,3-benzene-tricarboxylate on pyruvate metabolism in rat-liver mitochondria, *Eur. J. Biochem.*, 78, 183, 1977.
833. Stucki, J. W., Brawand, F., and Walter, P., Regulation of pyruvate metabolism in rat liver mitochondria by adenine nucleotides and fatty acids, *Eur. J. Biochem.*, 27, 181, 1972.
834. Stumpf, B., Boie, A., and Seubert, W., In vitro stimulation of gluconeogenesis from pyruvate and of $^{14}CO_2$-fixation by dexamethasone phosphate and vasopressin in the kidney cortex of adrenalectomised rats, *Biochem. Biophys. Res. Commun.*, 49, 164, 1972.
835. Solanki, K., Moser, U., and Walter, P., Inhibition of gluconeogenesis by ascorbate, iso-ascorbic acid and dehydro-ascorbate in rat liver hepatocytes, *Experientia (Basel)*, 37, 633, 1981.
836. Sumper, M. and Riepertinger, C., Structural relationship of biotin-containing enzymes. Acetyl CoA carboxylase and pyruvate carboxylase from yeast, *Eur. J. Biochem.*, 29, 237, 1972.
837. Sundaram, T. K., Physiological role of pyruvate carboxylase in a thermophilic *Bacillus*, *J. Bacteriol.*, 113, 549, 1973.
838. Sundaram, T. K., Cazzulo, J. J., and Kornberg, H. L., Anaplerotic CO_2 fixation in mesophilic and thermophilic bacilli, *Biochim. Biophys. Acta*, 192, 355, 1969.
839. Sundaram, T. K., Cazzulo, J. J., and Kornberg, H. L., Pyruvate holocarboxylase formation from the apoenzyme and D-biotin in *Saccharomyces cerevisiae*, *Arch. Biochem. Biophys.*, 143, 609, 1971.
840. Sundaram, T. K., Cazzulo, J. J., and Kornberg, H. L., Synthesis of pyruvate carboxylase from its apoenzyme and (+)-biotin in *Bacillus stearothermophilus.* Mechanism and control of the reaction, *Biochem. J.*, 122, 663, 1971.
841. Suomalainen, H. and Oura, E., Yeast nutrition and solute uptake, in *The Yeasts*, Vol. 2, Rose, A. H. and Harrison, J. S., Eds., Academic Press, New York, 1971, 3.
842. Sutnick, M., Grover, W., and Patel, M., Impairment of hepatic pyruvate metabolism in phenyl ketonuria, *Life Sci.*, 15, 1945, 1974.

843. Sutton, M. R., Fall, R. R., Nervi, A. M., Alberts, A. W., Vagelos, P. R., and Bradshaw, R. A., Amino acid sequence of *Escherichia coli* biotin carboxylase carrier protein (9100), *J. Biol. Chem.*, 252, 3934, 1977.

844. Sutton, R. and Pollak, J. K., The increasing adenine nucleotide concentration and the maturation of rat liver mitochondria during neonatal development, *Differentiation*, 12, 15, 1978.

845. Sweetman, L., Ultramicroassays of enzymes for rapid and inexpensive diagnosis of inherited disorders of metabolism, *Ind. J. Pediatr.*, 48, 133, 1981.

846. Sweetman, L., Two forms of biotin-responsive multiple carboxylase deficiency, *J. Inher. Metab. Dis.*, 4, 53, 1981.

847. Sweetman, L. and Burri, B. J., Abnormal holocarboxylase synthetases with elevated K_m values for biotin as the cause for one form of the inherited human disorder, biotin-responsive multiple carboxylase deficiency, *Proc. Int. Symp. Biotin-Dependent Enzymes*, Adelaide, 1982.

848. Sweetman, L., Packman, S., Yoshino, M., Cowan, M., Wara, D., Ammann, A., and Nyhan, W., Biotin responsive multiple carboxylase deficiency, *Pediatr. Res.*, 13, 426, 1979.

849. Swiatek, K. R., Development of gluconeogenesis in pig liver slices, *Biochim. Biophys. Acta*, 252, 274, 1971.

850. Tachiki, K. H. and Baxter, C. F., Role of carbon dioxide fixation, blood aspartate and glutamate in the adaptation of amphibian brain tissues to a hyperosmotic internal environment, *Neurochem. Res.*, 5, 993, 1980.

851. Tada, K., Sugita, K., Fujitani, K., Uesakai, T., Takada, G., and Omura, K., Hyperalaninemia with pyruvicemia in a patient suggestive of Leigh's encephalomyelopathy, *Tohoku J. Exp. Med.*, 109, 13, 1973.

852. Tada, K., Takada, G., Omura, K., and Itokawa, Y., Congenital lactic-acidosis due to pyruvate carboxylase deficiency. Absence of an inhibitor of thiamine pyro phosphate-ATP-phosphyryl transferase, *Eur. J. Pediatr.*, 127, 141, 1978.

853. Tada, K., Yoshida, T., Konno, T., Wada, Y., Yokoyama, Y., and Arakawa, T., Hyperalaninemia with pyruvicemia due to pyruvate carboxylase deficiency in the liver, *Tohoku J. Exp. Med.*, 97, 99, 1969.

854. Takada, G., Ohtake, M., Miyabayashi, S., and Tada, K., Lymphoblasts and diagnosis of pyruvate carboxylase deficiency, *Tohoku J. Exp. Med.*, 136, 103, 1982.

855. Tang, T. T., Good, T. A., Dyken, P. R., Johnsen, S. D., McCreadie, S. R., Sy, S. T., Lardy, H. A., and Rudolph, F. B., Pathogenesis of Leigh's encephalomyelopathy, *J. Pediatr.*, 81, 189, 1972.

856. Taylor, B. L., Barden, R. E., and Utter, M. F., Identification of the reacting form of pyruvate carboxylase, *J. Biol. Chem.*, 247, 7383, 1972.

857. Taylor, B. L., Frey, W. H., Barden, R. E., Scrutton, M. C., and Utter, M. F., The use of the ultracentrifuge to determine the catalytically competent forms of enzymes with more than one oligomeric structure. Multiple reacting forms of pyruvate carboxylase from chicken and rat liver, *J. Biol. Chem.*, 253, 3062, 1978.

858. Taylor, B. L., Routman, S., and Utter, M. F., The control of the synthesis of pyruvate carboxylase in *Pseudomonas citronellolis*. Evidence from double labeling studies, *J. Biol. Chem.*, 250, 2376, 1975.

859. Taylor, D. J., Crabtree, B., and Smith, G. H., The intracellular location of pyruvate carboxylase, citrate synthase and 3-hydroxyacyl-CoA dehydrogenase in lactating rat mammary gland, *Biochem. J.*, 171, 273, 1978.

860. Taylor, P. H., Nielsen, J., and Keech, D. B., Substrate activation of pyruvate carboxylase by pyruvate, *Biochem. Biophys. Res. Commun.*, 37, 723, 1969.

861. Taylor, P. H., Pyruvate Carboxylase and PEP Carboxykinase in Sheep, Ph.D. thesis, University of Adelaide, Australia, 1971, 88.

862. Taylor, P. H., Wallace, J. C., and Keech, D. B., Gluconeogenic enzymes in sheep liver. Intracellular localization of pyruvate carboxylase and phosphoenolpyruvate carboxykinase in normal, fasted and diabetic sheep, *Biochim. Biophys. Acta*, 237, 179, 1971.

863. Taylor, W. M., Prpic, V., Exton, J. H., and Bygrave, F. L., Stable changes to calcium fluxes in mitochondria isolated from rat livers perfused with α-adrenergic agonists and with glucagon, *Biochem. J.*, 188, 443, 1980.

864. Terheggen, H. G., Lowenthal, A., and Walter, P., Lacticacidosis due to pyruvate carboxylase deficiency, *Pediatr. Res.*, 11, 1016, 1977.

865. Teuful, H., Menahan, L. A., Shipp, J. C., Boning, S., and Wieland, O., Effect of oleic acid on the oxidation and gluconeogenesis from [1-^{14}C] pyruvate in the perfused rat liver, *Eur. J. Biochem.*, 2, 182, 1967.

866. Theodore, T. S. and Englesberg, E., Mutant of *Salmonella typhimurium* deficient in the carbon dioxide-fixing enzyme phosphoenolpyruvic carboxylase, *J. Bacteriol.*, 88, 946, 1964.

867. Thiers, R. E. and Vallee, B. L., Distribution of metals in subcellular fractions of rat liver, *J. Biol. Chem.*, 226, 911, 1957.

868. Thoene, J. G., Baker, H., Yoshino, M., and Sweetman, L., Biotin-responsive carboxylase deficiency associated with subnormal plasma and urinary biotin, *New Engl. J. Med.*, 304, 817, 1981.

869. Thoene, J. G., Lemons, R. M., Borgsko, K. Z., and Baker, H., Juvenile multiple carboxylase deficiency: evidence for impaired biotin absorption, *Pediatr. Res.*, 16, 179A, 1982.
870. Thoene, J. G., Lemons, R., and Baker, H., Impaired intestinal absorption of biotin in juvenile multiple carboxylase deficiency, *New Engl. J. Med.*, 308, 639, 1983.
871. Thoene, J., Sweetman, L., and Yoshino, M., Biotin-responsive multiple carboxylase deficiency, *Am. J. Hum. Genet.*, 31, 64A, 1979.
872. Thoene, J. and Wolf, B., Biotinidase deficiency in juvenile multiple carboxylase deficiency, *Lancet*, 398, 1983.
873. Thomas, A. P. and Halestrap, A. P., The role of mitochondrial pyruvate transport in the stimulation by glucagon and phenylephrine of gluconeogenesis from L-lactate in isolated rat hepatocytes, *Biochem. J.*, 198, 551, 1981.
874. Tilghman, S. M., Hanson, R. W., and Ballard, F. J., Hormonal regulation of phosphoenolpyruvate carboxykinase (GTP) in mammalian tissues, in *Gluconeogenesis: Its Regulation in Mammalian Species*, Hanson, R. W. and Mehlman, M. A., Eds., John Wiley & Sons, New York, 1976, 47.
875. Ting, I. P. and Osmond, C. B., Photosynthetic phosphoenolpyruvate carboxylase, *Plant Physiol.*, 51, 439, 1973.
876. Titheradge, M. A. and Coore, H. G., The mitochondrial pyruvate carrier: its exchange properties and its regulation by glucagon, *FEBS Lett.*, 63, 45, 1976.
877. Titheradge, M. A. and Haynes, R. C., Jr., The hormonal stimulation of ureogenesis in isolated hepatocytes through increases in mitochondrial ATP production, *Arch. Biochem. Biophys.*, 201, 44, 1980.
878. Titheradge, M. A. and Haynes, R. C., Jr., The control of uncoupler-activated ATPase activity in rat liver mitochondria by adenine nucleotide transport, *J. Biol. Chem.*, 255, 1471, 1980.
879. Titheradge, M. A., Stringer, J. L., and Haynes, R. C., Jr., The stimulation of mitochondrial uncoupler-dependent ATPase in isolated hepatocytes by catecholamines and glucagon and its relationship to gluconeogenesis, *Eur. J. Biochem.*, 102, 117, 1979.
880. Toews, C. J., Lowy, C., and Ruderman, N. B., The regulation of gluconeogenesis. The effect of pent-4-enoic acid on gluconeogenesis and on the gluconeogenic metabolite concentrations of isolated perfused rat liver, *J. Biol. Chem.*, 245, 818, 1970.
881. Tolbert, B., Ph.D. thesis, Case Western Reserve University, Cleveland, 1970.
882. Tommasi, M., Jouvet-Telinge, A., Kopp, N., Pialat, J., and Gilly, J., Alpers infantile cerebral polio dystrophy: a case with an abnormality of hepatic pyruvate carboxylase, *Ann. Anat. Pathol.*, 22, 337, 1977.
883. Topper, Y. J. and Hastings, A. B., A study of the chemical origins of glycogen by use of ^{14}C labelled carbon dioxide, acetate and pyruvate, *J. Biol. Chem.*, 179, 1255, 1949.
884. Topping, D. L., Clark, D. G., Storer, G. B., Trimble, R. P. and Illman, R. J., Acute effects of ethanol on the perfused rat liver: studies on lipid and carbohydrate metabolism, substrate cycling and perfusate amino acids, *Biochem. J.*, 184, 97, 1979.
885. Tosaka, O., Morioka, H., and Takinami, K., The role of biotin-dependent pyruvate carboxylase in L-lysine production, *Agric. Biol. Chem.*, 43, 1513, 1979.
886. Toshima, K., Kuroda, Y., Hashimota, T., Ito, M., Watanabe, T., Miyao, M., and Kunio, U., Enzymologic studies and therapy of Leigh's disease associated with pyruvate decarboxylase deficiency, *Pediatr. Res.*, 16, 430, 1982.
887. Trautman, R. and Crampton, C. F., Application of the Archibald Principle for the ultracentrifugal determination of the molecular weight in urea solutions of histone fractions from calf thymus, *J. Am. Chem. Soc.*, 81, 4036, 1959.
888. Triebwasser, K. C. and Freedland, R. A., The effect of glucagon on ureagenesis from ammonia by isolated rat hepatocytes, *Biochem. Biophys. Res. Commun.*, 76, 1159, 1977.
889. Trüper, H. G., CO_2-fixierung und Intermediarstoffwechsel bei *Chromatium okenii* Perty, *Arch. Mikrobiol.*, 49, 23, 1964.
890. Tsuchiyama, A., Oyanagi, K., Soguwa, H., Nakao, T., Ogawa, K., and Fujita, S., Normal activities of hepatic pyruvate dehydrogenase and pyruvate carboxylase in Leigh's syndrome, *Tohoku J. Exp. Med.*, 139, 67, 1983.
891. Utter, M. F., The role of CO_2 fixation in carbohydrate utilization and synthesis, *Ann. N.Y. Acad. Sci.*, 72, 451, 1959.
892. Utter, M. F., Pathways of phosphoenolpyruvate synthesis in glycogenesis, *Iowa St. J. Sci.*, 38, 97, 1963.
893. Utter, M. F., Metabolic roles of oxaloacetate, in *Citric Acid Cycle — Control and Compartmentation*, Lowenstein, J. M., Ed., Marcel Dekker, New York, 1969, 249.
894. Utter, M. F., Pyruvate carboxylase, in *Metabolic Regulation and Enzyme Action*, FEBS Symp., Vol. 19, Sols, A. and Grisolia, S., Eds., Academic Press, London, 1970, 91.
895. Utter, M. F., The biochemistry of manganese, *Med. Clin. N. Am.*, 60, 713, 1976.

896. Utter, M. F., Barden, R. E., and Taylor, B. L., Pyruvate carboxylase: an evaluation of the relationships between structure and mechanism and between structure and catalytic activity, *Adv. Enzymol.*, 42, 1, 1975.
897. Utter, M. F. and Fung, C.-H., Possible control mechanisms of liver pyruvate carboxylase, in *Regulation of Gluconeogenesis,* Söling, H. D. and Willms, B., Eds., Academic Press, New York, 1971, 1.
898. Utter, M. F. and Keech, D. B., Formation of oxaloacetate from pyruvate and CO_2, *J. Biol. Chem.*, 235, PC17, 1960.
899. Utter, M. F. and Keech, D. B., Pyruvate carboxylase. I. Nature of the reaction, *J. Biol. Chem.*, 238, 2603, 1963.
900. Utter, M. F., Keech, D. B., and Scrutton, M. C., A possible role for acetyl CoA in the control of gluconeogenesis, *Adv. Enz. Reg.*, 2, 49, 1964.
901. Utter, M. F., Keech, D. B., and Scrutton, M. C., A possible role for acetyl-CoA in the control of gluconeogenesis, in *Advances in Enzyme Regulation,* Vol. 2, Weber, G., Ed., Pergamon Press, New York, 1964, 49.
902. Utter, M. F. and Kolenbrander, H. M., Formation of oxaloacetate by CO_2 fixation on phosphoenolpyruvate, in *The Enzymes,* Vol. 6, Boyer, P. D., Ed., Academic Press, New York, 1972, 117.
903. Utter, M. F. and Kurahashi, K., Mechanism of action of oxalacetic carboxylase, *J. Biol. Chem.*, 207, 821, 1954.
904. Utter, M. F. and Scrutton, M. C., Pyruvate carboxylase, in *Current Topics in Cellular Regulation,* Vol. 1, Horecker, B. L. and Stadtman, E. R., Eds., Academic Press, New York, 1969, 253.
905. Vaartjes, W. J., den Breejen, J. N., Geelen, M. J. H., and Van den Bergh, S. G., Persistent changes in the initial rate of pyruvate transport by isolated rat liver mitochondria after preincubation with adenine nucleotides and calcium ions, *Biochim. Biophys. Acta,* 592, 28, 1980.
906. Valentine, R. C., Shapes and symmetries of biological molecules, *Proc. 4th Reg. Conf. (Europe) Electron Microscopy,* Tipografia Poliglotta Vaticana, Rome, 2, 3, 1968.
907. Valentine, R. C., Wrigley, N. G., Scrutton, M. C., Irias, J. J., and Utter, M. F., Pyruvate carboxylase. VIII. The subunit structure as examined by electron microscopy, *Biochemistry,* 5, 311, 1966.
908. Van Berkel, T. J. C., Kruijt, J. K., and Koster, J. F., Hormone-induced changes in pyruvate kinase. Effects of glucagon and starvation, *Eur. J. Biochem.*, 81, 423, 1977.
909. Van Berkel, T. J. C., Kruijt, J. K., Van den Berg, G. J., and Koster, J. F., Differences in the effect of glucagon and starvation upon L-type pyruvate kinase from rat liver, *Eur. J. Biochem.*, 92, 553, 1978.
910. Van Biervliet, J. P., Bruinvis, L., Van der Heiden, C., Ketting, D., Wadman, S. K., Willemsen, J. L., and Monnens, L. A., Report of a patient with severe chronic lactic-acidemia and pyruvate carboxylase deficiency, *Dev. Med. Child. Neurol.*, 19, 392, 1977.
911. Van Biervliet, J. P. G. M., Duran, M., Wadman, S. K., Monnuns, L. A. H., and Williams, J. C., A patient with pyruvate carboxylase deficiency, *Pediatr. Res.*, 10, 900, 1976.
912. van Marrewijk, W. J. A. and de Zwaan, A., Amino acid metabolism of *Astacus leptodactylus* Esch. IV. Role of carbon dioxide fixation and acetyl-CoA pathway in the biosynthesis of amino acid precursors, *Comp. Biochem. Physiol.*, 53B, 361, 1976.
913. Verhoeven, A. J., Hensgens, H. E. S. J., Meijer, A. J., and Tager, J. M., On the nature of the stimulation by glucagon of citrulline synthesis in rat-liver mitochondria, *FEBS Lett.*, 140, 270, 1982.
914. Visser, C. M. and Kellogg, R. M., Biotin. Its place in evolution, *J. Mol. Evol.*, 11, 171, 1978.
915. Von Brand, T., Carbohydrates. II. Metabolism of carbohydrates. IX. Terminal anaerobic processes, in *Biochemistry of Parasites,* Academic Press, New York, 1973, 136.
916. Von Glutz, G. and Walter, P., Compartmentation of acetyl-CoA in rat-liver mitochondria, *Eur. J. Biochem.*, 60, 147, 1975.
917. Von Glutz, G. and Walter, P., Regulation of pyruvate carboxylation by acetyl-CoA in rat liver mitochondria, *FEBS Lett.*, 72, 299, 1976.
918. Wagle, S. R., Studies on pyruvate carboxylase activity in alloxan diabetic and normal animals, *Biochem. Biophys. Res. Commun.*, 14, 533, 1964.
919. Wagle, S. R., Effects of biotin deficiency on pyruvate metabolism, *Fed. Proc. Fed. Am. Soc. Exp. Biol. Med.*, 121, 1519, 1966.
920. Wagner, G. and Wuthrich, K., Sequential resonance assignments in protein ^1H nuclear magnetic resonance spectra. Basic pancreatic trypsin inhibitor, *J. Mol. Biol.*, 155, 347, 1982.
921. Wagner, G. and Wuthoich, K., Amide proton exchange and surface conformation of the basic pancreatic trypsin inhibitor in solution. Studies with two-dimensional nuclear magnetic resonance, *J. Mol. Biol.*, 160, 343, 1982.
922. Wagner, P. and Yount, R. G., Stoichiometry of labelling of myosin's proteolytic fragments by a purine disulfide analog of adenosine triphosphate, *Biochemistry,* 14, 1900, 1975.
923. Wagner, P. and Yount, R. G., Subunit location of sulfydryl groups of myosin labeled with a purine disulfide analog of adenosine triphosphate, *Biochemistry,* 14, 1908, 1975.
924. Wakat, D. K. and Haynes, R. C., Jr., Glucocorticoid-stimulated utilization of substrates in hepatic mitochondria, *Arch. Biochem. Biophys.*, 184, 561, 1977.

925. Walajtys-Rode, E., Coll, K. E., and Williamson, J. R., Effects of branched chain alpha keto acids on the metabolism of isolated rat liver cells. II. Interactions with gluconeogenesis and urea synthesis, *J. Biol. Chem.*, 254, 11521, 1979.
926. Walker, P. R. and Bailey, E., Role of malic enzyme in lipogenesis, *Biochim. Biophys. Acta*, 187, 591, 1969.
927. Walker, P. R. and Bailey, E., Fat body pyruvate carboxylases of the adult male desert locust, *Comp. Biochem. Physiol.*, 36, 623, 1970.
928. Walker, I. D., Harris, J. I., Runswick, M. J., and Hudson, P., The subunits of rabbit-muscle phosphofructokinase. A search for sequence repetition, *Eur. J. Biochem.*, 68, 255, 1976.
929. Wallace, J. C. and Newsholme, E. A., A comparison of the properties of fructose 1,6-diphosphatase, and the activities of other key enzymes of carbohydrate metabolism, in the livers of embryonic and adult rat, sheep and domestic fowl, *Biochem. J.*, 104, 378, 1967.
930. Wallace, J. C., Young, M. R., Tolbert, B., and Utter, M. F., Studies of the relationships between the catalytic properties and physical characteristics of pyruvate carboxylases from diverse sources, *Proc. Aust. Biochem. Soc.*, 3, 1, 1970.
931. Walter, P., Pyruvate carboxylase: intracellular localization and regulation, in *Gluconeogenesis: Its Regulation in Mammalian Species*, Hanson, R. W. and Mehlman, M. A., Eds., John Wiley & Sons, New York, 1976, 239.
932. Walter, P. and Anabitarte, M., On the use of glutamate dehydrogenase as a mitochondrial marker enzyme for the determination of the intracellular distribution of rat liver pyruvate carboxylase, *FEBS Lett.*, 12, 289, 1971.
933. Walter, P. and Anabitarte, M., Intracellular distribution of pyruvate carboxylase in livers of normal and cortisol treated rats, *FEBS Lett.*, 37, 170, 1973.
934. Walter, P., Brawand, F., Moerikofer-Zwez, S., and Stoecklin, F. B., Regulation of gluconeogenesis in cell-free systems of rat liver, in *Biochemistry of Metabolic Processes*, Lennon, D. L. F., Stratman, F. W., and Zahlten, R. N., Eds., Elsevier, Amsterdam, 1983, 83.
935. Walter, P., Paetkau, V., and Lardy, H. A., Paths of carbon in gluconeogenesis and lipogenesis. III. The role and regulation of mitochondrial processes involved in supplying precursors of phosphoenolpyruvate, *J. Biol. Chem.*, 241, 2523, 1966.
936. Warnes, D. M., Seamark, R. F., and Ballard, F. J., The appearance of gluconeogenesis at birth in sheep. Activation of the pathway associated with blood oxygenation, *Biochem. J.*, 162, 627, 1977.
937. Warren, G. B. and Tipton, K. F., Pig liver pyruvate carboxylase: purification, properties and cation specificity, *Biochem. J.*, 139, 297, 1974.
938. Warren, G. B. and Tipton, K. F., Pig liver pyruvate carboxylase: the reaction pathway for the carboxylation of pyruvate, *Biochem. J.*, 139, 311, 1974.
939. Watford, M., Hod, Y., Chiao, Y-B., Utter, M. F. and Hanson, R. W., The unique role of the kidney in gluconeogenesis in the chicken. The significance of a cytosolic form of phosphoenolpyruvate carboxykinase, *J. Biol. Chem.*, 256, 10023, 1981.
940. Watson, J. A. and Lowenstein, J. M., Citrate and the conversion of carbohydrate into fat. Fatty acid synthesis by a combination of cytoplasm and mitochondria, *J. Biol. Chem.*, 245, 5993, 1970.
941. Weber, G., The molecular correlation concept: studies on the metabolic patterns of hepatomas, *Gann. Monogr.*, 1, 151, 1966.
942. Weber, G. and Morris, H. P., Comparative biochemistry of hepatomas. III. Carbohydrate enzymes in liver tumours of different growth rates, *Cancer Res.*, 23, 987, 1963.
943. Weber, K. and Osborn, M., The reliability of molecular weight determinations by dodecyl sulfate-polyacrylamide gel electrophoresis, *J. Biol. Chem.*, 244, 4406, 1969.
944. Weinberg, M. B. and Utter, M. F., Effect of thyroid hormone on the turnover of rat liver pyruvate carboxylase and pyruvate dehydrogenase, *J. Biol. Chem.*, 254, 9492, 1979.
945. Weinberg, M. B. and Utter, M. F., Effect of streptozotocin-induced diabetes mellitus on the turnover of rat liver pyruvate carboxylase and pyruvate dehydrogenase, *Biochem. J.*, 188, 601, 1980.
946. Weiss, G., Lamartiniere, C. A., Muller-Ohly, B., and Seubert, W., The intracellular location of pyruvate carboxylase in rat liver, *Eur. J. Biochem.*, 43, 391, 1974.
947. Weiss, G., Muller-Ohly, B., and Seubert, W., Control of pyruvate carboxylase in alloxan diabetic rats, *Biochem. Biophys. Res. Commun.*, 46, 225, 1972.
948. Weyler, W. L., Sweetman, L., Maggia, D. C., and Nyhan, W. L., Deficiency of propionyl-CoA carboxylase and methylcrotonyl-CoA carboxylase in a patient with methylcrotonylglycinuria, *Clin. Chim. Acta*, 76, 321, 1977.
949. Whitehead, C. C., The assessment of biotin status in man and animals, *Proc. Nutr. Soc.*, 40, 165, 1981.
950. Whitehead, C. C., Armstrong, J. A., and Waddington, D., Determination of the availability to chicks of biotin in feed ingredients by a bioassay based on the response of blood pyruvate carboxylase activity, *Br. J. Nutr.*, 48, 81, 1982.
951. Whitehead, C. C. and Bannister, D. W., Blood pyruvate carboxylase activity as a criterion of biotin status in chickens and turkeys, *Br. J. Nutr.*, 39, 547, 1978.

952. Whitehead, C. C. and Bannister, D. W., Biotin status blood pyruvate carboxylase activity and performance in broilers under different conditions of bird husbandry and diet processing, *Br. J. Nutr.*, 43, 541, 1980.
953. Whitehead, C. C. and Bannister, D. W., Aspects of metabolism related to the occurrence of skin lesions in biotin deficient chicks, *Br. Poultry Sci.*, 22, 467, 1981.
954. Whitehead, C. C., Bannister, D. W., and Cleland, M. E., Metabolic changes associated with the occurrence of fatty liver and kidney syndrome in chicks, *Br. J. Nutr.*, 40, 221, 1978.
955. Whitehead, C. C., Bannister, D. W., and D'Mello, J. P. F., Blood pyruvate carboxylase activity as a criterion of biotin status in young pigs, *Res. Vet. Sci.*, 29, 126, 1980.
956. Whitehead, C. C., Bannister, D. W., Evans, A. L., Siller, W. G., and Wight, P. A. A., Biotin deficiency and fatty liver and kidney syndrome in chicks given purified diets containing different fat and protein levels, *Br. J. Nutr.* 35, 115, 1976.
957. Whittaker, V. P., The application of subcellular fractionation techniques to the study of brain function, in *Prog. Biophys. Mol. Biol.*, Vol. 15, Butler, J. A. V. and Huxley, H. E., Eds., Pergamon Press, Oxford, 1963, 109.
958. Wilbur, D. O. and Patel, M. S., Development of mitochondrial pyruvate metabolism in rat brain, *J. Neurochem.*, 22, 709, 1974.
959. Williamson, J. R., Mechanism for the stimulation *in vivo* of hepatic gluconeogenesis by glucagon, *Biochem. J.*, 101, 11C, 1966.
960. Williamson, J. R., Control of gluconeogenesis by acetyl-CoA in rats treated with glucagon and anti-insulin serum, *Biochem. Biophys. Res. Commun.*, 24, 765, 1966.
961. Williamson, J. R., Effects of fatty acids, glucagon and anti-insulin serum on the control of gluconeogenesis and ketogenesis in rat liver, in *Advances in Enzyme Regulation*, Vol. 5, Weber, G., Ed., Pergamon Press, New York, 1967, 229.
962. Williamson, J. R., Calculation of metabolic concentrations in the cytosol and mitochondria of rat liver, in *Energy Level and Metabolic Control in Mitochondria*, Papa, S., Tager, J. M., Quagliariello, E., and Stater, E. C., Eds., Adriatica Editrice, Bari, 1969, 385.
963. Williamson, J. R., Browning, E. T., and Olson, M. S., Interrelations between fatty acid oxidation and the control of gluconeogenesis in perfused rat liver, in *Advances in Enzyme Regulation*, Vol. 6, Weber, G., Ed., Pergamon Press, New York, 1968, 67.
964. Williamson, J. R., Browning, E. T., and Scholz, R., Control mechanisms of gluconeogenesis and ketogenesis. I. Effects of oleate on gluconeogenesis in perfused rat liver, *J. Biol. Chem.*, 244, 4607. 1969.
965. Williamson, J. R., Browning, E. T., Scholz, R., Kreisberg, R. A., and Fritz, I. B., Inhibition of fatty acid stimulation of gluconeogenesis by (t)-decanoylcarnitine in perfused liver, *Diabetes*, 17, 194, 1968.
966. Williamson, J. R., Herczeg, B., Coles, H., and Danish, R., Studies on the ketogenic effect of glucagon in intact rat liver, *Biochem. Biophys. Res. Commun.*, 24, 437, 1966.
967. Williamson, J. R., Jakob, A., and Scholz, R., Energy cost of gluconeogenesis in rat liver, *Metabolism*, 20, 13, 1971.
968. Williamson, J. R., Kreisberg, R. A., and Felts, P. W., Mechanism for the stimulation of gluconeogenesis by fatty acids in perfused rat liver, *Proc. Natl. Acad. Sci. U.S.A.*, 56, 247, 1966.
969. Williamson, J. R. and Murphy, E., Calcium homeostasis and compartmentation in liver, *Adv. Exp. Med. Biol.*, 132, 671, 1980.
970. Williamson, J. R., Rostand, S. G., and Peterson, M. J., Control factors affecting gluconeogenesis in perfused rat liver. Effects of 4-pentenoic acid, *J. Biol. Chem.*, 245, 3242, 1970.
971. Wilson, J. E., Brain hexokinase. A proposed relation between soluble-particulate distribution and activity *in vivo*, *J. Biol. Chem.*, 243, 3640, 1968.
972. Wilson, J. E., Ambiquitous enzymes: variation in intracellular distribution as a regulatory mechanism, *Trends Biochem. Sci.*, 3, 124, 1978.
973. Wilson, J. E., Brain hexokinase, the prototype ambiquitous enzyme, in *Current Topics in Cellular Regulation*, Vol. 16, Horecker, B. L. and Stadtman, E. R., Eds., Academic Press, New York, 1980, 1.
974. Wimhurst, J. M. and Harris, E. J., The actions of avenociolide and ethanol on glucose metabolism and on related enzyme activities in the isolated perfused rat liver, *Biochim. Biophys. Acta*, 437, 51, 1976.
975. Wimhurst, J. M. and Manchester, K. L., Some aspects of the kinetics of rat liver pyruvate carboxylase, *Biochem. J.*, 120, 79, 1970.
976. Wimhurst, J. M. and Manchester, K. L., A comparison of the effects of diabetes induced with either alloxan or streptozotocin and of starvation on the activities in rat liver of the key enzymes of gluconeogenesis, *Biochem. J.*, 120, 95, 1970.
977. Wimhurst, J. M. and Manchester, K. L., Role of bivalent cations in the control of enzymes involved in gluconeogenesis, *FEBS Lett.*, 10, 33, 1970.
978. Wolf, B., An approach to the study of biotin metabolism in fibroblast culture, *Pediatr. Res.*, 14, 529, 1980.

979. Wolf, B. and Feldman, G. L., The biotin-dependent carboxylase deficiencies, *Am. J. Hum. Genet.*, 34, 699, 1982.
980. Wolf, B., Grier, R. E., Allen, R. J., Goodman, S. I., and Kien, C. L., Biotinidase deficiency: the enzymatic defect in late-onset multiple carboxylase deficiency, *Clin. Chim. Acta*, 131, 273, 1983.
981. Wolf, B., Grier, R. E., Parker, W. D., Goodman, S. I., and Allen R. J., Deficient biotinidase activity in late-onset multiple carboxylase deficiency, *New Engl. J. Med.*, 308, 161, 1983.
982. Wolf, B., Hsia, Y. E., Sweetman, L., Feldman, G., Boychuk, R. B., Bart, R. D., Crowell, D. H., Di Mauro, R. M., and Nyhan, W. L., Multiple carboxylase deficiency: clinical and biochemical improvement following neonatal biotin treatment, *Pediatrics*, 68, 113, 1981.
983. Wood, H. G. and Barden, R. E., Biotin enzymes, *Ann. Rev. Biochem.*, 46, 385, 1977.
984. Wood, H. G., Lochmüller, H., Riepertinger, C., and Lynen, F., Transcarboxylase. IV. Function of biotin and the structure and properties of the carboxylated enzyme, *Biochem. Z.*, 337, 247, 1963.
985. Woronick, C. L. and Johnson, M. J., Carbon dioxide fixation by cell-free extracts of *Aspergillus niger*, *J. Biol. Chem.*, 235, 9, 1960.
986. Wright, L. D., Cresson, E. L., Skeggs, H. R., Wood, J. R., Peck, R. L., Wolf, D. E., and Folkers, K., Isolation of crystalline biotin from yeast extract, *J. Am. Chem. Soc.*, 74, 1996, 1952.
987. Wuthrich, K., Wider, G., Wagner, G., and Braun, W., Sequential resonance assignments as a basis for determination of spatial protein structures by high resolution proton nuclear magnetic resonance, *J. Mol. Biol.*, 155, 311, 1982.
988. Yamazaki, R. K., Glucagon stimulation of mitochondrial respiration, *J. Biol. Chem.*, 250, 7924, 1975.
989. Yamazaki, R. K. and Graetz, G. S., Glucagon stimulation of citrulline formation in isolated hepatic mitochondria, *Arch. Biochem. Biophys.*, 178, 19, 1977.
990. Yamazaki, R. K., Mickey, D. L., and Story, M., Rapid action of glucagon on hepatic mitochondrial calcium metabolism and respiratory rates, *Biochim. Biophys. Acta*, 592, 1, 1980.
991. Yamazaki, R. K. and Utter, M. F., personal communication.
992. Yeung, D. and Oliver, I. T., Gluconeogenesis from amino acids in neonatal rat liver, *Biochem. J.*, 103, 744, 1967.
993. Yeung, D., Stanley, R. S., and Oliver, I. T., Development of gluconeogenesis in neonatal rat liver. Effect of triamcinolone, *Biochem. J.*, 105, 1219, 1967.
994. Yeung, R. T. T., Yeung, D. C. Y., and Au, K. S., Hypoglycaemia associated with lipid accumulation in primary hepatocellular carcinoma, *Cancer*, 32, 1482, 1973.
995. Yoshida, T., Tada, K., Konno, T., and Arakawa, T., Hyperalaninemia with pyruvicemia due to pyruvate carboxylase deficiency of the liver, *Tohoku J. Exp. Med.*, 99, 121, 1969.
996. Young, M. R., Tolbert, B., and Utter, M. F., Pyruvate carboxylase from *Saccharomyces cerevisiae*, in *Methods in Enzymology*, Vol. 13, Colowick, S. P. and Kaplan, N. O., Eds., Academic Press, New York, 1969, 250.
997. Yount, E. A. and Harris, R. A., Inhibition of gluconeogenesis by oxalate, *Biochim. Biophys. Acta*, 633, 122, 1980.
998. Yphantis, D. A., Equilibrium ultracentrifugation of dilute solutions, *Biochemistry*, 3, 297, 1964.
999. Zakharchuk, L. M., Ivanovskii, R. N., and Kondrat'eva, E. N., Acetate metabolism in *Ectothiorhodospira shaposhnikovii* during growth in the dark, *Mikrobiologiya*, 49, 383, 1980.
1000. Zaleski, J. and Bryla, J., Effects of oleate, palmitate and octanoate on gluconeogenesis in isolated rabbit liver cells, *Arch. Biochem. Biophys.*, 183, 553, 1977.
1001. Zielke, H. R., Sumbilla, C. M., Krause, B. L., and Ozand, P. T., Activity of gluconeogenic enzymes in fibroblasts from infants dying of the sudden infant death syndrome, *J. Cell. Biol.*, 87, 322A, 1980.
1002. Freytag, S. O. and Collier, K-J., Molecular cloning of a cDNA for human pyruvate carboxylase. Structural relationship to other biotin-containing carboxylases and regulation of mRNA content in differentiating preadipocytes, *J. Biol. Chem.*, 259, 12831, 1984.
1003. Khew. Goodall, Y- S., Pyruvate Carboxylase: Its Interactions with Acetyl - CoA, Ph.D. thesis, University of Adelaide, Australia, 1985, 76.

INDEX

A

C

G

H

M

N

Q

R

S